Global Congress on Process Safety 2016

Topical Conference at the 2016 AIChE Spring Meeting and 12th Global Congress on Process Safety

Houston, Texas, USA
10-14 April 2016

ISBN: 978-1-5108-2474-4

Printed from e-media with permission by:

Curran Associates, Inc.
57 Morehouse Lane
Red Hook, NY 12571

Some format issues inherent in the e-media version may also appear in this print version.

Printed by Curran Associates, Inc. (2016)

AIChE
120 Wall Street, FL 23
New York, NY 10005-4020

Phone: (800) 242-4363
Fax: (203) 775-5177

www.aiche.org

Additional copies of this publication are available from:

Curran Associates, Inc.
57 Morehouse Lane
Red Hook, NY 12571 USA
Phone: 845-758-0400
Fax: 845-758-2633
Email: curran@proceedings.com
Web: www.proceedings.com

TABLE OF CONTENTS

441682 Achieving Flare Risk Analysis Consistency through Auditing

Abstract:

A study was conducted to assess how PHAs in a large petrochemical complex evaluated the hazards of elevated flares. A total of twelve PHAs encompassing seventeen different elevated flare systems that utilized the layer of protection analysis (LOPA) methodology were evaluated. Based on this study a guideline was created that outlines typical hazard scenarios, initiating events, and potential safeguards that could be applied to prevent or mitigate the hazards.

The guideline generated from this study can be used to audit PHAs to determine if hazards and initiating events are adequately considered. The guideline can also be used to evaluate whether initiating event frequencies and risk reduction factors (credits) applied to safeguards were within an acceptable range. The guideline can be used not only as a tool for auditing existing PHAs but can be used as a reference for future PHAs in facilitating discussion on what hazard scenarios should be considered along with what safeguards may be applicable.

12TH GLOBAL CONGRESS ON PROCESS SAFETY

An Ounce of Observation, a Pound of Prevention – a Barrier-Based Approach to Identify Leading Indicators

Marisa Pierce
DNV GL
1400 Ravello Dr
Houston, Texas 77057
Marisa.pierce@dnvgl.com

Amar Ahluwalia
DNV GL
1400 Ravello Dr
Houston, Texas 77057
Amar.ahluwalia@dnvgl.com

Prepared for Presentation at
American Institute of Chemical Engineers
2016 Spring Meeting
12th Global Congress on Process Safety
Houston, Texas
April 11-13, 2016

—

An Ounce of Observation, a Pound of Prevention – a Barrier-Based Approach to Identify Leading Indicators

Marisa Pierce
DNV GL
1400 Ravello Dr
Houston, Texas 77057
Marisa.pierce@dnvgl.com

Amar Ahluwalia
DNV GL

Keywords: Leading Indicators, Barrier Risk Management, Bowtie, Human Factors, Downstream, Incident Investigation / Near Miss

Abstract

Currently, many incidents or near misses are investigated using Root Cause Analysis (RCA) or other techniques. DNV GL applied the barrier-based bowtie methodology in a novel manner for a customer in the nuclear industry so that they could determine the weak links in their work practice procedures for reactor isolation during maintenance. This paper discusses the application of the same bowtie methodology to evaluate well-known incidents that have occurred in the refining industry such as Ciniza Oil Refinery Fire and Explosion (2004), BP Texas City Refinery Fire and Explosion (2005), Valero-McKee LPG Fire (2007), Chevron Richmond Refinery Fire and Explosion (2012), and others. Additionally, the paper discusses the visual effectiveness of a bowtie to summarize the results of a review of multiple events - in order to identify common factors contributing to incidents and near misses and assess the priorities for strengthening or adding barriers. For the purposes of this paper, the term "event" will be used in place of "near miss", "incident" or "accident" for consistency. The first step is the development of a generic bowtie diagram showing the left-hand (prevention) side and the right-hand (mitigation) side for loss of containment for similar events based on information that is publically available, most notably through the U.S Chemical Safety and Hazard Investigation Board (CSB). Next, the generic bowtie diagram is annotated for a set of different events. This is accomplished through a review of the sequence of events for each incident to determine where/when the barrier failure(s) occurred. The failed barriers are then tallied and categorized to determine the patterns associated with these failures. Finally, a summary bowtie diagram is developed to show how many times each barrier was compromised across the studied events, and which barriers were effective in detecting the deviation.

The findings from application of this methodology to the nuclear power facility accomplished the following:

- identified the common factors contributing to events reported over a period of time
- revealed which barriers were effective in detecting or preventing the top event
- identified priorities for strengthening or adding barriers
- provided guidance for development of an action plan to strengthen existing barriers and add additional barriers to reduce the likelihood or consequences of future events

Overall, it is anticipated that the use of this same approach in reviewing undesired events in refining facilities will proactively give the users insight into trends that might have previously been unnoticed. The use of the methodology will also demonstrate which barriers are effective and help the user identify and focus on strengthening the barriers that previously failed. It can also help users identify systemic issues with culture, organization, or layout between different facilities or regions based on the trends.

1 Introduction

The BP America Texas City Refinery Fire and Explosion (2005), Valero - McKee LPG Fire (2007), Chevron Richmond Refinery Fire and Explosion (2012), and more recently the Exxon Mobil Torrance Refinery Explosion (2015) are some well-known examples of downstream major events that have resulted in fatalities and/or injuries, asset damage, and business interruption losses. The fact that such events still continue to happen serves as a wake-up call for the refining industry. These events are low frequency-high consequence, as compared to the high frequency, (relatively) low consequence events that often occur. Currently, undesired events are investigated using Taproot ®, "5 Whys" or other techniques.

Being high-consequence events, the root cause investigations for such events tend to have longer timelines, are comprehensive and may involve several investigation parties. There is still a possibility, and it is more than likely, that some important failure mechanisms are overlooked or not all plausible events have actually been identified. However, when these multiple source reports are analyzed combined, by contrast, it provides hard information as to how things can go wrong and bypass any sort of bias (ignorance, lack of understanding etc.). This type of analysis coupled with a barrier-based approach provides a holistic view of what went wrong and how often failures occur in a very visual means using bowties. Failure or weakening of barriers is a general cause of undesirable events and is a further reason to consider a barrier-based approach over conventional investigation techniques. The reports of such events can be analyzed to determine commonalities of root causes at facilities, departments, or organizations by understanding where technical or procedural barriers are failing. This provides one of the most effective means of improving process safety and truly embodies continual learning.

Lewis and Smith discussed the development of a bowtie and applications of the bowtie methodology in communicating risk and risk mitigation to different stakeholders at the 2010 Global Congress on Process Safety [1]. This paper more specifically discusses how bowtie diagrams were applied in a novel manner for a nuclear facility, so weak links in work practice procedures for reactor isolation during maintenance could be identified. The same barrier-based approach can be extended to loss of containment incidents available through public sources and databases to gather crucial learnings to prevent such incidents from happening in the future. Refining operations involve a number of risks that can lead to major events. It is, therefore, crucial to understand the causes that lead to such events. Uncontrolled releases of hydrocarbons are large risk contributors to downstream operations and have caused catastrophic events in the past.

This paper seeks to demonstrate a methodology that will assist operating companies in accomplishing the following goals with the application of a barrier-based approach:

- identify common factors contributing to incidents or near misses reported over a period of time
- reveal which barriers are effective in detecting or preventing the top event
- identify priorities for strengthening or adding barriers
- provide guidance for development of an action plan to strengthen existing barriers and add additional barriers to reduce the likelihood or consequences of future events

2 Methodology

The cornerstone of risk management is the development and maintenance of controls or barriers to prevent or mitigate risk of undesirable events. For the purposes of this paper, the term "event" will be used in place of "near miss", "incident" or "accident" for consistency. For this study, the risk events of interest are those that degrade or defeat barriers that are used to prevent or mitigate loss of containment events. Therefore, an important step in this study is a systematic identification of these barriers and how they interact to prevent or mitigate such events. The following four-step approach was used for this study, as illustrated in Figure 1.

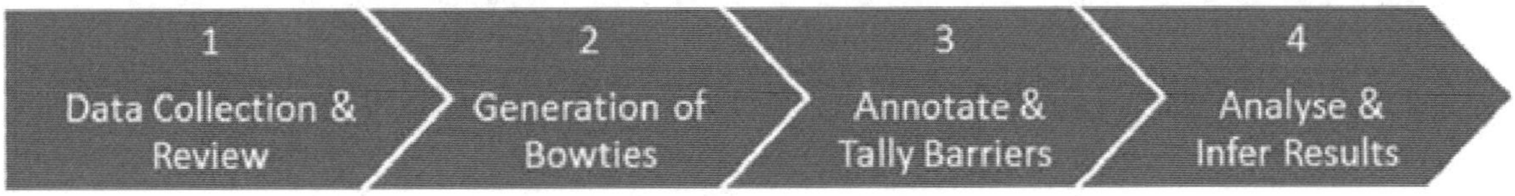

Figure 1. Four-Step Approach

2.1 Step One: Data Collection & Review

The first step was to collect a list of events available on CSB's website involving refineries. The next step was to select incidents based on the availability of information concerning the event and the date of the event's occurrence. Events within the last 15 years were considered for the study. The events that were selected were those whose investigations were completed (not currently ongoing) by the CSB. The reports were reviewed to obtain immediate root causes and findings or recommendations. Root causes were classified based on the type of barrier deficiency they were most aligned with in a bowtie diagram for loss of primary containment (LOPC) of hydrocarbons in refineries. Such barriers might include mechanical integrity, relief systems, corrosion prevention or emergency response procedure. After the root causes were tabulated, the findings or recommendations were reviewed to determine systemic issues in management system elements and human factors.

The findings were categorized based on elements of an effective process safety management system. The Center for Chemical Process Safety includes 20 elements and four pillars in their Risk-Based Process Safety (RBPS) management system framework which is shown in Table 1.

Table 1. Risk-Based Process Safety Pillars and Elements [3]

Commit to Process Safety	Process Safety Culture
	Compliance with Standards
	Process Safety Competency
	Workforce Involvement
	Stakeholder Outreach
Understand Hazards and Risks	Process Knowledge Management
	Hazard Identification and Risk Analysis
Manage Risk	Operating Procedures
	Safe Work Practices
	Asset Integrity and Reliability
	Contractor Management
	Training and Performing Assurance
	Management of Change
	Operational Readiness
	Conduct of Operations
	Emergency Management
Learn from Experience	Incident Investigation
	Measurement and Metrics
	Auditing
	Management Review and Continuous Improvement

For the purpose of our study we have chosen to use CCPS's RBPS management system elements.

2.2 Step Two: Generation of Bowties

The second step was the generation of the bowtie diagrams.

- Bowtie No. 1 (BT1): Generic LOPC bowtie for a refinery
- Bowtie No. 2 (BT2): Refinery-specific bowtie for each event reviewed

A generic bowtie was developed for the scenario of loss of primary containment (LOPC) of hydrocarbons at a refinery as shown in Figure 2. The orange circle in the center of the diagram is the "Top Event" that is the focus of the barrier analysis. The yellow/black crosshatched box at the top of the diagram is the Hazard or energy source that can lead to the Top Event if it is not controlled. The blue boxes on the left side are potential Causes that could cause the Top Event to occur. Similarly, the red boxes on the right side are potential Consequences that could result if the Top Event occurs. The blue rectangles on the pathways between the Causes and the Top Event are Barriers that can intervene in the chain of events to prevent the Cause from leading to the Top Event. Similarly, Barriers are shown on the right hand side of the diagram that could reduce the Consequences that are possible if the Top Event occurs. Because of this arrangement, the left-hand side of the Bowtie diagram is commonly called the Prevention side while the right-hand side is referred to as the Mitigation side. The causes (left hand side) considered include:

- Operating outside of normal operating parameters
- High temperature hydrogen attack (HTHA)
- Corrosion
- Erosion
- Fatigue
- "Breaking containment"/Opening of pipes and equipment (planned)
- Start-up/shutdown operations
- Adverse weather
- External impact
- Loss of structural support

It should be noted that the causes considered are not reflected in their entirety in Figure 2. The figure is meant to show a generic bowtie and bowtie BT1 could not be fully captured without compromising the legibility of the font in the diagram. The consequences (right hand side) considered include:

- Fire and explosion
- Toxic release
- Environmental release

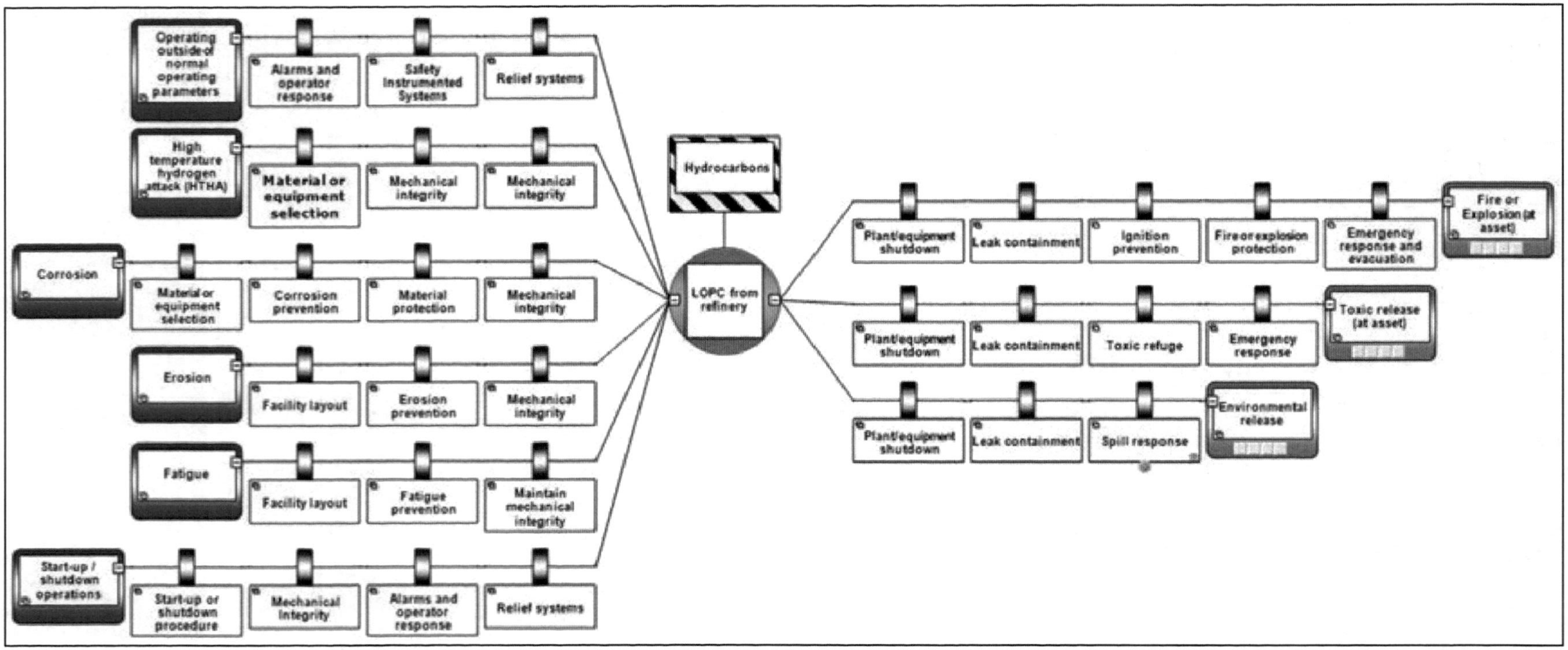

Figure 2. Generic LOPC Bowtie (BT1) for a Refinery with Typical Barriers

—

The preventive barriers (left side) and the mitigating barriers (right side) considered for the study are shown in Table 2. It should be noted that the barriers considered for this exercise are not reflected in their entirety in Figure 2, since this figure is meant to show a generic bowtie and the bowtie could not be fully captured without compromising the legibility of the font in the diagram. For each incident, the BT1 was used to develop BT2 considering the barriers that were compromised and the barriers that succeeded.

Table 2. Barrier List

Barrier Description	Barrier Type
Mechanical Integrity	Preventive
Process Control	Preventive
Alarms and Operator Response	Preventive
Safety-Instrumented Systems	Preventive
Relief Systems	Preventive
Corrosion Prevention	Preventive
Material Protection	Preventive
Facility Layout	Preventive
Erosion Prevention	Preventive
Fatigue Prevention	Preventive
Isolation of affected sub-systems and intervention	Preventive
Re-instatement of Plant	Preventive
Start-up or Shutdown Procedure	Preventive
Equipment and Pipe Protection	Preventive
Special Operating Procedures for Adverse Weather Conditions	Preventive
Structural Integrity	Preventive
Material or Equipment Selection	Preventive
Plant Shutdown	Mitigating
Leak Containment	Mitigating
Ignition Prevention	Mitigating
Passive Fire/ Explosion Protection	Mitigating
Active Fire Suppression	Mitigating
Emergency Response and Evacuation	Mitigating
Toxic Refuge	Mitigating
Spill Response	Mitigating

BT1 was then populated for each case study based on the findings to generate BT2 Bowties. BT2 for each incident shows the barriers that were compromised and those barriers that succeeded. Only the CSB findings that were related to barriers on the bowties were considered. A generic bowtie was used to represent an entire refinery for ease. However, it may be more effective for the operating company to develop or generate specific bowties for each process unit. The bowtie generated for BP Texas City Refinery Fire and Explosion (2005) is shown in Figure 3, Figure 4 and Figure 5. In each figure, the barriers that were compromised have a yellow box with the number of occurrences of barrier failure and the following description "Low (1 Occurrence)". This is how BT2 for each refinery event was developed.

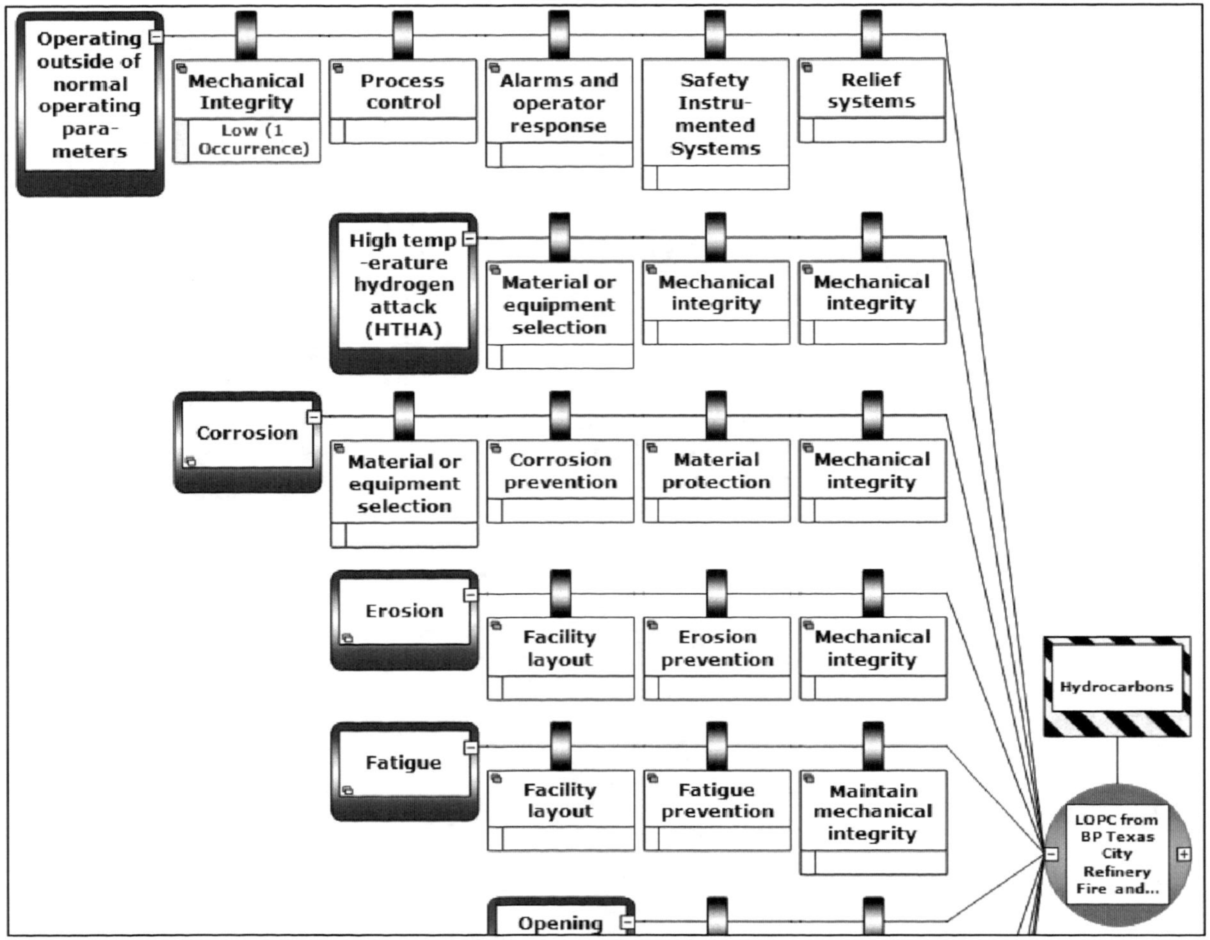

Figure 3. Upper Left Hand Side of Bowtie (BT2) for BP Texas City Refinery Fire and Explosion (2005)

—

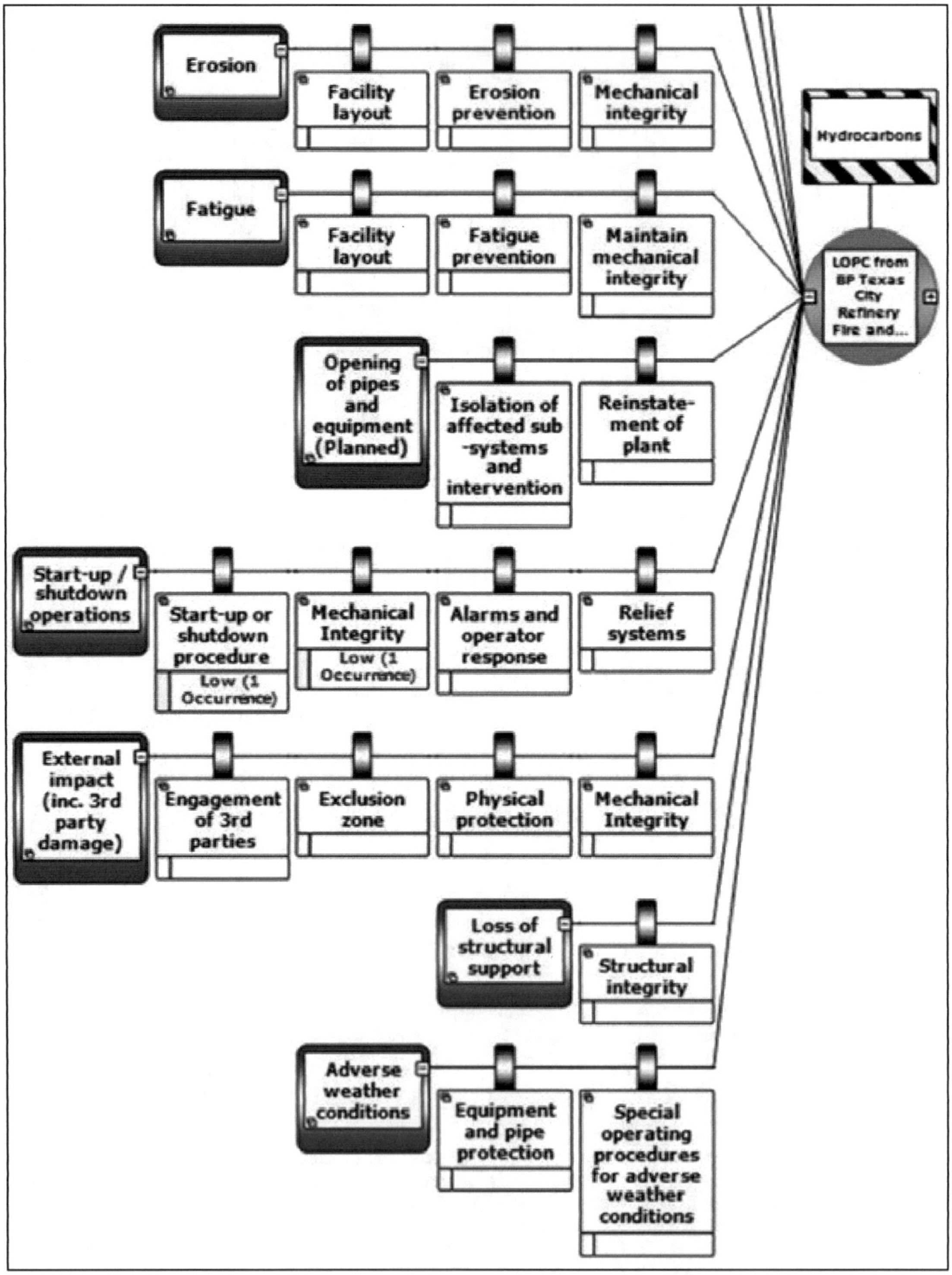

Figure 4. Bottom Left Hand Side of Bowtie (BT2) for BP Texas City Refinery Fire and Explosion (2005)

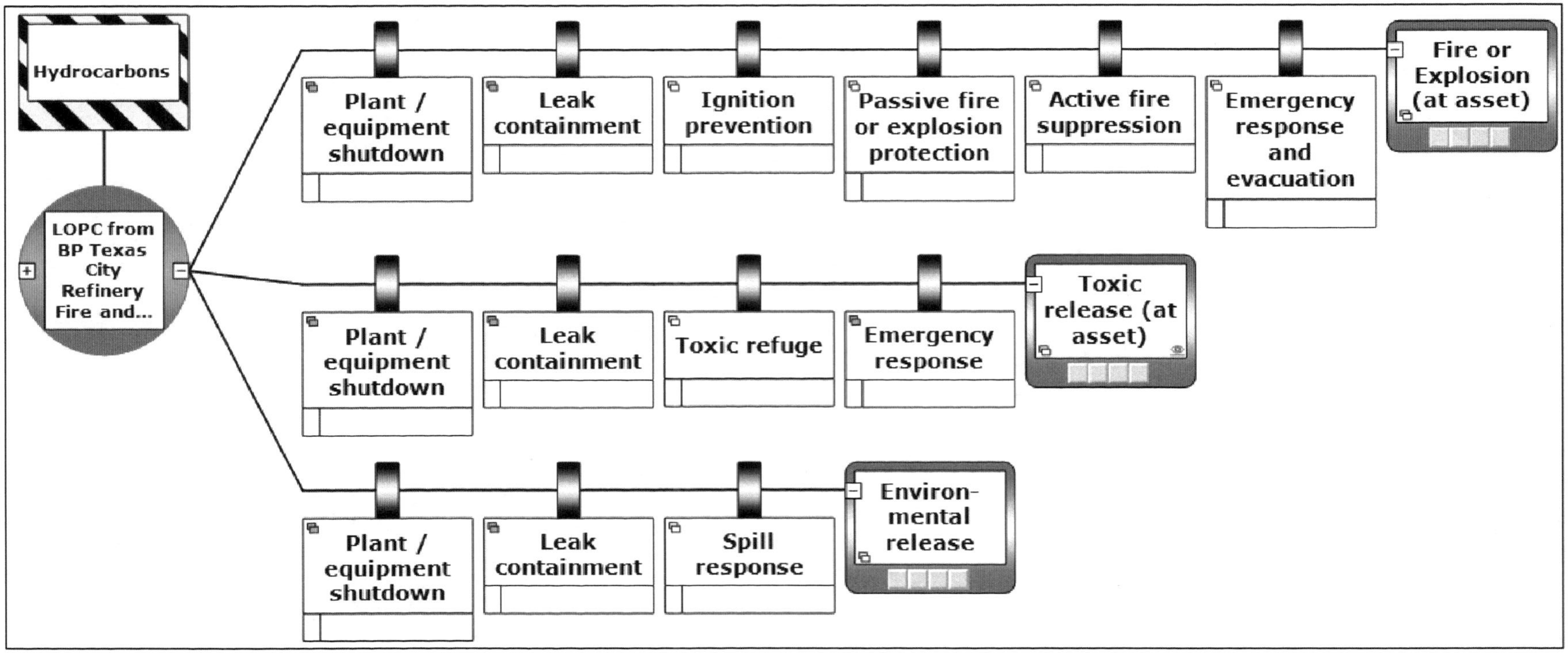

Figure 5. Right Hand Side of Bowtie (BT2) for BP Texas City Refinery Fire and Explosion (2005)

—

2.3 Step Three: Annotation and Tally of Barriers

All of the event-specific bowties (BT2) were annotated to develop Bowtie No. 3 (BT3) to show the summation of barriers that had been compromised in all case studies. "Compromised barriers" are defined as:

- Barriers that were in place but the action associated with the barrier was compromised
- Barrier had failed or degraded or
- Barrier was missing (not implemented)

The annotations were based on the findings of the CSB's investigation results as it related to a barrier on the bowtie diagram, and are at the discretion of the assessor. The annotated version BT3 was created to show the number of events where the barriers were compromised and the number of events where the barriers performed as intended. Scores were applied to the barriers that failed. The scores were the number of times that the barrier was compromised in the case studies reviewed and were determined based on the assessor's acceptability criteria. The criticality was assigned based on the number of times the barrier failed. The criticality level and associated number of barrier failures were selected arbitrarily and can be modified based on the operating company's needs and as acceptability criteria change. The following values were used for the scores in this assessment and are shown in Table 3.

Table 3. Barrier Criticality Levels and Descriptions

Barrier Criticality Level	No. of times Barrier Compromised	Criticality Color
None	0	Green
Low	1	Yellow
Medium	2	Orange
High	3 or More	Red

2.4 Step Four: Analyze and Infer Results

Bowtie BT3 was studied to determine where/when the barrier failure(s) occurred as well as when they succeeded in reducing the severity of the conseqences. The barriers (explicitly mentioned in the CSB reports) that were degraded were classified in categories based on CCPS's 20 Elements of Risk-Based Process Safety. The information on the systemic management issues would be compared amongst all of the events selected to see patterns across industry.

3 Scope

The scope of this review was focused on refinery events that occurred in the U.S. in the last 15 years (from 2001), where the information was available to the public through the CSB reports. The events that were selected were those events whose investigations were completed (not currently ongoing) by the CSB. The investigations that were reviewed are shown in Table 4.

Table 4. Refinery Event Case Studies Reviewed

Year	Event	Geographic Location	Operating Company	Area of the Refinery	Results
2004	Ciniza Oil Refinery Fire and Explosion	Jamestown, NM	Giant Industries	Hydrofluoric Alkylation Unit	6 injuries
2005	BP Texas City Refinery Fire and Explosion	Texas City, TX	BP	Isomerization Unit	15 fatalities 180 injuries
2005	BP Texas City Refinery (Hydrogen) Fire	Texas City, TX	BP	Resid Hydrotreater Unit	1 injury (minor)
2007	Valero - McKee LPG Fire	Sunray, TX	Valero - McKee	Propane Deasphalting Unit	4 injuries
2009	Citgo Refinery Hydrofluoric Acid Release and Fire	Corpus Christi, TX	Citgo	Hydrogen Fluoride Alkylation Unit	2 injuries
2009	Silver Eagle Refinery Flash Fire and Explosion	Woods Cross, UT	Silver Eagle Refining	Mobil Distillate Dewaxing Unit	4 injuries
2010	Tesoro Catastrophic Heat Exchanger Rupture	Anacortes, WA	Tesoro	Catalytic Reformer/ Naphtha Hydrotreater Area	7 fatalities
2012	Chevron Richmond Refinery Fire and Explosion	Richmond, CA	Chevron	Crude Unit #4	6 injuries (minor)

4 Case Studies

4.1 Ciniza Oil Refinery Fire and Explosion (2004)

On April 8, 2004, mechanics at the Giant Industries Ciniza refinery were in the process of taking an isostripper recirculation pump offline in order to repair a leaking seal. The isostripper pumps had a history of repeated seal leaks. The operator attempted to close the inlet plug valve on the suction with a valve wrench. The valve wrench and not the position indicator on the valve was used to determine the position of the valve. In reality, the suction valve was open. The operator had verified that the pressure had been bled off from the pump housing, because flow from the pump housing through a vent hose subsided. It was discovered later that the vent hose was plugged and the pump was not depressured. As the pump casing was being separated, alkylate at conditions of 150 psig and 350 F was released injuring the nearby personnel. Not long afterwards, there were a series of explosions. A total of 6 people were injured [4]. The CSB noted 6 failures during the event. Three of the findings, shown in Table 5, are related directly to barriers that failed. While barriers might have succeeded in reducing the likelihood in this event, there is no indication as to which barriers were especially effective.

Table 5. Compromised Barriers, CSB's Findings and Relevant RBPS Management System Elements for Ciniza Oil Refinery Fire and Explosion [4]

CSB's Findings	Barrier Compromised	Barrier Type	Relevant RBPS Management System Elements
Mechanical Integrity The multiple pump seal failures were addressed as they occurred but the personnel did not identify the root causes of the seal failures in order to prevent them. (Section 6.1, p. 5)	Mechanical Integrity	Preventive	Asset Integrity and Reliability
Corrosion and Scale Formation Giant Industries Management did not investigate why excessive iron fluoride generation caused the mechanical seals to fail repeatedly. Mechanical seals failed immediately due to the erosion from materials that formed from corrosion. (Section 6.2, p. 7)	Mechanical Integrity	Preventive	Asset Integrity and Reliability
	Corrosion Prevention	Preventive	
	Erosion Prevention	Preventive	
Lockout/Tagout and Isolation Mechanics and operators did not properly verify that the pump was isolated or free of liquid before locking and tagging out the pump. (Section 6.6, p. 10)	Isolation of affected sub-systems and intervention	Preventive	Safe Work Practices

4.2 BP Texas City Refinery Fire and Explosion (2005)

On March 23, 2005, there were explosions and subsequent fires at the BP Texas City Refinery that killed 15 people and injured another 180. The event occurred during the startup of an isomerization1 (ISOM) unit when a raffinate splitter tower was overfilled. The pressure relief devices opened resulting in a release of flammable liquid from a blowdown stack that was not equipped with a flare. All of the fatalities occurred in or near office trailers located close to the blowdown drum [5]. The CSB noted 7 technical findings during the event. Three of the findings, shown in Table 6, are related directly to barriers that failed. For the first finding, there were two underlying factors out of seven that were directly related to barriers. While barriers might have succeeded in reducing the likelihood in this event, there is no indication as to which barriers were especially effective.

Table 6. Compromised Barriers, CSB's Findings and Relevant RBPS Management System Elements for BP Texas City Refinery Fire and Explosion [5]

CSB's Findings	Underlying Factors	Barrier Compromised	Barrier Type	Relevant RBPS Management System Elements
Raffinate splitter tower was overfilled, leading to high pressure, which activated relief valves that discharged flammable liquid to the blowdown system. (Section 1.5, pp. 5 - 6)	The tower level indicator failure	Mechanical Integrity	Preventive	Asset Integrity and Reliability
	Outdated and ineffective startup procedures	Startup procedures	Preventive	Operating Procedures
Size of blowdown drum was insufficient to contain the liquids sent to it by the pressure relief valves. Relief system safety study was performed. (Section 1.5, pp. 7)	N/A	Relief Systems	Preventive	Hazard Identification and Risk Analysis
Blowdown drum and stack were not connected to a flare during major modifications which was not in compliance with BP's own safety standards. (Section 1.5, pp. 7)	N/A	Relief Systems	Preventive	Compliance with Standards

4.3 BP Texas City Refinery (Hydrogen) Fire (2005)

On July 28, 2005, the BP Texas City Refinery experienced a catastrophic rupture of an 8" elbow on the outlet of a heat exchanger in the Residual Hydrotreater Unit (RHU) due to High Temperature Hydrogen Attack (HTHA). Highly-flammable hydrogen at more than 600 °F was released from the elbow and ignited which led to a fireball and a fire that burned for two hours. The rupture caused one minor injury of a BP employee who was assisting with the unit shutdown [6].

The RHU at the BP Texas City Refinery contains two heat exchangers in series used to preheat hydrogen gas before it enters a furnace to be heated further and then to a reactor, where hydrogen combines with nitrogen and sulfur compounds within the reactor feedstock to form ammonia and hydrogen sulfide. Three months before the event, a maintenance contractor installed a carbon steel elbow instead of a low-alloy steel elbow which was specified by BP's procedures for hydrogen at temperatures above 450 F. The contractor had replaced the outlet elbow with the elbow intended for the inlet pipe (which was transporting hydrogen below 450 F). The material of construction of the elbow and the failure mode of HTHA were confirmed by post-incident metallurgical analysis. The CSB's findings are shown in Table 7 and are related directly to barriers that failed [6]. While barriers might have succeeded in reducing the likelihood in this event, there is no indication as to which barriers were especially effective.

Table 7. Compromised Barriers, CSB's Findings and Relevant RBPS Management System Elements for Tesoro Catastrophic Heat Exchanger Rupture [6]

CSB's Findings	**Barrier Compromised**	**Barrier Type**	**Relevant RBPS Management System Elements**
The rupture of the elbow was the result of its being made of carbon steel, which was not adequate for the service for which it was used. The reason that the carbon steel was put in place was in part due to BP not requiring the maintenance contractor to use special measures to prevent inadvertent switching of the elbows or positive material identification post-assembly inspection (p. 6).	Equipment or material selection	Preventive	Contractor Management

4.4 Valero - McKee LPG Fire (2007)

On February 16, 2007, liquid propane under high pressure was released in the Propane De-Asphalting (PDA) unit of Valero's McKee Refinery near the town of Sunray, Texas. The resulting propane vapor cloud was ignited, leading to a subsequent fire that injured workers, damaged unit piping and equipment, and collapsed a major pipe rack [7]. The fire grew rapidly and threatened surrounding units, including a Liquefied Petroleum Gas (LPG) storage area. Four workers were seriously burned, including a contractor. The release occurred in a control station that had not been isolated from the process. As a result, water in the propane accumulated in low points that were in the control station. During a period when the ambient temperature was below freezing, the water froze, expanded and cracked the pipe. The ice sealed the crack until the temperatures increased, at which time the ice melted, the opening in the pipe was open to the atmosphere and propane was released. The fire led to collapse of a nearby pipe rack and release of chlorine from nearby storage cylinders exposed to radiant heat. Plant personnel were not ordered to evacuate until 15 minutes after the fire erupted [7]. The CSB noted 12 technical findings during the event. Three of the findings, shown in Table 8, are related directly to barriers that failed. While barriers might have succeeded in reducing the likelihood in this event, there is no indication as to which barriers were especially effective.

Table 8. Compromised Barriers, CSB's Findings and Relevant RBPS Management System Elements for Valero - McKee LPG Fire [7]

CSB's Findings	Barrier Compromised	Barrier Type	Relevant RBPS Management System Elements
Freeze protection practices did not ensure that process units were systematically reviewed to identify and mitigate freezing hazards. (p. 7)	Equipment and Pipe Protection	Preventive	Hazard Identification and Risk Analysis
The field operators were not able to access manual isolation valves and local pump controls to isolate high-pressure propane source because they were in close proximity to the fire. No remotely operable shut-off valves (ROSOV) were installed in the PDA to enable control room operators to isolate the leak which was in direct violation of Valero's standards (p. 7).	Plant/Equipment Shutdown	Mitigating	Compliance with Standards
Pipe rack was installed on a non-fireproofed structural support. Flame impingement led to the collapse of the pipe rack. (p. 8)	Structural integrity	Preventive	Hazard Identification and Risk Analysis

4.5 Citgo Refinery Hydrofluoric Acid Release and Fire (2009)

On July 19, 2009, Citgo's Corpus Christi East Refinery experienced a release of hydrocarbons from a break at two threaded connections as a result of the violent shaking of the process recycle pipe. This occurred due to blockage in flow from a control valve failure. The manual bypass around the control valve was not accessible after the release. The hydrocarbons were ignited by equipment in the neighboring unit. The fire caused subsequent equipment failures and releases of hydrogen fluoride (HF) in the Alkylation Unit. As a result, two employees were injured. While the refinery had an HF water mitigation system, 30 pounds out of the 51 pounds released were not captured by the mitigation system. Additionally, Citgo nearly used up its stored water supply for the fire suppression and HF water mitigation systems. Before the water supply was completely exhausted, the Refinery Terminal Fire Company began pumping salt water from the Corpus Christi ship channel into the refinery's water system using a barge, specially equipped for firefighting. However, multiple failures were noted during the salt water transfer. The CSB found that Citgo had never conducted safety audits of its HF alkylation units, which was not in accordance with best practices found in American Petroleum Institute's Recommended Practices (API RP). According to API RP 751, *Safe Operation of Hydrofluoric Acid Alkylation Units*, refineries should audit their HF operations once every 3 years. The CSB's findings, shown in Table 9, are related directly to barriers that failed. While barriers might have succeeded in reducing the likelihood in this event, there is no indication as to which barriers were especially effective. While there were HF mitigation and fire suppression systems in place, they were still considered to be compromised since there was a fire which led to the HF release [8].

Table 9. Compromised Barriers, CSB's Findings and Relevant RBPS Management System Elements for Citgo Refinery Hydrofluoric Acid Release and Fire [8]

CSB's Findings	Barrier Compromised	Barrier Type	Relevant RBPS Management System Elements
HF Water Mitigation System was assumed to be available to minimize the consequences of an HF release per the Alkylation Unit's PHA when in reality it was not fully available. Additionally, no safety audits were performed on the HF Water Mitigation System in accordance with best practices (API RP 751).	Leak Containment	Mitigative	Hazard Identification and Risk Analysis
			Compliance with Standards

4.65 Silver Eagle Refinery Flash Fire and Explosion (2009)

On November 4, 2009, the Silver Eagle Refinery in Woods Cross, Utah, experienced a catastrophic pipe rupture in a 10" inch pipe at the bottom of a reactor in the Mobil Distillate Dewaxing Unit. The ruptured pipe released hydrogen which caught fire immediately and exploded sending a blast wave that damaged 100 homes in the surrounding vicinity. According to a metallurgical failure study and analysis performed for the CSB, the line ruptured as a result of sulfidation corrosion which caused the pipe walls to thin over time. Four employees suffered minor injuries during the event [9]. The CSB's findings, shown in Table 10, are related directly to barriers that failed. While barriers might have succeeded in reducing the likelihood in this event, there is no indication as to which barriers were especially effective.

Table 10. Compromised Barriers, CSB's Findings and Relevant RBPS Management System Elements for Silver Eagle Refinery Fire and Explosion [9]

CSB's Findings	Barrier Compromised	Barrier Type	Relevant RBPS Management System Elements
Third party investigation performed for the CSB identified accelerated corrosion in the component that failed. No corrosion inspections were performed.	Mechanical Integrity	Preventive	Asset Integrity and Reliablity
An inherently safer, more corrosion resistant metallurgy was not implemented more broadly in crude unit high temperature service.	Material or equipment selection	Preventive	Hazard Identification and Risk Analysis

4.7 Tesoro Catastrophic Heat Exchanger Rupture (2010)

On April 2, 2010, the Tesoro Anacortes Refinery experienced a catastrophic rupture of a heat exchanger in the Catalytic Reformer / Naphtha Hydrotreater (NHT) unit due to High Temperature Hydrogen Attack (HTHA). Highly-flammable hydrogen and naphtha at more than 500 °F were released from the ruptured heat exchanger and ignited which led to an explosion and an intense fire that burned for more than three hours. The rupture killed seven Tesoro employees who were working near the heat exchanger at the time of the event.

The NHT unit at the Tesoro Anacortes Refinery contained two parallel groups, or banks, of three heat exchangers used to preheat process fluid before it entered a reactor, where impurities were treated for subsequent removal. At the time of the event, the Tesoro workers were in the process of putting one group of heat exchangers in service. The heat exchanger that ruptured was in the other group of heat exchangers. The refinery had a history of leaks and fires during this operation [10]. The CSB noted eight key technical findings and 11 organizational findings during the event. One of the technical findings and two of the organizational findings, shown in Table 11, are related directly to barriers that failed. While barriers might have succeeded in reducing the likelihood in this event, there is no indication as to which barriers were especially effective.

Table 11. Compromised Barriers, CSB's Findings and Relevant RBPS Management System Elements for Tesoro Catastrophic Heat Exchanger Rupture [10]

CSB's Findings	Barrier Compromised	Barrier Type	Relevant RBPS Management System Elements
The rupture of the heat exchanger was the result of its being made of carbon steel, which was not adequate for the service for which it was used. (Section 1.2, pp. 3 - 4)	Equipment or material selection	Preventive	Hazard Identification and Risk Analysis
Successful identification of HTHA is highly dependent on the specific techniques employed and the skill of the inspector. While early inspections could have indicated that there was HTHA in the early phases, it was not sufficiently reliable to ensure mechanical integrity and prevent HTHA equipment damage. (Section 1.2.2, p. 6)	Mechanical Integrity	Preventive	Asset Integrity and Reliability
Tesoro procedures did not prohibit or effectively limit the use of additional personnel during the nonroutine hazardous startup of the NHT heat exchangers. (Section 1.2.2, p. 6)	Operating Procedures	Preventive	Operating Procedures

4.8 Chevron Richmond Refinery Fire and Explosion (2012)

On August 6, 2012, the Chevron Refinery in Richmond, California, experienced a catastrophic pipe rupture in one of the crude units. The event occurred from piping referred to as the "4-sidecut" stream, which is one of several process streams exiting the refinery's Crude Unit Atmospheric Column [11]. At the time of the event, light gas oil was flowing through the 8-inch line. The ruptured pipe released flammable, high temperature light gas oil, which then partially vaporized into a large vapor cloud. Approximately two minutes after the process fluid was released, it ignited. The line ruptured as a result of sulfidation corrosion which caused the pipe walls to thin over time. Six employees suffered minor injuries during the event and subsequent emergency response efforts [11]. Four of the CSB's findings, shown in Table 12, are related directly to barriers that failed. While barriers might have succeeded in reducing the likelihood in this event, there is no indication as to which barriers were especially effective.

Table 12. Compromised Barriers, CSB's Findings and Relevant RBPS Management System Elements for Chevron Richmond Refinery Fire and Explosion [11]

CSB's Findings	Barrier Compromised	Barrier Type	Relevant RBPS Management System Elements
Chevron identified accelerated corrosion in the 52-inch component in a 2002 inspection. Yet, no Condition Monitoring Location (CML) was added to ensure future monitoring. The 52-inch component was not inspected again. (p.21)	Mechanical Integrity	Preventive	Asset Integrity and Reliablity
The Chevron Energy Technology Company (ETC) 100 percent component inspection recommendation for high-risk piping systems, which was established in 2009, was not implemented at Richmond. (p. 26)			
An inherently safer, more corrosion resistant metallurgy was not implemented more broadly in crude unit high temperature service as a result of an incent that occurred at the Richmond Refinery in 2007. (p. 28)	Material or equipment selection	Preventive	Hazard Identification and Risk Analysis
In 1991, the 4-sidecut equipment used for hydrogen sulfide and reduced sulfidation corrosion was taken out of service. No management of change was implemented to assess the impact of its removal. (p. 33)			Management of Change
The CSB found that in addition to the sulfur content increase, the 4-sidecut draw temperature increased from 625 °F in 1992 to 680 °F in 2002 which appeared to increase the corrosion rate in the carbon steel pipe on 4-sidecut. (pp. 34 - 36)			

5 Results

The barriers for the events were tallied and shown in an annotated form in BT3. BT3 illustrates how a bowtie can be used to annotate barrier failure across multiple events. Scores were applied to the barriers that failed (as given in Table 3 in Section 2.3) to present barrier criticality levels shown in the bowtie below in Figure 6, Figure 7 and Figure 8 . The scores were the number of times that the barrier was compromised in the case studies reviewed and were determined based on the assessor's acceptability criteria. The criticality was assigned based on the number of times the barrier failed.

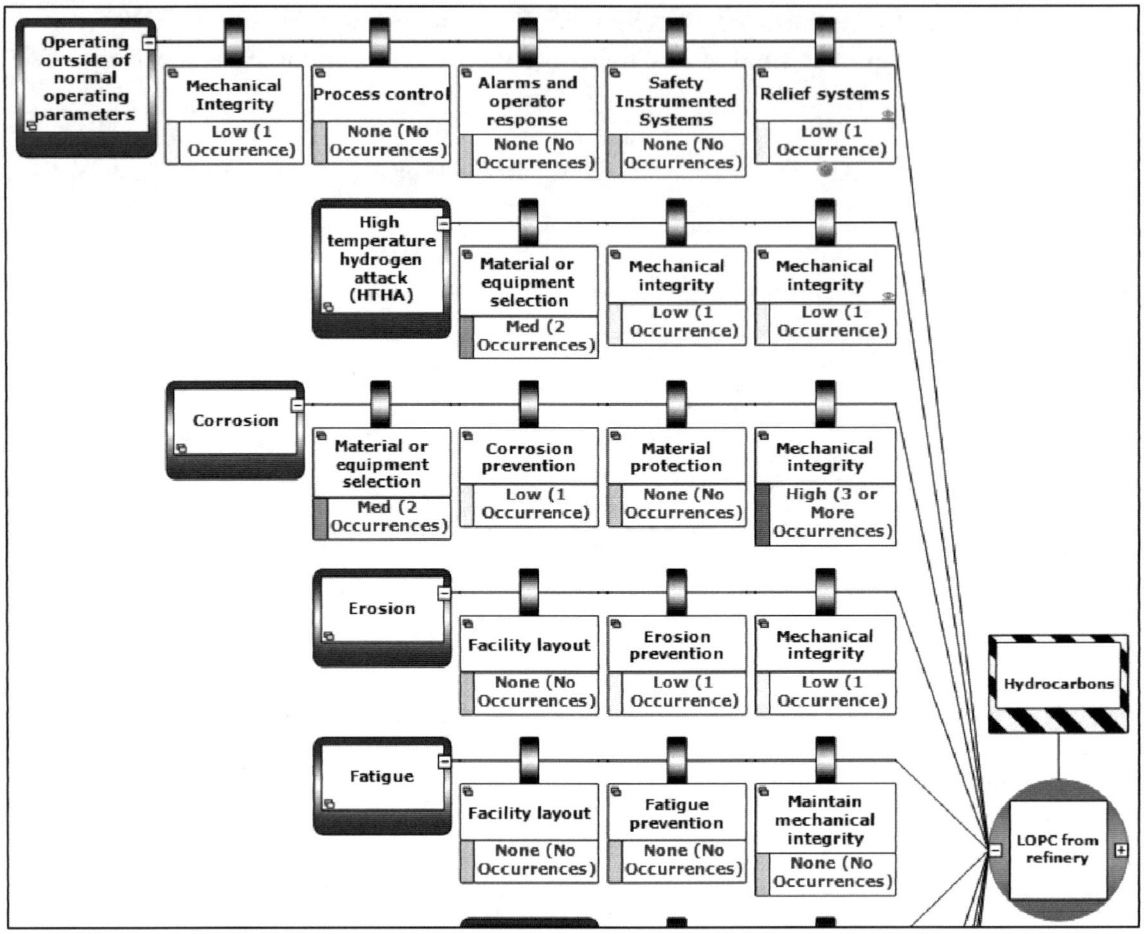

Figure 6. Upper Left Hand Side of Annotated Bowtie (BT3) for Compromised Barriers Based on 8 CSB Incident Investigation Reports

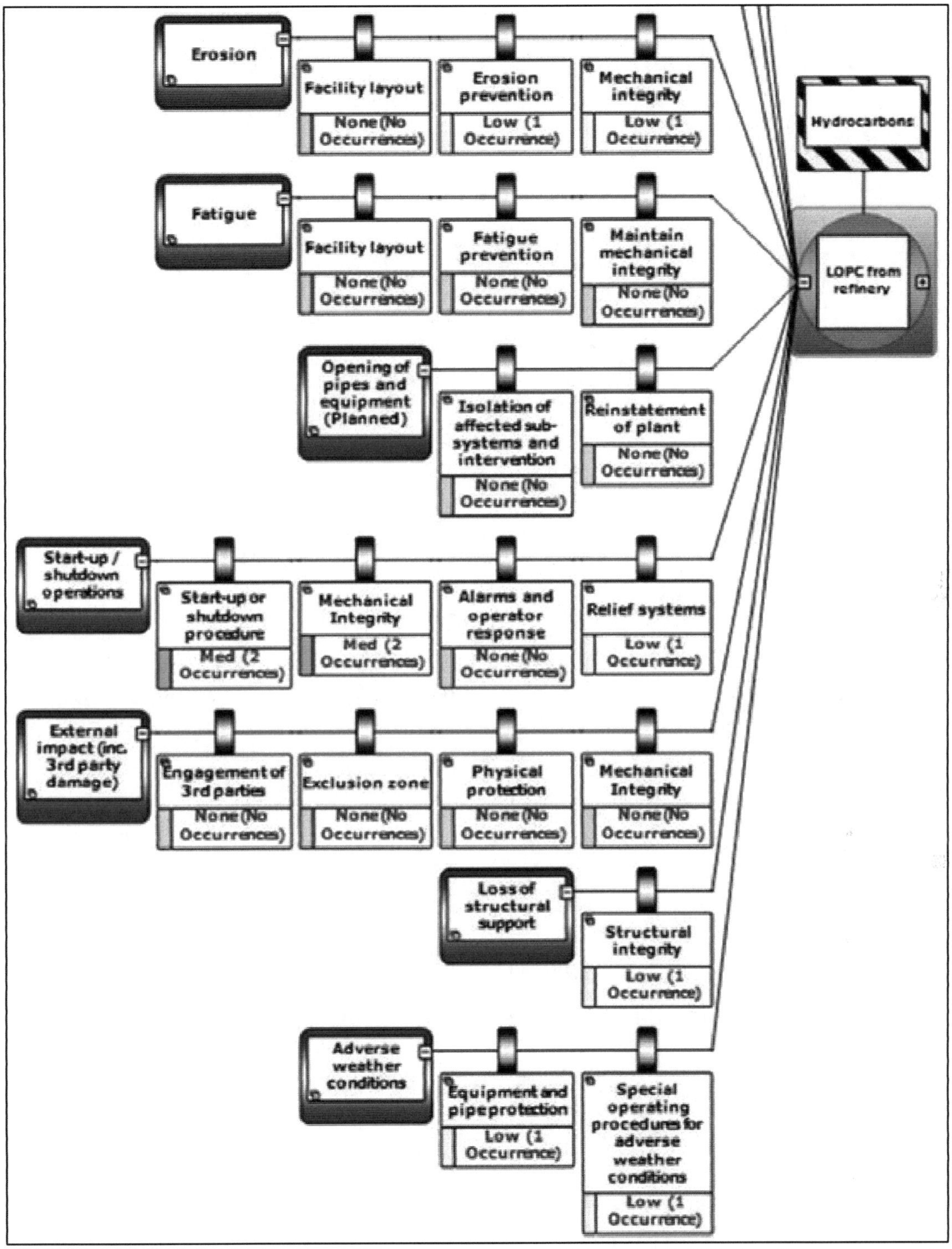

Figure 7. Bottom Left Hand Side of Annotated Bowtie (BT3) for Compromised Barriers Based on 8 CSB Incident Investigation Reports)

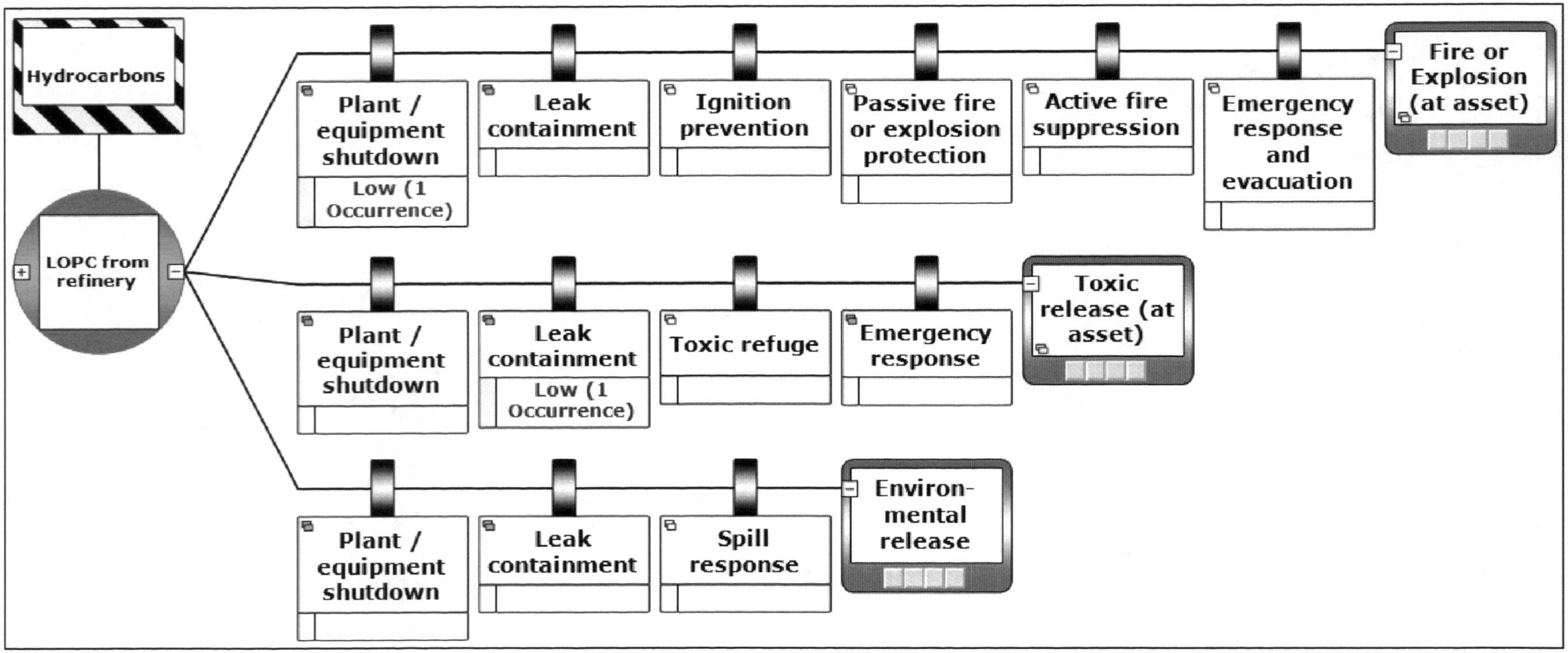

Figure 8. Right Hand Side of Annotated Bowtie (BT3) for Compromised Barriers Based on 8 CSB Incident Investigation Reports

—

The results of the annotation for compromised barriers for the event-specific bowties (based on BT2 Bowties) are shown in Table 13. Only the barriers that were compromised are shown in the table. They are included specifically with the cause or consequence they addressed. The barriers that failed the most often were associated with startup procedures and corrosion causes, and the specific barriers compromised include startup or shutdown procedures and mechanical integrity.

Table 13. Compromised Barriers

Barrier Compromised	Related Cause or Consequence	Number of Times Barrier Compromised
Mechanical integrity	Corrosion	3
Startup or Shutdown Procedure	Start-up or shutdown operations	2
Mechanical integrity	Start-up or shutdown operations	2
Material or equipment selection	High temperature hydrogen attack (HTHA)	2
Mechanical integrity	Operating outside of normal operating parameters	1
Mechanical integrity	High temperature hydrogen attack (HTHA)	1
Mechanical Integrity	Erosion	1
Relief systems	Operating outside of normal operating parameters	1
Relief systems	Start-up or shutdown operations	1
Material or equipment selection	Corrosion	1
Corrosion prevention	Corrosion	1
Erosion prevention	Corrosion	1
Plant shutdown	Fire or explosion	1
Leak containment	Toxic release	1

Figure 9 shows which types of barriers failed when considering the bowties from an overall, higher level perspective. Most of the barriers that failed were associated with Mechanical Integrity (45%), with Material or Equipment Selection (15%), with Startup or Shutdown Procedures (10%), Relief Systems (10%) and other barrier failures all represented equally in the limited data analyzed.

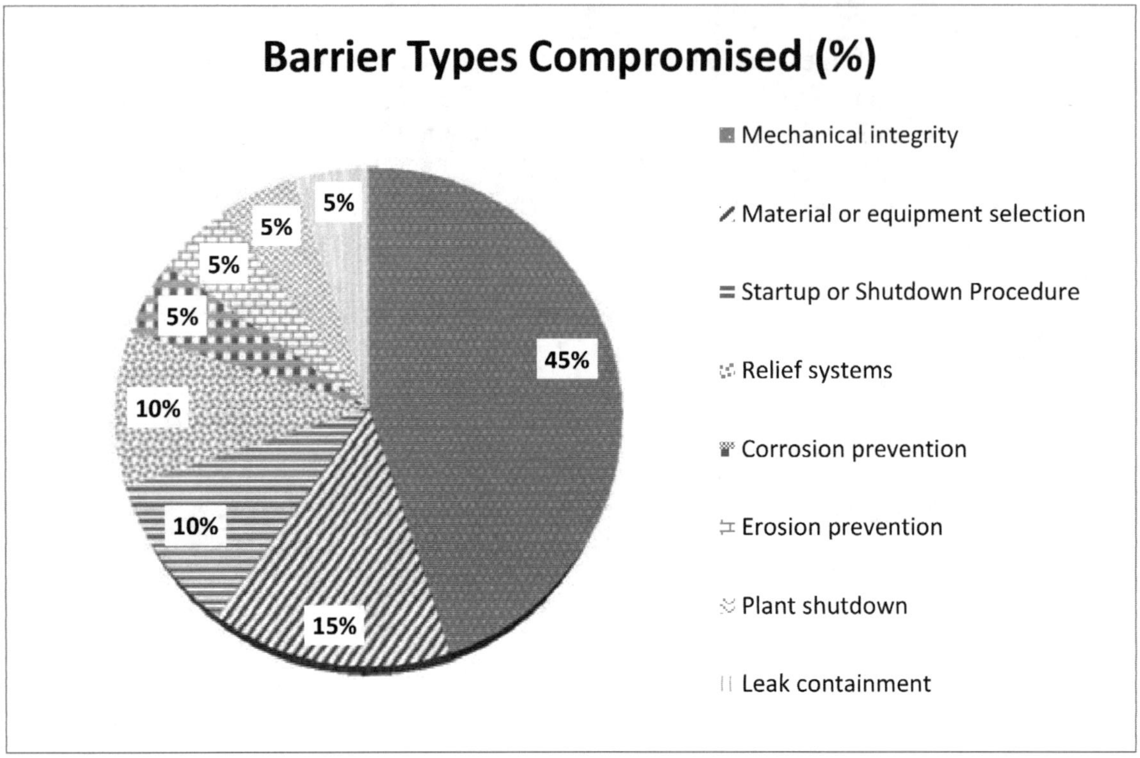

Figure 9. Distribution of Barriers Compromised

The general findings of all 8 case studies related to technical or applicable organizational gaps along with immediate root causes were compared to determine if there were patterns or themes associated with deficiencies. The deficiencies were identifed and characterized based on the Risk-Based Process Safety (RBPS) Elements that the authors determined to be most fitting for that deficiency. The deficiencies were then annotated and tabulated inTable 14.

Table 14. Top Deficient RBPS Management System Elements as Found in 8 Case Studies

Relevant RBPS Management System Elements	No. of Times Observed in Case Studies	Ciniza Oil Refinery Fire and Explosion (2004)	BP Texas City Refinery Fire and Explosion (2005)	BP Texas City Refinery (Hydrogen) Fire (2005)	Valero - McKee LPG Fire (2007)	Citgo Refinery Hydrofluoric Acid Release and Fire (2009)	Silver Eagle Refinery Flash Fire and Explosion (2009)	Tesoro Catastrophic Heat Exchanger Rupture (2010)	Chevron Richmond Refinery Fire and Explosion (2012)
Hazard Identification and Risk Analysis	6		X		X	X	X	X	X
Asset Integrity and Reliability	5	X			X		X	X	X
Compliance with Standards	3		X		X	X			
Operating Procedures	2		X					X	
Safe Work Practices	1	X							
Management of Change	1								X
Contractor Management	1			X					

—

In five out of the eight case studies, deficiencies associated with *Hazard Identification and Risk Analysis* were identified and for four out of the eight cases, *Asset Integrity and Reliability* were the second-highest deficient areas. Based on the findings, it is evident that *Hazard Identification and Risk Analysis* and *Asset Integrity and Reliability* could be improved in refineries. The other common deficiencies associated with the findings and root causes were associated with *Compliance with Standards* and *Operating Procedures*. *Compliance with Standards* failures were related to the standards developed by the operating company not being followed. There were a few additional findings related to *Safe Work Practices*, *Management of Change* and *Contractor Management*. The findings presented in this paper are very similar to the results delivered in a presentation by OSHA for "Refinery NEP Frequently Cited PSM Elements" as of 2011. These elements, shown in Table 15, list in order from highest to lowest the number of citations [12].

Table 15. Ranking of Violated PSM Elements in 2011 [12]

Ranking from highest to lowest number of Citations	PSM Element
1	Mechanical Integrity
2	Process Safety Information
3	Operating Procedures
4	Management of Change
5	Incident Investigation
6	Contractors
7	Compliance Audits
8	Training
9	Emergency Planning and Response
10	Employee Participation

BT3 can be further reviewed to determine patterns with common causes. From the cases reviewed, it appears that the barriers that were compromised were associated with Corrosion and Startup or Shutdown operations as shown in Table 16. This could help an operating company see patterns across different units or facilities.

Table 16. Causes of Incidents

Related Causes	Number of Times Cause Was Associated with Compromised Barrier
Corrosion	6
Start-up or shutdown operations	5
High temperature hydrogen attack (HTHA)	3
Operating outside of normal operating parameters	2
Erosion	1

6 Findings

The application of the barrier methodology that was used for a nuclear power facility illustrated the following in the context of 8 refinery events reviewed:

- It helped identify the common factors contributing to events reported over a period of time. Based on the specific events that were reviewed, it showed that *Hazard Identification and Risk Analysis* and *Asset Integrity and Reliability* were compromised more than other management system elements.
- The methodology helped identify priorities for strengthening or adding barriers. In all of the cases reviewed, it was evident that the refineries that were studied needed to address mechanical integrity-related and startup procedure-related barriers.
- The methodology could also be further reviewed to determine patterns with common causes. From the cases reviewed, it appears that the barriers that were compromised the most were associated with Corrosion and Startup or Shutdown operations.

There are other ways the methodology can be used that were not covered in this paper. When information is available on barriers that succeeded in halting the event or reducing the consequences of the event, the bowtie can be used to show patterns for effectiveness. The methodology can be applied more specifically to each unit in the refinery (such as fluid catalytic cracker, catalytic reformer, amine treatment, etc.), which could be useful in helping operating companies to compare the barriers being compromised between units within refineries and even for refineries at different geographic locations. This approach utilizing bowties/a barrier methodology could also potentially be insightful in seeing patterns in management system gaps.

7 References

[1] Lewis, Steve and Kris Smith. Lessons Learned from Real World Application of the Bow-tie Method. *Global Congress on Process Safety,* March 2010, San Antonio, Texas. Unpublished conference paper,. http://risktecsolutions.co.uk/ (accessed on March 17, 2016).

[2] Process Safety Management ofHhighly Hazardous Chemicals, 29 C.F.R §1910.119, Subpart H, 2013. Washington, DC: US Federal Government, http://www.ecfr.gov.

[3] Center for Chemical Process Safety of the American Institute of Chemical Engineers. *Guidelines for Risk Based Process Safety.* New York: Wiley, 2007. Print.

[4] Case Study: Giant Industries' Ciniza Oil Refinery. No. 2004-08-I-NM, US Chemical Safety and Hazard Investigation Board. Washington, DC, USA (October 2005), http://www.csb.gov/ (accessed February 14, 2016).

[5] Refinery Explosion and Fire Investigation Report. No. 2005-04-I-TX, US Chemical Safety and Hazard Investigation Board. Washington, DC, USA (March 2007), http://www.csb.gov/ (accessed February 14, 2016).

[6] Safety Bulletin: BP Texas City Refinery Fire Positive Material Verification: Prevent Errors During Alloy Steel Systems Management. No. 2005-04-B, US Chemical Safety and Hazard Investigation Board. Washington, DC, USA (October 12, 2006), http://www.csb.gov/ (accessed March 17, 2016).

[7] LPG Fire at Valero - McKee Refinery Investigation Report. No. 2007-05-I-TX, US Chemical Safety and Hazard Investigation Board. Washington, DC, USA (July 2008), http://www.csb.gov/ (accessed February 14, 2016).

[8] Urgent Recommendations (for the CITGO Refinery Hydrofluoric Acid Release and Fire): US Chemical Safety and Hazard Investigation Board. Washington, DC, USA (December 9, 2009), http://www.csb.gov/ (accessed March 17, 2016).

[9] "CSB Releases Analysis Showing Cause of Rupture and Hydrogen Blast in 2009 Silver Eagle Refinery Accident in Woods Cross; Pipe Walls Thinned Due to Corrosion That Went Uninspected for Years." US Chemical Safety and Hazard Investigation Board. Washington, DC, USA (April 10, 2014), http://www.csb.gov/ (accessed March 17, 2016).

[10] Catastrophic Rupture of Heat Exchanger Investigation Report. No. 2010-08-I-TX, US Chemical Safety and Hazard Investigation Board. Washington, DC, USA (May 2014), http://www.csb.gov/ (accessed February 14, 2016).

[11] Chevron Richmond Refinery Fire Interim Investigation Report. US Chemical Safety and Hazard Investigation Board. Washington, DC, USA (April 2013), Internet. http://www.csb.gov/(accessed February 17, 2016).

[12] OSHA's Refinery & Chemical National Emphasis Programs. US Chemical Safety and Hazard Investigation Board. Washington, DC, USA (March 2012), Internet. http://www.ppsa.org/ (accessed March 18, 2016).

442312 Assessing the Risk from Helicopter Operation at Offshore North WEST JAVA FIELD

Abstract:

PHE ONWJ as the operator of Offshore North West Java field uses helicopter for medical evacuation, project requirements, and management visit. Due to operational requirements, the helicopter usage in PHE ONWJ increases significantly; at present it is almost doubled compared to that in 2010. PHE ONWJ acknowledged the inherent risks associated with helicopter operations, and has put effort to understand these risks and to manage them better.

To kick-off this initiative, a comprehensive helicopter risk assessment study which covers two risk aspects of the helicopter operation has been done to assess:
- risk to helicopter passengers, and
- risk to the installations, including personnel onboard.

Correspondingly, two separate studies have been carried out: helicopter transportation risk study to analyze the risk to helicopter passengers, and helicopter impact risk study with focus on the risk to personnel and facilities onboard the installations. The study was conducted utilizing a well-known methodology and data published by OGP and DNV GL combined with brainstorming session with Event Tree Analysis (ETA) being used as the main tool to identify and analyze all foreseeable accidental scenarios, likelihoods and consequences. The result of individual risk was calculated as IRPA while societal risk was presented in F-N Curve. These risk parameters were subsequently evaluated against relevant PHE ONWJ's risk criteria.

The study shows that risks to helicopter passengers and installations are within tolerable levels. It is also found out that the risk from helicopter impact contributes about 1-2% of the overall risk at offshore installations, depending on the actual number of flights to/from each installation and its lay out. It is concluded that the helicopter flights management and the offshore installations conditions are suitable for PHE ONWJ to sustain the current operational requirements.

441376 Passive Fire Protection (PFP) Optimization for Oil and Gas Facilities

Abstract:

The structural integrity assessment of offshore topside structures focuses on protecting pressurized hydrocarbon piping and vessels from damage and fracture, and providing safe escaping routes for the personnel for a specific amount of time. In order to achieve these goals, it is critical to apply a sufficient quantity of Passive Fire Protection (PFP) on the topside structure steel members. On the other hand, excessive use of PFP on the structure results in considerable additional cost and extra added dead load to the structure. The main objective of this paper is to demonstrate that by employing finite element analysis (FEA) which accounts for material and geometric nonlinearities as well as load re-distribution, PFP application can be optimized.

Simplified and conservative approaches are available to estimate the extent and amount of the PFP on the offshore structures. However, the main concern with simplified approaches is that they can lead to over-application of the PFP resulting in substantial increase in the topsides weight. With the use of advanced engineering analysis and knowledge of critical load paths, an optimized PFP scheme can be developed that would maintain the structural integrity of the structure and also provide sufficient escape time for personnel during fire. Ductility level analysis will be used to demonstrate that by using proper heat transfer simulation and FEA, amount of PFP application can be optimized. Detailed analysis will be carried out using general-purpose finite element software package. Temperature dependent thermal and mechanical material properties will be taken into account in the analysis. Methodology and analysis of the results of PFP optimization will be discussed in this paper.

442387 Particle Size Reduction on the 20L Sphere, a Computational Approach

Abstract:

Dust explosions might occur when a combustible dust is dispersed near an ignition source in a confined facility. This means that this kind of event is a main concern for any industry that handle or store metallic or organic particle material as part of their process. Nowadays, the 20L sphere has become the standard device for characterizing parameters associated with flammable dust explosions like the maximum pressure (Pmax), the maximum rate of pressure rise (dP/dt max), the minimum explosive concentration (MEC), and the deflagration index (Kst). The 20L sphere contains an explosion chamber, a dispersion system, an ignition system pressure measurement and an automatic control system. For several years, there have been generated some questions about the conditions in which the standard test is performed in order evaluate some parameters. One of those questions is associated with a reduction of the particle size due to fragmentation phenomenon in the 20 L sphere. When this type of phenomenon occur, a mischaracterization of the flammable dust takes place, as the Pmax and dP/dt max tends to radically increase with particle size reduction. Also, depending of the cohesive characteristics of each material, agglomeration phenomenon may also occur, increasing the particle size. According to the above, this study aims to assess the agglomeration and fragmentation phenomenon during the 20L standard test. For this purpose a CFD simulation was developed, where the particle interaction was modeled using DEM, and cohesion models were implemented to represent agglomeration and fragmentation of particles. Four different time frames were defined for the study: an initial time (1 ms), high turbulence time (20 ms), the explosion time according to ASTM standard (60 ms) and a time of low turbulence (100 ms). Also, for each time five different areas were stablished to evaluate and quantify the fragmentation and agglomeration. Each area was defined according to the particles flow around the sphere, which was study in a previous work. An analysis of the 20L sphere standard test was presented. The phenomenon of fragmentation and agglomeration in dust explosions have an important effect on the severity of the explosion because it affects particle size distribution. This implies that values such as the MEC and Kst, calculated during the test, are related to different conditions than the ones needed. The results allowed to determine the degree of agglomeration and fragmentation in the different areas. It was obtained that the areas with high turbulence presented a higher degree of fragmentation than the areas with low turbulence.

12TH GLOBAL CONGRESS ON PROCESS SAFETY

LPS 50th LOSS PREVENTION SYMPOSIUM

Leveraging Data from a Quantitative Risk Assessment to Determine Blast and Fire Risk to Safety Critical Equipment in the Chemical and Refining Industries

Phillip R. Hodge
Baker Engineering and Risk Consultants, Inc.
11011 Richmond Ave. Suite 700
Houston, TX 77042
PHodge@BakerRisk.com

Thomas Anderson
Baker Engineering and Risk Consultants
3330 Oakwell Court, Suite 100
San Antonio, TX 78218
TAndererson@BakerRisk.com

David Black
Baker Engineering and Risk Consultants
11011 Richmond Avenue, Suite 700
Houston, TX 77042
DBlack@BakerRisk.com

Murtaza Gandhi
Baker Engineering and Risk Consultants
11011 Richmond Avenue, Suite 700
Houston, TX 77042
MGandhi@BakerRisk.com

Anay Raibagkar
BakerRisk Canada
5575 North Service Road, Suite 103
Burlington, Ontario, Canada L7L 6M1
ARaibagkar@BakerRisk.com

Prepared for Presentation at
American Institute of Chemical Engineers
2016 Spring Meeting
12th Global Congress on Process Safety
Houston, Texas
April 11-23, 2016

Leveraging Data from a Quantitative Risk Assessment to Determine Blast and Fire Risk to Safety Critical Equipment in the Chemical and Refining Industries

Phillip R. Hodge
Baker Engineering and Risk Consultants, Inc.
11011 Richmond Ave. Suite 700
Houston, TX 77042
PHodge@BakerRisk.com

T. Anderson, M. Gandhi, D. Black, A. Raibagkar

Keywords: Facility Siting Study, Quantitative Risk Analysis, Safety Critical Equipment, Blast, Fire, Consequence Modeling

Abstract

In the event of a loss of containment, refineries and chemical plants depend on safety critical equipment to isolate the release, prevent escalation, and respond to potential injuries. Even when considering potential catastrophic releases, these systems are often assumed to be available. However, safety critical equipment, including the power and electronics required to activate these systems, may be damaged by fires and explosions; therefore, proper analysis is necessary to determine the availability of safety critical systems in the aftermath of an event. Thankfully, the information to make this determination is readily available in a plant's facility siting study or quantitative risk assessment. While current facility siting methodologies often focus chiefly on loss of life, they can be readily modified to include the risk to safety critical equipment.

This paper outlines a method of screening safety critical equipment in the chemical and refining industries for blast and thermal vulnerability and risk using data available in a facility siting study or quantitative risk assessment. The methodology addresses the specific failure modes and vulnerabilities of safety systems and develops a case study for a chemical facility. The case study identifies safety critical equipment warranting further study or requiring damage mitigation, and outlines solutions to improve the availability of these systems in the event of a fire or explosion in the facility.

1 Introduction

Quantitative Risk Assessments (QRAs) used for facility siting are typically complex and robust studies. QRAs represent a significant time investment for a facility because of the large quantity of input data and detailed modeling required. However, facilities can look to maximize the value from their QRAs by expanding upon the typical outputs from the study. For instance, a QRA typically focuses only on loss of life, but by altering the vulnerability modeling in the QRA, the study can look at other types of losses with minimal incremental investment.

A key candidate for this extension is Safety Critical Equipment (SCE). SCE can refer to a wide range of items within a plant such as deluge systems, pressure relief devices, etc. Often these systems depend on elements not typically classified as safety critical, for example motor controls, electrical transformers or switchgears, generators, firewater tanks, cable trays, and pipe racks. These safety systems are often designed to mitigate consequences and prevent escalation. As such, it is important that these protective layers are reliable, available, and operate as intended. Often the assumption is made that this equipment is available to safely shut down a unit to prevent escalation after an event. A modified QRA can help to validate this assumption and improve the availability of SCE after an event.

This paper presents a methodology for leveraging a QRA to analyze the risk to indoor SCE in a process facility. The methodology addresses the specific failure modes and vulnerabilities of safety systems and develops a case study for a chemical facility. The case study identifies safety critical equipment warranting further study or requiring damage mitigation, and outlines solutions to improve the availability of these systems in the event of a fire or explosion in the facility.

2 Methodology

This section outlines the methodology used to determine the risk to SCE in a process facility and assumes that information from a comprehensive QRA or similar study is available. The analysis and case studies presented in this paper utilize Baker Engineering and Risk Consultants, Inc.'s (BakerRisk®) proprietary dispersion, fire, and blast modelling software, SafeSite$_{3G}^{©}$, to model thousands of fire and blast scenarios. Other methods of conducting the hazard analysis and QRA are available [1], but, whatever method is used, it is important to consider the range of possible conditions (magnitude, duration, wind direction, weather conditions, etc.) that would impact each building housing SCE. Figure 1 below shows a flowchart of a typical QRA methodology. This paper focuses on the vulnerability portion of the assessment.

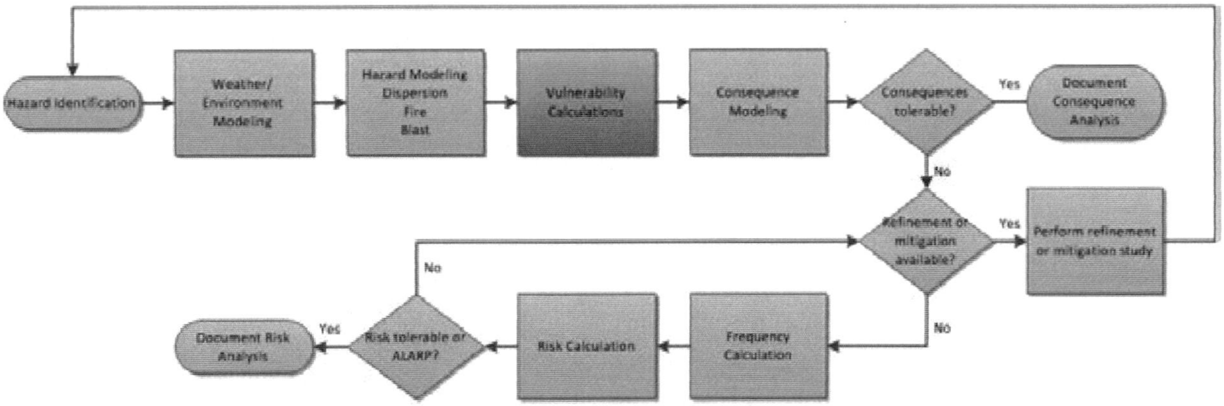

Figure 1. Flowchart of a Typical QRA Methodology

Blast Equipment Vulnerability

Blast equipment vulnerability (EV) represents the likelihood that a piece of equipment within a building will sustain damage to the point of losing functionality as a result of overall building damage experienced during an explosion. Blast EV is dependent on the predicted building damage level (BDL), the location of the equipment within the building floor plan, the sensitivity of the equipment to sudden movement or impact, and the equipment support conditions. The BDL is dependent on the construction of the building and the pressure and impulse of the blast wave hitting the building. Various methods exist for calculating the BDL. The case studies utilize the method described by Baker [2].

The support conditions for the equipment are influenced by the age of the supports, the potential for slippage, and the strength of the attachment. The primary factor, however, is the location of the attachment. Equipment is, in general, sensitive to sudden movement and excessive vibrations. Surface (wall or ceiling) mounted equipment would be subjected to the response of the surface to which they are mounted. Therefore, it is anticipated that the EV values of surface-mounted equipment would be significantly greater than that of non-surface-mounted equipment for the same predicted BDL. Based on equipment mounting, Figure 2 and Figure 3 show a building with high EV and low EV, respectively. Table 1 describes the five BDLs and the potential equipment damage associated with each BDL. Based on evidence collected through numerous industrial accident investigations, BakerRisk® has determined that there is a significant increase in potential EV for surfaced-mounted equipment. As such, EV values have been developed for both types of mounting and are reported in Table 2.

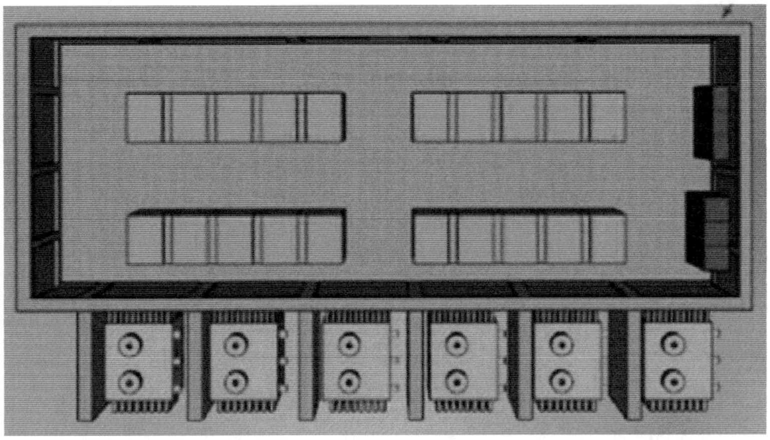

Figure 2. Example Substation Layout with High EV

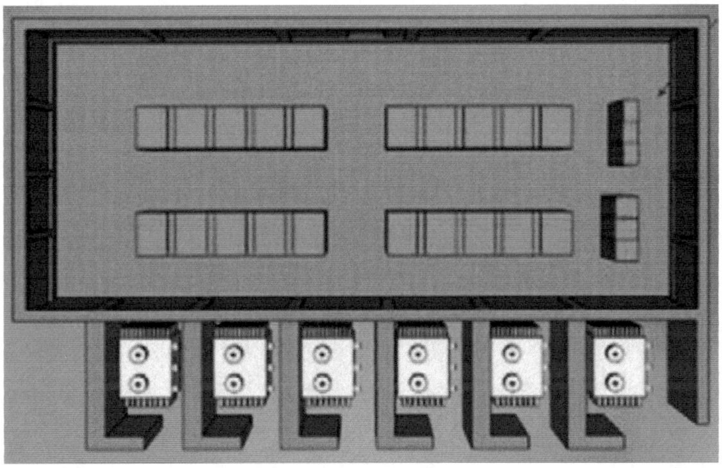

Figure 3. Example Substation Layout with Lower EV

Table 1. BDL Descriptions and Potential Equipment Damage

BDL	Potential Building Damage	Potential Equipment Damage	
		Surface-mounted	Not surface-mounted
BDL 1 Minor Damage	Walls sustain the onset of visible damage. Repairs are necessary for cosmetic reasons only.	A very low probability of equipment failure is predicted when exterior walls sustain the onset of visible damage and equipment is mounted to these surfaces.	No loss of equipment functionality is anticipated.
BDL 2 Moderate Damage	Localized damage. Walls facing the blast sustain moderate damage, while other walls and the roof sustain minor to moderate damage. Building can be repaired and reused.	A moderately high probability of equipment failure is predicted when an exterior wall sustains moderate damage.	A very low probability of equipment failure is predicted when exterior walls or the roof sustains moderate damage and equipment is mounted off exterior surfaces.
BDL 2.5 Heavy Damage	Widespread building damage. Walls facing the blast fail or sustain major damage, while other walls and the roof sustain moderate damage. Building repair may not be practical.	A high probability of equipment failure is predicted when an exterior wall fails or sustains major damage.	A moderate probability of equipment failure is predicted when an exterior wall fails or sustains major damage due to damage to key pieces of equipment resulting from debris.
BDL 3 Major Damage	Walls facing the blast fail, while other walls have compromised structural integrity. This may cause eventual collapse of the building. Building repair is not practical.	The equipment is predicted to completely lose functionality due to exterior wall failure.	A high probability of equipment failure is predicted when exterior walls fail due to damage to the majority of equipment resulting from debris.
BDL 4 Building Collapse	Primary and secondary structural members fail or sustain major damage resulting in building collapse.	The equipment is predicted to completely lose functionality due to building collapse.	The equipment is predicted to completely lose functionality due to building collapse.

Table 2. Relationship Between BDL and EV

BDL	EV	
	Surface-mounted	Not surface-mounted
1	1%	0%
2	70%	1%
2.5	90%	50%
3	100%	90%
4	100%	100%

Thermal Equipment Vulnerability

In general, a heat flux value of 25 kW/m^2 is used as a guideline for the onset of damage for process equipment [3]. However, this value does not represent damage to the more vulnerable electrical systems controlling the equipment. Most SCE will include electronic controls either in the form of power management or remote control of start-up, which will likely be the most sensitive part of the equipment.

The key metrics to determine the thermal EV of the SCE are the thermal load on the exterior of the building, the duration of the flame, the thermal resistance of the building, the air mixing within the building, the location of the equipment within the building, and the failure mode of the electronics within the building. Values for the thermal resistance of generic metal and concrete masonry unit (CMU) buildings are provided in Table 3 below. A one dimensional transient heat transfer analysis was used to calculate the temperature rise in the building. For a screening-level analysis, the electronics have been assumed to be located on the wall impacted by the jet fire, the air inside the building is assumed to be perfectly mixed to maintain conservatism, and the HVAC system is assumed to fail quickly and provide minimal cooling to the exposed building. However, more rigorous modelling can be used to remove conservatism, if needed.

The recommended screening-level thermal EVs are presented in Table 4 below. The number, type, and make-up of the electrical controls used in a typical chemical processing facility can range between a few hundred to thousands; therefore, it would not typically be cost-effective to assess the vulnerability of each type of equipment separately. Moreover, the vulnerability of the equipment can range from 0% to 100% over a wide range of temperatures. For a screening level study, a value of 50°C is recommended as a threshold value to model a 100% failure of all electronic controls with an EV of 0% for building temperatures below 50°C. This binary form of vulnerability assessment is conservative as 50°C represents a lower bound value for most electrical devices [4]. For a more detailed analysis, these assumptions could be altered to reflect the specifics of the electronics.

Table 3. Heat Transfer Analysis Properties

Type	K (W/m·K)	Rho (Kg/m^3)	C_p (J/kg·K)
Metal	0.05	28	1,700
CMU	1.6	114	920

Table 4. Screening Study Thermal EVs

Temperature Inside the Building	EV
Above 50° C	100%
Below 50° C	0%

Risk Determination

Once the EV determination is complete, consequence modelling should be done to determine the impact of the failure. The consequence of losing a given piece of SCE will vary based on the nature of the SCE and the magnitude of the event. Consequences can be calculated as simple failures of the SCE for screening studies, or detailed event trees can be constructed to more adequately represent the loss of the SCE. Loss of containment failure rates can be determined via a simple parts count approach or detailed fault trees as warranted by the scope of the study. The case studies presented in this paper use a parts count approach to determine the blast or fire scenario frequency and a combination of different approaches to arrive at the consequence of the event.

3 Case Study

The following case study illustrates the concepts presented in the methodology. During the design phase of a chemical plant, the project safety engineers requested that a QRA be completed for the site. Of the 181 buildings on site, 83 were deemed to be functionally occupied and were assessed for typical occupant vulnerabilities. The remaining 98 buildings included analyzer shelters, remote instrument enclosures, unit substations, motor control centers (MCC), and deluge buildings. Typically, these buildings would be excluded from the study; however, the project elected to utilize their QRA to assess these buildings for equipment vulnerability.

Initial Screening Study

Of the buildings analyzed, 46 of the buildings were predicted to experience negligible risk (<1E-5 SCE failures per year). Thirty-five of the buildings did not house SCE and were deemed to have negligible consequences upon failure. Of the remaining 17 buildings, 4 had risk in excess of 1E-3 failures per year. The project identified these buildings as candidates for further analysis and potential mitigation. Table 5 shows the 4 identified buildings and the mitigation measures used to bring their risk below 1E-3 SCE failures per year.

Table 5. Greenfield Screening Study EV Risk Summary

Building	SCE Failures/year			Mitigation Plan
	Blast	Fire	Total	
Unit 5 Deluge Building	5.0E-5	4.2E-3	4.2E-3	Thermal insulation added
Unit 3 MCC	1.5E-3	4.4E-5	1.6E-3	Wall mounting removed
Electrical Substation	1.7E-3	1.1E-3	2.8E-3	Building relocated
Fire Water Pump House	6.2E-3	4.7E-5	6.2E-3	Building redesigned

The Unit 5 Deluge Building was by necessity located close to a source of high fire risk. To ensure that the deluge system would be available in the event of a fire near the building, thermal insulation was added to the building. This increased the time to reach a critical temperature inside the building substantially and mitigated the risk to tolerable levels. In addition, the project changed the electrical and communication cable layout to bury the cables and bring them into the building underground to provide additional thermal protection.

The Unit 3 MCC was controlling a unit with high explosion hazards. The design of the MCC had mounted cabinets and controls on the wall. A larger pre-engineered metal building was purchased for the MCC so that controls could be moved off of the wall. This reduced the risk to tolerable levels.

Relocating the Electrical Substation

The electrical substation was of special interest to the project. It was designed to provide power to four units. During the HAZOP for the units, a loss of power scenario was identified for one of the units that would result in a large hydrocarbon release and significant damage to the process equipment. There were no significant impacts to the other three units on a loss of power. A detailed fault tree was performed to determine the frequency of loss of power. A backup generator was added to the project to improve the availability of the system should power from the neighboring co-generation (CoGen) facility be lost. However, a single point of failure was identified for the system: the power from both the CoGen plant and the backup generator was routed through the substation. As such, excessive blast risk to this building was deemed intolerable.

Because the project had identified this building as a potential hazard early in the design of the facility, there was opportunity to move the substation to a safer location. To aid in this effort, data was utilized from the QRA to provide iso-vulnerability contours for the building using both surface-mounted (gray line) and non-surface-mounted (black line) options. The contours are shown in Figure 4 below.

The generated contours showed that the current location of the substation was inadequate, but a viable alternative was not available. The project determined that the unit in question could be removed from the analysis. They did this on the grounds that, if a major accident occurred within this unit, it would no longer be a concern if the substation remained operational. The other three units could be shut down safely without the substation. A refinement was done to remove the unit from the analysis, which generated the contours in Figure 5. The facility then used these contours to determine an acceptable location for the substation using equipment that was not wall or ceiling mounted. Moving the substation further away also significantly decreased the fire risk. Risk contours were also generated, but the project elected to use a consequence-based approach.

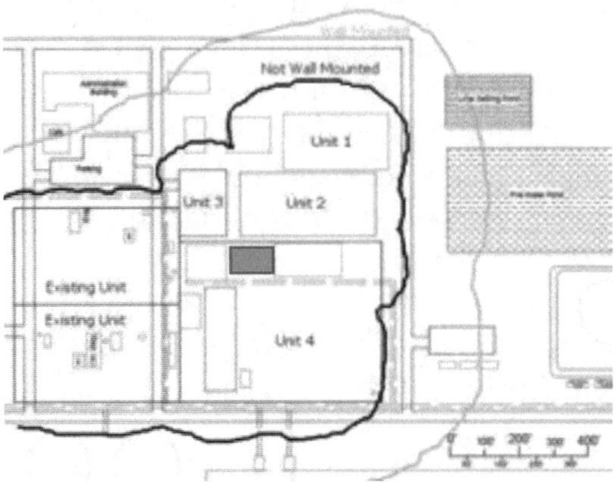

Figure 4. Four Unit Iso-Vulnerability Blast Contours

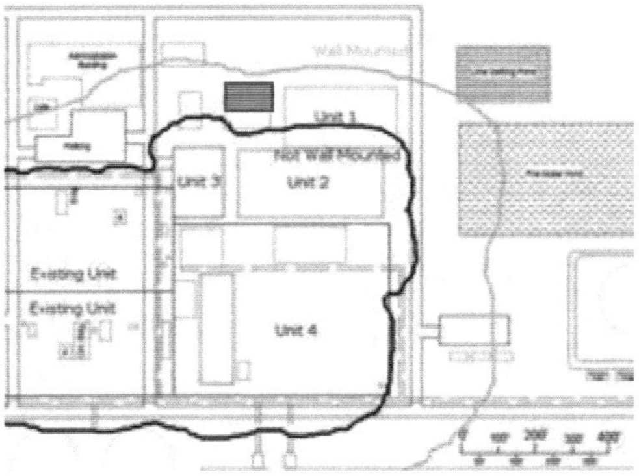

Figure 5. Three Unit Iso-Vulnerability Blast Contours

Upgrading a Fire Water Pump House

Due to concerns raised by their environmental impact study, the project had located the fire water pump house in the center of the plant surrounded by units handling highly flammable and explosive materials. The project was concerned that an explosion could damage the fire water pump, making it unavailable to respond to the event. Since the building could not be as easily relocated as the substation had been, the project elected to strengthen the building design. The overall blast response for the building was improved using standard dynamic design techniques but with a risk-based design criteria. The original and adjusted pressure-impulse curves for the building are shown in Figure 6 and Figure 7. By increasing the blast capacity of the building and removing surface-mounted equipment, the overall blast EV risk for the fire water pump was decreased to acceptable levels.

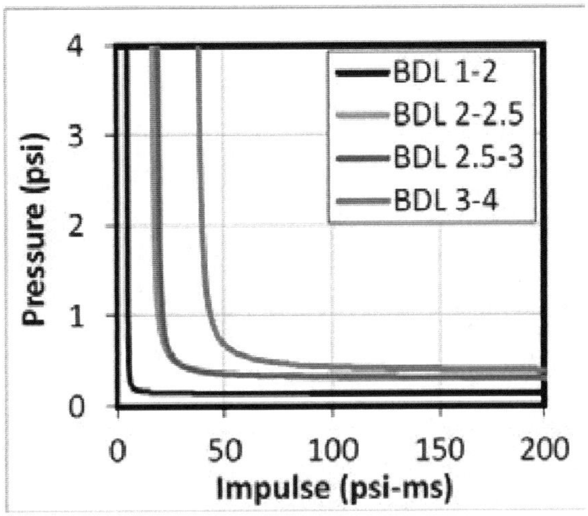

Figure 6. Original Pressure-impulse Curves

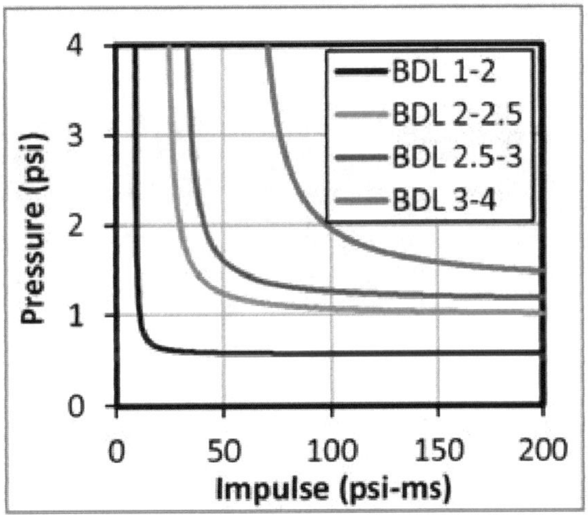

Figure 7. Adjusted Pressure-impulse Curves

Refined Consequence Analysis of the Fire Water Pump Diesel Tanks

During the redesign of the fire water pump house, the project raised the possibility of the diesel tanks used to fuel the fire water pumps being damaged by a blast. The QRA EV screening study had identified the maximum load that was predicted to be applied to the diesel tanks. At this point, the project elected to perform a more detailed consequence assessment of the maximum predicted load to determine how it affected the storage tanks and their support structure. Figure 8 below shows the effective stress on the tanks as they are loaded by a blast wave. Based on this analysis, it was determined the tanks were not predicted to fail under the maximum loads calculated to impact them.

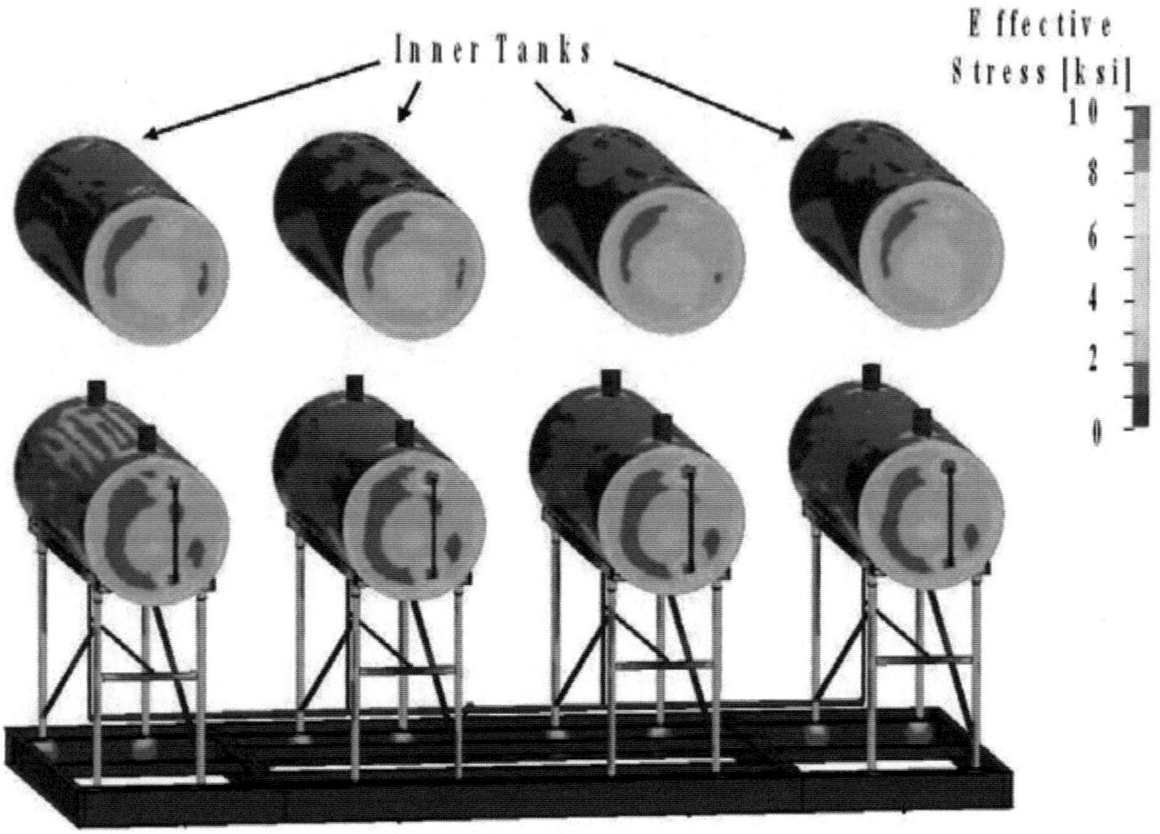

Figure 8. Outdoor Storage Tanks

4 Conclusions

As demonstrated in the case study, a QRA can be effectively leveraged to provide additional information about the risk to safety critical equipment within a process plant. The study primarily depends on a modification to the vulnerability assessment portion of the QRA procedure which will allow for screening studies to be conducted. When the screening study identifies short-comings, the same methodology can be utilized to help identify mitigation measures. In addition, more detailed studies can be selectively run to refine the consequence or failure rate modeling for the significantly smaller number of SCE locations identified. In general, moving equipment off of walls potentially exposed to a blast is an effective way to lower the predicted blast EV. Increasing the thermal resistance of a building and ensuring that cable trays are not exposed to significant fire hazards are methods of reducing the thermal EV. When possible, these considerations should be done early in the design phase of a project when it is still feasible to move equipment and reduce hazards to improve the safety of the plant. When relocation is no longer an available option, building upgrades can be implemented to improve the building response to blast and fire hazards.

5 References

[1] CCPS. *Guidelines for Chemical Process Quantitative Risk Analysis* 2nd Edition, Center for Chemical Process Safety, American Institute of Chemical Engineers, New York, NY, 2000.

[2] Q. A. Baker. "Explosion Risk and Structural Damage Assessment Code (ERASDAC)," 30th Department of Defense Explosive Safety Seminar, 2002.

[3] T. F. Barry. "An Introduction to Quantitative Risk Assessment in Chemical Process Industries," *The SFPE Handbook of Fire Protection Engineering*, 3rd Edition, National Fire Protection Association, Quincy, MA, 1995.

[4] J. L. Scheffey, L. A. Jona, T. A. Toomey, R. Byrd, and F. W. Williams. "Analysis of Quick Response Fire Fighting Equipment on Submarines-Phase II, Full Scale Doctrine and Tactics Tests," NRL Memorandum Report 6632, 1990.

442208 Utilizing Accelerating Rate Calorimetry for Insulation Fire Hazard Identification and Management

Abstract:

Auto-oxidation of process fluids may occur as a result of a leak or spill on insulation. Consequently, this can result in smoldering or fires, even when the initial temperature of the fluid is well below its autoignition temperature. The primary cause of an insulation fire is that heat gains from oxidation of the absorbed material on the high surface area of the insulation exceed any heat losses via conductive or convective cooling. Since closed cell insulation systems are very expensive, it is important to evaluate the most appropriate and cost-effective insulation system. This paper will outline a calorimetric and kinetic modeling approach to predict the Temperature of No Return (TNR) for a given thickness of insulation. The data will aid in determining the best type of insulation system and development of safe operating discipline to prevent insulation fires.

12TH GLOBAL CONGRESS ON PROCESS SAFETY

Compliance with Proposed Regulatory Requirements on Damage Mechanisms

Michael Tia
Kazarians & Associates, Inc.
100 West Broadway, Suite 970,
Glendale, California
Michael.Tia@Kazarians.com

Prepared for Presentation at
American Institute of Chemical Engineers
2016 Spring Meeting
12th Global Congress on Process Safety
Houston, Texas
April 11-13, 2016

Compliance with Proposed Regulatory Requirements on Damage Mechanisms

Michael Tia
Kazarians & Associates, Inc.
100 West Broadway, Suite 970,
Glendale, California
Michael.Tia@Kazarians.com

Keywords: California Accidental Release Prevention Program, Process Safety Management, Damage Mechanism Reviews, Corrosion

Abstract

As part of the proposed improvements to the California Process Safety Management (8 CCR, Division 1, Chapter 4, 5189.1) and Accidental Release Prevention Program (19 CCR Division 21, Chapter 4.5), California petroleum refineries will be required to conduct a Damage Mechanism Review (DMR) of all covered processes. The DMR requirement of the regulations includes a strict review as well as documentation of the damage mechanisms that may affect the covered processes.

The proposed DMR requirements can be summarized as identification of potential damage mechanisms, determination of existing safeguards, determination of vulnerabilities and recommendation of measures to reduce the risk associated with the mechanism. To comply with the specifics of the regulations, an approach is discussed in this article. The intent of the approach is to generate a documented analysis that provides the required information for the specific damage mechanisms in a process.

In the first step, existing engineering and metallurgy documents are collected and catalogued. In the next step, the Process Hazard Analysis of the covered process and any other similar risk analyses are reviewed. In that process, the specifics of the damage mechanism and existing mitigative measures (i.e., inspection program) are identified and recorded. Doing so will establish potential vulnerabilities and identify whether recommended improvements are warranted. In the last step, a separate report is generated based on the analysis completed in the second step to document the identified vulnerabilities, existing mitigative measures, and recommendations. The report is organized such that the basis of the vulnerabilities and recommendations are clearly identified. The report will provide operations and other affected personnel with a single living document summarizing the damage mechanism related concerns associated with the process that can be updated on a periodic basis.

1. Introduction

Stemming from several catastrophic releases of hazardous chemicals [References 1 and 2], federal and California regulatory agencies have begun to consider improvements to update the Process Safety Management and Risk Management Plan regulations. Specifically, the California Governor's Office of Emergency Services (CalOES) and California's Occupational Health and Safety Administration (CalOSHA) have already issued several revisions of proposed modifications to the California Accidental Release Prevention Program (CalARP) and PSM regulations [see References 3 and 4 for current version] respectively.

A critical element of the proposed regulation is to establish a requirement for petroleum refiners in California to conduct a Damage Mechanism Review (DMR) of equipment in a covered system and revalidation of the DMR on a five year cycle. As DMRs are to be conducted to ensure the ongoing integrity of the covered system and associated equipment, this requirement of the regulations is defined under the Mechanical Integrity section of both regulations.

The list of DMR requirements in the proposed regulations effectively covers all the issues that one could raise related to the impact of damage mechanisms on process piping and equipment. This article provides guidance and an approach for addressing the damage mechanism review clause of the proposed regulation for California refineries, which can also be applied to non-regulated systems.

2. Regulatory Requirements and Suggested Compliance

The proposed requirements of the California Accidental Release Prevention (CalARP) Program and California Process Safety Management (PSM) regulations related to Damage Mechanism Review are provided in Appendix A. The latest revisions of both the CalARP and PSM regulations contain nearly identical requirements for Damage Mechanism Review as shown in the side-by-side comparison in Appendix A.

Note, the regulations have not been finalized and published for implementation. Therefore, changes to these regulatory requirements may occur. The intent of this article is to provide some guidance on initiating the process for achieving compliance with the regulations. Although the regulations have not been finalized, implementation of all elements of the proposed regulations may prove beneficial in improving process safety, as well as providing useful and structured knowledge in maintaining the ongoing integrity of process equipment. Additionally, the occurrence of damage mechanisms in the process industry, petrochemical and other, will continue to manifest itself. As such, it may be in the best interest of chemical processing and storage facilities to ensure the ongoing mechanical integrity of their system by implementing aspects of the propose requirements.

Each element of the proposed regulations will be discussed further below.

California Accidental Release Prevention Program (CalARP)	Process Safety Management
(f) Damage Mechanism Review	**(k) Damage Mechanism Review**
(1) The owner or operator shall complete a Damage Mechanism Review (DMR) for each process for which a damage mechanism exists. Where no DMR is performed, the owner or operator shall document the rationale for the determination that no damage mechanism exists. The owner or operator shall determine and document the priority order for conducting the DMR based on process operating history, PHA schedule and inspection records. No less than 50 percent of the initial DMRs shall be completed within three (3) years of the effective date of this Article, and the remainder within five (5) years of the effective date of this Article. If the employer has conducted and documented a DMR for a process unit within five (5) years prior to the effective date of this Section, and that DMR includes the elements identified in paragraph (f)(6), that DMR may be used to satisfy the employer's obligation to complete an initial DMR under this paragraph.	*(1) The employer shall complete a Damage Mechanism Review (DMR) for each process for which a damage mechanism exists. Where no DMR is performed, the employer shall document the rationale for the determination that no damage mechanisms exist. The employer shall determine and document the priority order for conducting DMRs based on the process operating history, the PHA schedule and inspection records.* *(2) The employer shall complete no less than fifty (50) percent of initial DMRs within 3 years and all remaining DMRs within five (5) years of the effective date of this Section. If the employer has conducted and documented a DMR for a process unit up to five (5) years prior to the effective date of this Section, and that DMR includes the elements identified in (k)(8), that DMR may be used to satisfy the employer's obligation to complete an initial DMR under this subsection.*

The regulation requires that a Damage Mechanism Review (DMR) be conducted for each process that is governed under the California PSM and CalARP regulations. Similar to the practice that many facilities with multiple covered processes utilize for conducting Process Hazard Analyses, DMRs may be distributed within a five year interval. For example, imagine a facility operates 30 individual processes. The DMRs can be spaced such that six (6) DMRs are conducted every year, satisfying the initial regulatory requirement for "50 percent of the initial DMRs shall be completed within three (3) years" and setting the cycle of DMR revalidations on a five (5) year cycle. It is typical for facilities with multiple covered processes to establish such a schedule to distribute workload and personnel demand among the five year interval.

Furthermore, depending on the corrosion/damage mechanism related analyses and documentation that a facility has already completed, many elements of the regulations may have already been satisfied or require a minimal amount of effort to ensure compliance. For example, facilities that maintain a Risk Based Inspection (RBI) Program [Reference 5], which establishes inspection intervals based on risk prioritization, may have satisfied several elements of the regulations. Since the RBI process requires identifying damage mechanisms and failure modes, assessing probability and consequence of failure, and risk management with inspection activities and process control, and identifying safe operating limits such as Integrity Operating Windows [Reference 6], required elements described in the proposed regulations may have already been addressed. Moreover, facilities may have developed corrosion diagrams or other similar documents identifying the major damage mechanisms of concern within a process unit. In some cases, facilities may have conducted thorough reviews of damage mechanisms in their Process Hazards Analyses. Reference 7 provides a paper discussing the process for addressing damage mechanism hazards as part of Process Hazard Analyses. Based on the extent in which a facility has addressed damage mechanisms, the effort to ensure compliance with the proposed regulations may not be substantial.

California Accidental Release Prevention Program (CalARP)	**Process Safety Management**
(2) A DMR shall be revalidated at least once every five (5) years.	*(3) A DMR shall be revalidated at least once every five years following its initiation.*

Similar to the requirements for Process Hazard Analyses, Damage Mechanism Reviews must also be revalidated on a five year cycle. The intent of this requirement is to ensure that DMRs are maintained accurate and reflect any system modifications as well as industry and site learnings. Moreover, both the revised PSM and CalARP regulations require that a PHA address "all applicable Damage Mechanism Review Reports" [References 3 and 4]. Therefore, the authors of this article recommend that the DMR of a process to be completed or revalidated prior to its PHA revalidation. Maintaining the DMR and PHA cycles for a process unit or system within a year of each other will ensure that there is an accurate and up-to-date DMR prior to the PHA session. This will allow for the PHA team to further review and validate the hazards and risks associated with identified damage mechanisms.

California Accidental Release Prevention Program (CalARP)	**Process Safety Management**
(3) A DMR shall also be conducted where a damage mechanism exists on new processes or a major change, prior to initial approval of the change.	*(4) Where a damage mechanism exists, a DMR shall also be conducted on new processes and as part of a major change, prior to approval of the change.*

To comply with this requirement, the facility's Management of Change (MOC) process can be updated to include a review of the change for potential introduction of new damage mechanisms or the potential to exacerbate the effects or likelihood of an existing damage mechanism. It is common for the MOC process to include a PHA or risk review of the change. The MOC should require that a DMR be performed of the change prior to or in conjunction with the MOC PHA.

Companies required to implement PSM and RMP regulations have typically implemented an MOC procedure and associated form(s). Typically these forms are checklist types to ensure that each element of the regulation as well as each foreseeable effect from the change (e.g., environmental effect) are reviewed and analyzed. The form should include checklist questions addressing potential damage mechanisms and the requirement to perform a DMR review for the change if necessary. Lastly, if a DMR is not required the form should include a section to provide the basis for this conclusion.

California Accidental Release Prevention Program (CalARP)	**Process Safety Management**
(4) As part of an incident investigation pursuant to Section 2762.9, where a damage mechanism is identified as a contributing factor, the owner or operator shall review the most recent DMR(s) that are relevant to the investigation. If a DMR has not been performed on the processes that are relevant to the investigation, a DMR shall be completed prior to completing the incident investigation.	*(5) Where a damage mechanism is identified as a contributing factor in an incident investigation, pursuant to subsection (o), the employer shall review the most recent DMRs that are relevant to the investigation. If a DMR has not been performed on the processes that are relevant to the investigation, a DMR shall be completed as part of the incident investigation.*

Common to incident investigations by outside investigators (see References 1 and 2 as examples of investigation reports by the CSB), investigators will request the PHA for the system with which the incident occurred. The purpose of this request is to determine whether the PHA team adequately addressed the specific incident. From the Process Hazard Analysis standpoint, the investigator is trying to determine whether the PHA was complete and whether appropriate controls were identified and if so, did any control fail? The Damage Mechanism Reviews will be used to similar effect, which will ensure that proper controls (e.g., inspection frequencies and types) are identified to reduce the potential for a catastrophic event.

California Accidental Release Prevention Program (CalARP)	**Process Safety Management**
(5) The DMR for a process unit shall be available to the team performing a PHA for that process unit.	*(6) The DMR for a process unit shall be available to the team performing a PHA for that process unit.*

The PHA facilitator should review the DMR and ensure the identified damage mechanisms and associated controls (e.g., inspection type and frequency) are addressed as part of the PHA. Specifically, the PHA team should conduct a Damage Mechanism Hazard Review during the PHA, which follows the system owner's PHA protocol for consequence development, identification of controls (safeguards), and risk ranking for each damage scenario. Reference 7 provides a method for integrating and addressing damage mechanisms as part of the PHA process. It is important to note that damage mechanism related discussions during PHA sessions must include a corrosion and materials expert to ensure accuracy of the discussion.

Since the DMR must be available to the PHA Team, it may be convenient that DMR and PHA revalidations are conducted within a similar timeline of each other. One suggestion would be to conduct the DMR revalidation within one year prior to the PHA revalidation. In doing so, this will ensure that the DMR is up-to-date prior to PHA revalidation and maintain the DMR and PHA cycles within similar timeframes.

California Accidental Release Prevention Program (CalARP)	**Process Safety Management**
(6) The DMR shall be performed by a team with expertise in engineering, operation of the processes under review, equipment and pipe	*(7) The DMR shall be performed by a team with expertise in engineering, equipment and pipe inspection, damage and failure mechanisms, and*

California Accidental Release Prevention Program (CalARP)	Process Safety Management
inspection, and damage and failure mechanisms. The team shall also include one member knowledgeable in the specific DMR method being used. The owner or operator shall provide for employee participation in this process, pursuant to Section 2762.10.	*the operation of the process or processes under review. The team shall include one member knowledgeable in the specific DMR methodology being used. The employer shall provide for employee participation in this process, pursuant to subsection (q).*

DMR Facilitator's Role - The DMR Facilitator's role is to guide the discussion using a standard and consistent methodology and ensure that there is detailed documentation of the discussion. Additionally, the DMR Facilitator is responsible for ensuring that all aspects of the regulations are satisfactorily completed.

Similar to PHA process, DMR is expected to be conducted in a group brain-storming session. Such sessions require a facilitator. Ideally, the facilitator should be well versed in the science of damage mechanisms, which effectively requires metallurgical knowledge. However, since facilitating a discussion session is not commonly done by such experts, the authors recommend that PHA facilitator to lead DMR discussions because they are trained on leading brain-storming sessions.

Materials and Corrosion Expert's and Inspectors Role - The Materials and Corrosion Expert provide the knowledge needed in identifying the damage mechanisms that may affect the system. They provide the materials, corrosion and inspection experience and the knowledge to accomplish the tasks provided in the technical requirements of the regulations (see below for the technical requirements)

Engineering and Operation Personnel - The Engineer and Operator provide the documents and knowledge that define the engineering and operational aspects of the process. The Engineer's and Operator's role is to provide information regarding normal operating process parameters (i.e., temperature, flow, pressure, composition), high/low operating limits, potential upset conditions (can be obtained in the PHA), and system conditions during alternative modes of operation to the materials and corrosion expert.

Additionally, the *Engineer and Operator* should note any unique or abnormal corrosion issues that have been observed. The DMR brain-storming session allows for members of engineering and operations to voice any observed concerns and receive an appropriate response for those identified issues. This may include any observed external corrosion on equipment, potential process deadlegs and mixpoints (injection points), cavitation heard in the system, abnormal levels of fouling in the system and upset conditions that may result in potential damage mechanisms.

California Accidental Release Prevention Program (CalARP)	Process Safety Management
(7) The DMR for each process shall include: *(A) Assessment of Process Flow Diagrams (PFDs);*	*(8) The DMR for each process shall include:* *(A) Assessment of Process Flow Diagrams (PFDs);*

California Accidental Release Prevention Program (CalARP)	Process Safety Management
(B) Identification of all potential damage mechanisms; *(C) Determination that the materials of construction are appropriate for their application and are resistant to potential damage mechanisms;* *(D) Qualitative assessment of the rate at which damage is likely to occur;* *(E) Methods to prevent or mitigate damage;* *(F) Review of operating parameters to identify operating conditions that could accelerate damage or that could minimize or eliminate damage;*	*(B) Identification of all potential damage mechanisms, pursuant to (k)(9);* *(C) Determination that the materials of construction are appropriate for their application and are resistant to potential damage mechanisms;* *(D) Methods to prevent or mitigate damage; and,* *(E) Review of operating parameters to identify operating conditions that could accelerate or otherwise worsen damage, or that could minimize or eliminate damage.*

The above section of the regulation establishes the technical requirements of the DMR. As stated previously, if the facility has already implemented an RBI or conducted a similar analysis to identify damage mechanisms (i.e., utilize the PHA process to identify damage mechanisms or develop other corrosion control documents), at least of portion of the requirements may have been met.

Indicated in the team requirement section of the proposed regulations, the DMR team must be composed of "*engineering, operation of the processes under review, equipment and pipe inspection, and damage and failure mechanisms*". Resultantly, a team will be required to complete the tasks identified in this section of the regulations. The required tasks are discussed further below:

(A) Assessment of Process Flow Diagrams (PFDs)
Many facilities may already have addressed the requirement for assessing the covered system PFDs by establishing some form of metallurgy and corrosion diagrams utilizing the PFDs. The following items are typically included on these diagrams:

1. Process material in the piping and equipment (e.g., crude, diesel, stove oil, etc.),
2. Normal operating parameters (flow, pressure, and temperature),
3. Metallurgy of equipment (piping and all major equipment, including identification of cladding/overlay in vessels or piping),
4. Damage mechanisms applicable to each segment of piping or equipment, and
5. Predicted corrosion rates (if applicable)

If not already developed, the DMR team should assess the PFDs and P&IDs and consider identifying the above listed items on the PFD. If similar materials and corrosion PFDs have already been developed, then the full DMR team should review these existing PFDs for accuracy.

(B) Identification of all potential damage mechanisms;
The materials and corrosion expert and the inspector are the responsible party for identifying damage mechanisms. Typical damage mechanisms in refining and hydrocarbon processing can be found in API 571 [Reference 8]. Many, if not all, of these damage mechanisms can also be found in other industries. The materials and corrosion engineer's primary role is to review the Piping & Instrumentation Diagrams (P&IDs), PFDs, and other system related Process Safety

Information (PSI) to identify all applicable damage mechanisms. The process engineer and member of operations will provide information regarding the process and normal operating parameters, such as flow, pressure, temperature, composition, which will assist the materials and corrosion engineer in establishing these damage mechanisms. Applicable damage mechanism should be identified on the system PFD as described in item (A) above.

(C) Determination that the materials of construction are appropriate for their application and are resistant to potential damage mechanisms;
The DMR team must be provided with the P&IDs and metallurgy related PSI in order to appropriately address potential damage mechanisms. While assessing the P&IDs and other PSI, metallurgical information should be notated on PFDs (PFD item 3 above). Industry standards and practices should by followed when determining whether the materials of construction are appropriate. During the review, the team will review piping and equipment, identify process material, and establish high/low operating parameters. At this point, it will be based on the materials and corrosion engineer and inspector or outside expert knowledge to determine whether certain materials must be upgraded. If it was deemed that materials upgrade is required, a recommendation must be generated and provided to management. In addition to the DMR, the PHA may also be a source to review damage mechanisms and proper metallurgy. Since PHAs have been an established requirement since the initial implementation of the RMP and PSM regulations in the 1990s, issuance of this type of recommendations during that analysis may be appropriate.

(D) Qualitative assessment of the rate at which damage is likely to occur; (Note, this is not identified in the PSM Regulation)
The DMR team may assess the corrosion rate in a variety of ways from purely qualitative to quantitative means. Facility management, in consultation with the regulatory authority must establish how these rates will be defined. Additionally, the materials and corrosion engineer will be the primary person responsible for establishing these rates. If possible, it may be beneficial to establish quantitative values using corrosion predictive software and models to postulate the rate of corrosion (i.e., establish rates in milli-inch per year).

(E) Methods to prevent or mitigate damage;
The methods to prevent or mitigate damage can be obtained from the system's mechanical program (e.g., inspection types/frequencies provided in an RBI, Mechanical Integrity (MI) program established based on industry or manufacture recommended practices, etc). Note, during the damage mechanism hazard review segment of the PHA, or even during scheduled system inspection, the effectiveness of the maintenance protocols should be considered and reviewed to ensure that the maintenance protocol is satisfactorily maintaining the integrity of the system component. That is, the corrosion rate is as expected.

(F) Review of operating parameters to identify operating conditions that could accelerate damage or that could minimize or eliminate damage;
During this stage of the DMR, the DMR team should review any alternative modes of operation (e.g., startup, shutdown, catalyst regeneration, complete unit recirculation) to determine whether any of these operating modes can result in additional damage mechanisms that may occur. One

such example is exceeding Minimum Pressurization-Temperature (MPT) curve of a vessel during startup or shutdown, which may result in brittle fracture concerns.

One source of potential initiating events that can assist in establishing these conditions is the PHA for the unit. As currently required by the CalARP, RMP and PSM regulations, PHAs must address "all modes of operation" [see Reference 4 for an example]. Therefore, the PHA can be a good source to obtain information regarding these different modes of operation.

Additionally, the PHA will have examined upset conditions that may exacerbate the rate of or result in new damage mechanisms that may not have manifested during normal operating condition. This issue stresses the importance of ensuring that PHAs and DMRs act as complements of each other. The PHA will systematically examine single point failures (e.g. operator closes a valve, pump fails, etc.) and determine the consequences of the failure. A suggestion when conducting a PHA is to encourage a damage mechanism/corrosion hazard review during the analysis to examine potential damage mechanisms that may arise due to upset conditions. Additionally, damage mechanisms that result from issues such as deadlegs may not be easily assessed using the PFDs and must be identified during field walks or by the operations and engineering representatives. Because of this issue, the PHA and DMR teams must identify noted deadlegs.

California Accidental Release Prevention Program (CalARP)	Process Safety Management
(8) For purposes of this section, damage mechanisms include, but are not limited to: *(A) Mechanical loading failures, such as ductile fracture, brittle fracture, mechanical fatigue, and buckling;* *(B) Erosion, such as abrasive wear, adhesive wear, and fretting;* *(C) Corrosion, such as uniform corrosion, localized corrosion, and pitting;* *(D) Thermal-related failures, such as creep, metallurgical transformation, and thermal fatigue;* *(E) Cracking, such as stress-corrosion cracking; and* *(F) Embrittlement, such as high-temperature hydrogen attack.*	*(9) For purposes of this subsection, damage mechanisms include, but are not limited to:* *(A) Mechanical loading failures, such as ductile fracture, brittle fracture, mechanical fatigue and buckling;* *(B) Erosion, such as abrasive wear, adhesive wear and fretting;* *(C) Corrosion, such as uniform corrosion, localized corrosion and pitting;* *(D) Thermal-related failures, such as creep, metallurgical transformation and thermal fatigue;* *(E) Cracking, such as stress-corrosion cracking; and,* *(F) Embrittlement, such as high-temperature hydrogen attack.*

A materials and corrosion expert is required during the entire DMR process to identify all possible damage mechanisms that may occur. An assessment of the PFD, which should include metallurgy and process information can be used as a starting point for identifying the possible damage mechanisms within a process unit. As a resource, many damage mechanisms in the refining and hydrocarbon processing can be found in API 571 [Reference 8]. API 571 addresses more than 70 damage mechanisms and can be used as a starting point to gain an understanding of damage mechanisms that may occur.

Also, it must be noted that API 571 is not comprehensive of all damage mechanisms. The damage mechanisms identified in API 571 are not unique to refining or hydrocarbon processing and all of them can be found in other process industries. It is the discretion of the materials and corrosion expert to determine which damage mechanisms are applicable and to use the appropriate publications and reference books.

California Accidental Release Prevention Program (CalARP)	Process Safety Management
(9) DMRs shall include an assessment of previous experience with the process including the inspection history and all damage mechanism data; a review of industry-wide experience with the process; and applicable standards, codes and practices.	*(10) DMRs shall include an assessment of previous experience with the process, including the inspection history and all damage mechanism data; a review of industry-wide experience with the process; and all applicable standards, codes and practices.*

The DMR team should take into consideration past experience, inspection results, industry knowledge, practices, standard, and codes when completing the technical requirements of the regulation. A review of past turnaround inspection results and past corrosion control plans may be beneficial to the DMR team.

California Accidental Release Prevention Program (CalARP)	Process Safety Management
(10) At the conclusion of the analysis, the team shall prepare a written DMR report that includes: *(A) The process unit(s) reviewed;* *(B) Damage mechanisms analyzed;* *(C) Results of the analyses conducted according to subsection (f)(7) above;* *(D) Recommendations for temporary mitigation;* *(E) Recommendations for prevention.*	*(11) At the conclusion of the analysis, the team shall prepare a written DMR report, which shall include the following:* *(A) The process unit and damage mechanisms analyzed;* *(B) Results of all analyses conducted, pursuant to (k)(8);* *(C) Recommendations for temporarily mitigating damage; and,* *(D) Recommendations for preventing damage.*

As stated in the regulation, a report shall be developed. Along with the following items identified by the regulations, the report should also include the following:

1. Purpose of the analysis
2. Scope of the analysis
3. Overview of the process
4. DMR methodology and approach applied
5. DMR team members and associated expertise
6. DMR meeting dates

Developing a detailed and well documented report will allow readers (e.g., auditors, operators, etc.) to easily understand the thought process behind the DMR. Additionally, a detailed report will provide the basis for recommendations issued during the DMR process, which will assist management addressing the issues posed.

California Accidental Release Prevention Program (CalARP)	Process Safety Management
(11) The report shall be provided to and, upon request, reviewed with all operating, maintenance, and other personnel, whose work assignments are within the process unit covered in the DMR.	*(12) The report shall be provided to and, upon request, reviewed with employees whose work assignments are within the process unit described in the DMR.*

Employees in the process unit addressed by the DMR should be informed that a DMR has been completed and the location of the latest DMR report. This can be accomplished through several means.

One of the required participants of the DMR is a member of operations. During the preparation phase of the DMR, request that the unit superintendent or manager inform all operators of the plans to complete or revalidate a DMR. Allow operators to voice any potential mechanical integrity issues that they may have noted or may be concerned about so that the DMR team can review and respond to the concerns (e.g., provide an operator questionnaire to be completed by select representatives of operations). Consider making the questionnaire anonymous if deemed necessary.

Following completion of the DMR, consider providing a summary presentation and/or report to members of operations, maintenance, and any other personnel that may be affected by the DMR. During the presentation or report, identify the location(s) where the DMR can be found and the person(s) that can be contacted for further information regarding the DMR.

California Accidental Release Prevention Program (CalARP)	Process Safety Management
(12) The owner or operator shall follow the corrective action work process documented in subsections 2762.16(d) and (e) when resolving the DMR team's findings and recommendations, determining corrective action for implementation, tracking to completion, and documentation of closeout.	*(13) The employer shall implement all recommendations in accordance with subsection (x).*

Similar to the requirements to track PHA, compliance audit, incident investigation, and other recommendations generated as part of PSM and RMP Program implementation, DMRs recommendations must also be tracked. DMRs recommendations should be tracked in a similar fashion as those recommendations, be it through a database, or other consolidated location.

Following the DMR session, recommendations owners should be established to ensure recommendations are completed within the required timeframe.

California Accidental Release Prevention Program (CalARP)	Process Safety Management
(13) DMR reports shall be retained for the life of the process unit.	*(14) DMR reports shall be retained for the life of the process unit.*

Pursuant to the requirement identified in the regulations, DMRs must be kept for the life of the system. Both a hard copy and an electronic copy should be maintained. The electronic copy should be kept in the facility's electronic database as part of PSI or mechanical integrity related documents. At least one copy of the DMR should be kept by the facility's Process Safety/Risk Management Department for ease of access in the event the DMR is required for auditing or review purposes.

3. Documentation

For each DMR consider developing a binder containing the final DMR report, recommendations from the review, and any supporting documents used in preparation of the DMR. The supporting documents should include the PFDs that were accessed, which should depict metallurgy and damage mechanisms on appropriate sections of the system.

Along with the DMR documentation, ensure that damage mechanism related documentation such as RBI reports, information regarding materials of construction, specification breaks, corrosion control plans, integrity operating windows (IOWs) or other set upper/lower operating limits and other technical information related to damage mechanisms are maintained as Process Safety Information for the system. As such, these documents must undergo the same rigors and practices that are applied to PSI established as part of the PSM and RMP regulations (e.g., P&IDs).

4. Limitations and Benefits

The process delineated in this paper provides guidance based on the author's interpretation of the proposed regulation and our history of implementing PSM and RMP programs for a multitude of industries and facilities. The author's understanding of the requirements, which is based on their experience with the PSM and RMP regulations, as well as their experience with the various local and other regulatory agencies, have been used to develop this article.

Additionally, based on the jurisdiction that governs a specific facility, the facility will still have to adhere to the specific requests and requirements of those governing offices. The purpose of this article is to begin providing some guidance to ensure that facilities that will fall under the jurisdiction of the regulation, begin the process of generating DMRs if not already done so.

5. Conclusion

With the continuing evolution of the regulatory environment to pursue modernization of the PSM and RMP regulations, the stress of further process safety improvements have occurred at not only California refineries, but other hazardous material handling industries within the state and beyond. Executive Order 13650 [Reference 9] was issued in response to recent catastrophic incidents at chemical handling facilities in the United States. The Chemical Safety Board on December 1, 2014 announced "to Modernize U.S. Process Safety Management Regulations is the Board's newest Most Wanted Safety Improvement" [Reference 10]. Within the same statement, the CSB states "that PSM and RMP, although written as performance-based regulations, appear to function primarily as reactive and activity based regulatory frameworks". It should be a primary focus of hazardous material handlers to ensure proactive process safety actions are taken. One crucial example is to ensure that the covered system maintains its mechanical integrity by conducting reviews such as DMRs to validate the continued integrity of the system.

Although a new requirement by a regulatory authority may seem as a daunting task, facilities within California have typically implemented some form of corrosion program to monitor, manage, mitigate, the effects of damage mechanism on process equipment. With that being said, at least parts of the requirements presented in the proposed regulations have already been addressed by current established practices.

No process system is immune to the effects of damage mechanisms. Although the proposed regulation is specific to petroleum refiners in the California, it may be beneficial for other hazardous materials handling facilities to consider implementing parts of the proposed regulations as part of their process safety practices. Initial steps such as requiring PHAs to specifically address damage mechanism related hazards, identifying the damage mechanisms applicable to the system and appropriate mechanisms in place to reduce the potential effects, ensuring safe high/low operating limits are established and operators are aware of these value may prove to be beneficial in protecting employees and the surrounding environment from releases of hazardous materials.

6. References

[1] *Investigation Report - Catastrophic Rupture of Heat Exchanger (Seven Fatalities) - Tesoro Anacortes Refinery - Anacortes, Washington, April 2010.* U.S. Chemical Safety and Hazard Investigation Board, Report 2010-08-I-WA, May 2014.

[2] *Final Investigation Report – Chevron Richmond Refinery – Pipe Rupture and Fire - Chevron Richmond Refinery #4 Crude Unit – Richmond, California, August 6, 2012.* U.S. Chemical Safety and Hazard Investigation Board, Report 2012-03-I-CA, January 2015.

[3] *Proposed General Industry Safety Order (GISO) §5189.1 - Process Safety Management for Petroleum Refineries*, Version 5.0 - September 14, 2015. State of California, Department of Industrial Relations, Division of Occupational Safety and Health

[4] *California Accidental Release Prevention (CalARP) Program.* Draft CalARP Program Regulations, September 24, 2015, California Governor's Office of Emergency Services

[5] *Risk-Based Inspection.* API Recommended Practice 580, 2nd Edition, November 2009.

[6] *Integrity Operating Windows.* API Recommended Practice 584, May 2014

[7] M. Tia, J. Waytashek, and M. Kazarians, *Integration of Damage Mechanism Review with Process Hazard Analyses*, presented at the 2015 Mary Kay O'Connor Process Safety Symposium, October 2015.

[8] *Damage Mechanisms Affecting Fixed Equipment in the Refining Industry.* American Petroleum Institute, API Recommended Practice 571, Second Edition, April 2011

[9] *Improving Chemical Facility Safety and Security.* Executive Order 13650, August 1, 2013

[10] *CSB Board Members Identify Modernization of Process Safety Management Regulations as the Agency's Second "Most Wanted Safety Improvement.* December 1, 2014, Available at http://www.csb.gov

Appendix A - Proposed DMR Regulations

California Accidental Release Prevention Program (CalARP)	Process Safety Management
CCR Title 19, Division 2, Chapter 4.5, 2762.5 (Mechanical Integrity), September 14, 2015 [Reference 3]	GISO §5189.1, Process Safety Management for Petroleum Refineries, Version 5. - September 14, 2015 [Reference 4]
(f) Damage Mechanism Review	**(k) Damage Mechanism Review**
(1) The owner or operator shall complete a Damage Mechanism Review (DMR) for each process for which a damage mechanism exists. Where no DMR is performed, the owner or operator shall document the rationale for the determination that no damage mechanism exists. The owner or operator shall determine and document the priority order for conducting the DMR based on process operating history, PHA schedule and inspection records. No less than 50 percent of the initial DMRs shall be completed within three (3) years of the effective date of this Article, and the remainder within five (5) years of the effective date of this Article. If the employer has conducted and documented a DMR for a process unit within five (5) years prior to the effective date of this Section, and that DMR includes the elements identified in paragraph (f)(6), that DMR may be used to satisfy the employer's obligation to complete an initial DMR under this paragraph.	*(1) The employer shall complete a Damage Mechanism Review (DMR) for each process for which a damage mechanism exists. Where no DMR is performed, the employer shall document the rationale for the determination that no damage mechanisms exist. The employer shall determine and document the priority order for conducting DMRs based on the process operating history, the PHA schedule and inspection records.* *(2) The employer shall complete no less than fifty (50) percent of initial DMRs within 3 years and all remaining DMRs within five (5) years of the effective date of this Section. If the employer has conducted and documented a DMR for a process unit up to five (5) years prior to the effective date of this Section, and that DMR includes the elements identified in (k)(8), that DMR may be used to satisfy the employer's obligation to complete an initial DMR under this subsection.*
(2) A DMR shall be revalidated at least once every five (5) years.	*(3) A DMR shall be revalidated at least once every five years following its initiation.*
(3) A DMR shall also be conducted where a damage mechanism exists on new processes or a major change, prior to initial approval of the change.	*(4) Where a damage mechanism exists, a DMR shall also be conducted on new processes and as part of a major change, prior to approval of the change.*
(4) As part of an incident investigation pursuant to Section 2762.9, where a damage mechanism is identified as a contributing factor, the owner or operator shall review the most recent DMR(s) that are relevant to the investigation. If a DMR has not been performed on the processes that are relevant to the investigation, a DMR shall be completed prior to completing the incident investigation.	*(5) Where a damage mechanism is identified as a contributing factor in an incident investigation, pursuant to subsection (o), the employer shall review the most recent DMRs that are relevant to the investigation. If a DMR has not been performed on the processes that are relevant to the investigation, a DMR shall be completed as part of the incident investigation.*
(5) The DMR for a process unit shall be available to the team performing a PHA for that process unit.	*(6) The DMR for a process unit shall be available to the team performing a PHA for that process unit.*

California Accidental Release Prevention Program (CalARP)	Process Safety Management
(6) The DMR shall be performed by a team with expertise in engineering, operation of the processes under review, equipment and pipe inspection, and damage and failure mechanisms. The team shall also include one member knowledgeable in the specific DMR method being used. The owner or operator shall provide for employee participation in this process, pursuant to Section 2762.10.	*(7) The DMR shall be performed by a team with expertise in engineering, equipment and pipe inspection, damage and failure mechanisms, and the operation of the process or processes under review. The team shall include one member knowledgeable in the specific DMR methodology being used. The employer shall provide for employee participation in this process, pursuant to subsection (q).*
(7) The DMR for each process shall include: (A) Assessment of Process Flow Diagrams (PFDs); (B) Identification of all potential damage mechanisms; (C) Determination that the materials of construction are appropriate for their application and are resistant to potential damage mechanisms; (D) Qualitative assessment of the rate at which damage is likely to occur; (E) Methods to prevent or mitigate damage; (F) Review of operating parameters to identify operating conditions that could accelerate damage or that could minimize or eliminate damage;	*(8) The DMR for each process shall include: (A) Assessment of Process Flow Diagrams (PFDs); (B) Identification of all potential damage mechanisms, pursuant to (k)(9); (C) Determination that the materials of construction are appropriate for their application and are resistant to potential damage mechanisms; (D) Methods to prevent or mitigate damage; and, (E) Review of operating parameters to identify operating conditions that could accelerate or otherwise worsen damage, or that could minimize or eliminate damage.*
(8) For purposes of this section, damage mechanisms include, but are not limited to: (A) Mechanical loading failures, such as ductile fracture, brittle fracture, mechanical fatigue, and buckling; (B) Erosion, such as abrasive wear, adhesive wear, and fretting; (C) Corrosion, such as uniform corrosion, localized corrosion, and pitting; (D) Thermal-related failures, such as creep, metallurgical transformation, and thermal fatigue; (E) Cracking, such as stress-corrosion cracking; and (F) Embrittlement, such as high-temperature hydrogen attack.	*(9) For purposes of this subsection, damage mechanisms include, but are not limited to: (A) Mechanical loading failures, such as ductile fracture, brittle fracture, mechanical fatigue and buckling; (B) Erosion, such as abrasive wear, adhesive wear and fretting; (C) Corrosion, such as uniform corrosion, localized corrosion and pitting; (D) Thermal-related failures, such as creep, metallurgical transformation and thermal fatigue; (E) Cracking, such as stress-corrosion cracking; and, (F) Embrittlement, such as high-temperature hydrogen attack.*
(9) DMRs shall include an assessment of previous experience with the process including the inspection history and all damage mechanism data; a review of industry-wide experience with the process; and applicable standards, codes and practices.	*(10) DMRs shall include an assessment of previous experience with the process, including the inspection history and all damage mechanism data; a review of industry-wide experience with the process; and all applicable standards, codes and practices.*

California Accidental Release Prevention Program (CalARP)	Process Safety Management
(10) At the conclusion of the analysis, the team shall prepare a written DMR report that includes: *(A) The process unit(s) reviewed;* *(B) Damage mechanisms analyzed;* *(C) Results of the analyses conducted according to subsection (f)(7) above;* *(D) Recommendations for temporary mitigation;* *(E) Recommendations for prevention.*	*(11) At the conclusion of the analysis, the team shall prepare a written DMR report, which shall include the following:* *(A) The process unit and damage mechanisms analyzed;* *(B) Results of all analyses conducted, pursuant to (k)(8);* *(C) Recommendations for temporarily mitigating damage; and,* *(D) Recommendations for preventing damage.*
(11) The report shall be provided to and, upon request, reviewed with all operating, maintenance, and other personnel, whose work assignments are within the process unit covered in the DMR.	*(12) The report shall be provided to and, upon request, reviewed with employees whose work assignments are within the process unit described in the DMR.*
(12) The owner or operator shall follow the corrective action work process documented in subsections 2762.16(d) and (e) when resolving the DMR team's findings and recommendations, determining corrective action for implementation, tracking to completion, and documentation of closeout.	*(13) The employer shall implement all recommendations in accordance with subsection (x).*
(13) DMR reports shall be retained for the life of the process unit.	*(14) DMR reports shall be retained for the life of the process unit.*

A Dynamic Duo of MOC Metrics – Using Quality and Age Scores to Drive Continuous Improvement of Management of Change Process

Jean M. Cronin
Cabot Corporation
3603 S. Saginaw Road, Midland, MI
jean.cronin@cabotcorp.com

Matthew J. Allen
Cabot Corporation
157 Concord Road, Billerica, MA
matthew.allen@cabotcorp.com

Prepared for Presentation at
American Institute of Chemical Engineers
2016 Spring Meeting
12th Global Congress on Process Safety
Houston, Texas
April 11-23, 2016

A Dynamic Duo of MOC Metrics – Using Quality and Age Scores to Drive Continuous Improvement of Management of Change Process

Jean M. Cronin
Cabot Corporation
3603 S. Saginaw Road, Midland, MI
jean.cronin@cabotcorp.com

Matthew J. Allen
Cabot Corporation

Keywords: management of change, MOC, metrics, continuous improvement

Abstract

Most chemical companies that have implemented a management of change (MOC) process are committed to continuously improving the process. Knowing how to measure and communicate MOC improvement progress can be difficult. Many common MOC metrics are labor intensive to generate, measure the mechanics of the process rather than the outcome, focus on the individual parts of the process rather than the whole, and/or are difficult to understand easily.

Cabot Corporation (Cabot) has implemented an MOC audit process that uses a simplified scoring method to measure the quality of individual MOC requests. Scores are determined for three stages of an MOC request: before approval, before startup, and before completion. These scores are used to calculate the overall quality score for the MOC request. The individual MOC scores can then be rolled up into an MOC quality score for the site.

In addition, Cabot developed an easily calculated single-number site MOC age score to measure whether MOC requests are being completed in a timely manner. The age score complements the quality score because time and quality are sometimes seen as trade-offs.

This paper will describe the methods used to calculate each metric. It will also address how to adapt the metrics to other businesses and/or goals. In addition, it will provide some implementation suggestions based on Cabot's experience.

These two metrics give each site the ability to easily track their MOC process. Although they do not communicate everything about the MOC process, the metrics are good indicators of the overall health of the process. Due to other concurrent improvement efforts, it is difficult to determine the direct impact of improving the MOC scores on Cabot's overall process safety performance. However, the business that originally implemented the metrics has experienced nearly 40% fewer process safety events (year over year) since the MOC metrics were launched.

1 Background

Most chemical companies, including Cabot Corporation (Cabot), that have implemented a management of change (MOC) process are committed to continuously improving the process. At Cabot, the MOC process has evolved from a paper-based system managed by each site to a corporate electronic-based MOC (eMOC) system used by all sites. After both the corporate MOC standard and eMOC system had been in place for several years, corporate audits and site self-assessments indicated that the Fumed Metal Oxides (FMO) business had an extremely high compliance with the standard and demonstrated an excellent MOC culture. It became apparent that new metrics were needed to drive further improvement.

However, knowing how to measure and communicate MOC improvement progress can be challenging. Many common MOC metrics are labor intensive to generate (such as % of changes that used the site's MOC procedure prior to making the change), measure the mechanics of the process rather than the outcome (such as average time from initiation to close out), focus on the individual parts of the process rather than the whole (such as % of temporary MOCs closed out on time), and/or are difficult to understand easily. As a result, in 2012 the FMO business developed an MOC audit process that uses a simplified scoring methodology to measure the quality of individual MOC requests. In 2013, FMO designed and implemented a complimentary metric to assess a site's performance in managing the age or lifecycle of their MOCs.

2 MOC Quality Score

2.1 MOC Quality Score Design

The MOC quality metric is designed to generate a single MOC quality score for each MOC that is audited. From the individual MOC scores, an overall site or business MOC quality score can be calculated. A simplified scoring system with minimum of 0 and maximum of 1 was chosen to help manage the time needed to complete the quality audits. Each element of an MOC is scored on this basis. A standard spreadsheet template was generated to ensure consistency and to ease implementation.

2.1.1 MOC Quality Score Elements

The MOC quality audit is intended to determine scores for three stages of an MOC request: before approval, before startup, and before completion. These scores are used to calculate the quality score for the whole MOC request. The individual MOC scores can then be rolled up into an overall MOC quality score for the site or business.

For each of the three stages of an MOC, the key elements for that stage were identified. Table 1 lists the various elements of the Cabot MOC process identified for each stage. By separating the elements by stage, it is easier to identify where in the MOC process issues are occurring.

Table 1. MOC Quality Score Elements by Stage

Before Approval Stage	Before Startup Stage	Before Completion Stage
Change category	Emergency MOC	Temporary MOC
Technical basis of change	PSSR	Documents updated (final)
Facility risk ranking	Communication with affected employees	Action tracking
Design safety review	Training	
PHA (process hazard analysis) needed	Documents updated (red line)	
PHA documentation		
PSSR required		
Documents to be updated list		

2.1.2 MOC Quality Scoring System

Cabot chose a simplified scoring system was chosen to help manage the time needed to complete the quality audits. Each element is generally scored as 0, 0.5, or 1.0, but two of the elements (PSSR required and temporary MOC) only have scoring options of 0 and 1.0. In addition to managing audit time, the simplified system makes it easier to ensure that different sites are evaluating the criteria in a similar way. With only three (or two) score choices, the differences between 0 and 0.5 and between 0.5 and 1.0 are much more easily understood.

2.1.3 MOC Quality Score Criteria

One of the fundamental philosophies of the MOC quality score is to align the maximum scores with agreed upon best practices, not just with compliance. In several instances, the FMO business has chosen to implement MOC practices that go above and beyond the regulatory requirements and the corporate MOC standard. Unless these higher practices are met, the maximum MOC quality score of 1.0 cannot be achieved.

For elements with three score options, the maximum score of 1.0 is only given if all of the criteria for compliance and best practice, if applicable, are met. The minimum score of 0 is given if the minimum criteria are not met, and the intermediate score of 0.5 is given if the minimum criteria are met but all of the compliance/best practice criteria is not met.

In addition to the score criteria, for some elements, the spreadsheet tool also includes notes that give more information and/or examples of the criteria to aid the auditor.

Table 2 shows the scoring criteria for several different elements.

Table 2. MOC Quality Scoring Criteria

<table>
<tr><th>Element</th><th>Score</th><th>Score Criteria</th><th>Score Criteria Notes</th></tr>
<tr><td rowspan="3">Technical Basis of Change</td><td>1.0</td><td>"Technical basis of change" and/or "Summary of change" sections contain all specific information on change.</td><td rowspan="3">It is acceptable to refer to "Summary of Change" section if all required information is contained there.

Examples of specific vs. generic information:

- Valve changed to model #VW23 vs. valve changed to new model.

- Piping material changed to carbon steel with grade and material compatibility information vs. piping material changed to carbon steel (no grade and/or material compatibility information).</td></tr>
<tr><td>0.5</td><td>"Technical basis of change" and/or "Summary of change" sections contain some but not all specific information.</td></tr>
<tr><td>0</td><td>"Technical basis of change" and/or "Summary of change" sections contain only generic information.</td></tr>
<tr><td rowspan="2">Emergency MOC</td><td>1.0</td><td>Appropriate emergency approvers are documented on MOC, and MOC request has been submitted for full approval within 2 working days of raising MOC request.</td><td rowspan="2">None</td></tr>
<tr><td>0.5</td><td>Appropriate emergency approvers are documented on MOC.</td></tr>
</table>

Element	Score	Score Criteria	Score Criteria Notes
	0	Emergency approvers are not documented.	
	NA	Not applicable if request is not an emergency MOC.	
Temporary MOC	1.0	Change reversed or extended/made permanent via new MOC request and documented as such prior to expiry date.	None
	0	Change not handled correctly before expiry date.	
	NA	Not applicable if request is not a temporary MOC	
Action tracking	1.0	All post start up actions (MOC, PHA, PSSR, etc.) entered into action tracking database with due dates and responsible parties assigned.	None
	0.5	Some but not all post start up actions entered into action tracking database with due dates and responsible parties assigned.	
	0	Post start up actions not entered into action tracking database or entered without due dates and/or responsible parties assigned.	

2.1.4 Determining Number of MOC Requests to Score

Cabot developed a formula to determine how many MOC requests should be audited. MOC quality audits were originally done monthly, and therefore, the number of requests to be audited was based on the number of requests that had been modified during the previous month. However, requests that were cancelled or that were in a status where scope was still being developed (send to responsible person, awaiting submission, DCS impact review) were not included. MOC requests in all other statuses were included meaning that an MOC was eligible for quality scoring before it was complete or even implemented.

The formula calculating how many MOC requests to audit was based on starting at a minimum audit size of 3 for a population of 5 MOCs. The audit size then increases with MOC population, but the required number of audit reduces by 35% each time the population size doubles to keep the audit size manageable. The maximum audit size was set at 10 per month (equivalent to population of 119 MOCs).

$$MOC\ audit\ size = 1.63 * Number\ of\ modified\ MOC's^{0.38}$$

$$\{MOC\ audit\ size\text{: minimum } of\ 3\ and \text{ maximum } of\ 10\}$$

2.1.5 Calculating MOC Quality Scores

The scores for each of the three MOC stages (before approval, before startup, and before completion) are calculated by averaging the scores assigned for each individual element in that stage. Any element that is 'not applicable' is ignored for the purposes of these calculations.

The quality score for the whole MOC request is calculated by averaging the scores assigned for all of the individual elements, not by averaging the scores for the three MOC stages. This results in each element having the same weight in the overall MOC quality score. Again, any element that is 'not applicable' is ignored for this calculation.

The MOC quality scores for each stage and the overall MOC quality score for the site are calculated by averaging the scores for each of the audited MOC requests. Using this method rather than averaging all of the scores for the individual elements was felt to give a more representative view of the site MOC quality with each MOC given equal weight regardless of how many individual elements were not applicable.

The MOC quality scores for each stage and the overall MOC quality score for the FMO business are calculated by determining the average for each of the site scores weighted by the number of MOCs audited at each site. This ensures that each MOC carries the same weight in the business MOC quality scores.

2.2 MOC Quality Score Implementation

The MOC quality metric was developed in 2012 by the FMO PSM (process safety management) leaders group which consists of the PSM leaders from each of the five FMO sites and the PSM coordinator for the FMO business. Once the criteria were defined, the sites did monthly audits for three months to test the system. Several small adjustments were made to the audit spreadsheet during this period to address errors and/or to make the file easier for the auditors to use (conditional formatting to highlight cells to complete, fixed entry choices, etc.).

During the next six months, the sites continued to perform monthly audits for tracking purposes only. One of the early lessons was that the sites needed refresher MOC training. Previous training had focused on how to use the electronic MOC system and on the requirements of the corporate MOC standard. A new training package was developed and rolled out which contained more guidance on meeting the expectations for a high-quality MOC request, including best practices that went above and beyond compliance with the corporate standard.

Following nine months of audits, the audit tool was felt to be stable and useful enough to set a goal for the FY14 continuous improvement plan (CIP). An overall improvement of 8% was achieved in FY14. The MOC quality score remains a metric tracked on the CIP through FY16. Figure 1 shows the improvement in overall MOC quality for the sites, and figure 2 shows the improvement in the four quality scores at the business level.

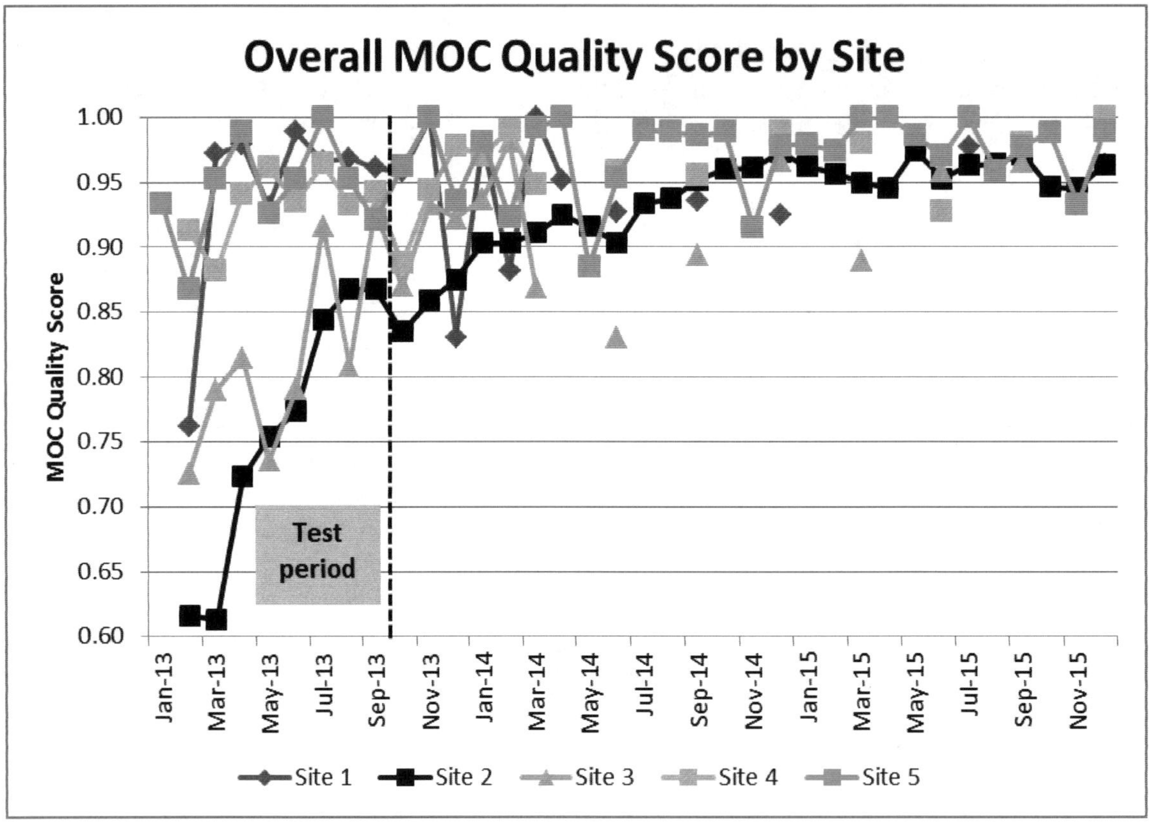

Figure 1. Overall MOC Quality Score by Site

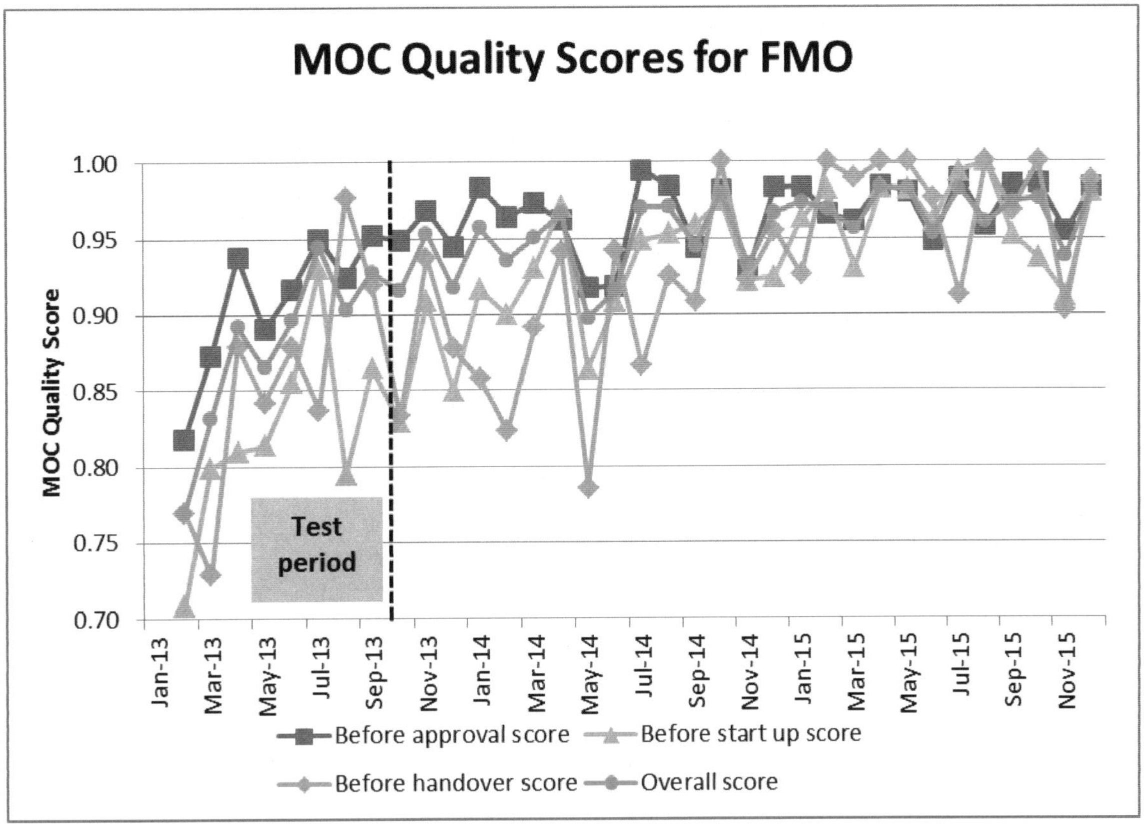

Figure 2. MOC Quality Scores for FMO

It should be noted that the various FMO sites chose to drive improvement using different methods depending on the maturity of their PSM culture, the specific areas needing improvement, how many people are involved in the MOC process, and the overall site culture. However, sites periodically reviewed their improvement plans with the other FMO sites to enable sharing of good practices. For example, the largest FMO site chose to schedule a regular MOC review meeting to ensure quality of MOCs before they were submitted. Other smaller sites chose to have their nominated PSM leader be a reviewer or approver of all MOCs. At the youngest site, targeted MOC refresher training was held several times as specific areas with low scores were identified. All sites' improvement efforts included one-on-one feedback to the person responsible for the MOC.

2.3 Improvements to MOC Quality Score Process

2.3.1 Quarterly MOC Quality Audits

Over time, several of the sites achieved a stable and high level of performance. At this point, the PSM group decided to allow the sites to perform MOC quality audits on a quarterly basis instead of on a monthly basis. The maximum audit size was increased to 15 per quarter, but the calculation of the MOC requests to audit was otherwise unchanged for those sites performing quarterly audits.

As of January 2016, three of the five sites are undertaking quarterly audits. The two sites who continue to perform monthly MOC quality audits do so for different reasons.

One of the sites performing monthly audits is the largest and most complex FMO site, and as such, their volume of MOCs is significantly greater than the other sites. Carrying out MOC quality audits on a monthly basis allows the PSM leader to manage his auditing time more easily with the time commitment spread over three months.

The second site undertaking monthly audits is the youngest FMO site, and as such, their PSM culture, while robust, is still in need of frequent, regular feedback to maintain performance. The site chooses to perform monthly audits to ensure that standards continue to be maintained at a high level.

2.3.2 Aid to Generate a Representative Sample to Audit

After some of the sites moved to quarterly audits, it became more difficult to choose a representative sample of MOCs to audit. Previously, the site PSM leaders rather easily filtered through the file with all of the MOCs modified in the previous month to pick the MOCs to audit. However, now the list was longer and was not so easily reviewed.

A PivotTable was added to the spreadsheet containing the list of MOCs modified during the previous month/quarter. The PivotTable contains filters for site, month, and whether the MOC is an emergency and/or temporary MOC. It provides the number of MOCs by responsible person and by change type versus the MOC stage. The site PSM leader can now easily identify individuals, types of changes, and stages that have more MOCs and thus should be represented in the audit. The filters ensure that the audit includes MOCs audited include emergency and/or temporary MOCs as needed.

3 MOC Age Score

During the initial trial period for the MOC quality score, the FMO PSM leaders group realized that the MOC quality score was missing a critical element: MOC lifecycle time. The group decided to implement a separate but complimentary score, the MOC age score. Since quality and time are often viewed as competing priorities, the group felt that tracking the two metrics would lead to better overall MOC performance.

3.1 MOC Age Score Design

The MOC age score is designed to give an overview of the relative ages of the open MOCs at a site. However, the group felt that simply calculating the average days open was too simplistic and didn't put enough emphasis on older open MOCs.

One plant already had some goals on their site scorecard to reduce the number of old, open MOCs, and these goals were incorporated into the MOC age score. The intent was to trial the age score and then to adjust the goals if needed. The original goals are still in place at this time, but the scoring system is flexible enough to accommodate updated goals if needed in the future.

To achieve the maximum MOC age score of 1, the site must have no open MOCs more than a year old and must have less than 20% of open MOCs more than 6 months old. The site that had these goals on their existing scorecard had chosen the 20% figure in recognition that MOCs for work to be done during a planned/scheduled outage may be raised significantly before the work is actually done.

In addition, the group felt that there should be a higher penalty for an open MOC over 2 years old versus an open MOC over 1 year old. Similarly, an open MOC over 1 year old should result in a greater penalty than an open MOC over 6 months old.

The resulting MOC age score calculation is shown below.

$$MOC\ age\ score = 1.0 - X * 1.5 - Y * 1.0 - (Z - 20\%) * 0.5$$

X = % of open MOCs over 2 years old

Y = % of open MOCs over 1 years old but less than 2 years old

Z = % of open MOCs over 6 months old but less than 1 year old

3.2 MOC Age Score Implementation

The MOC age metric was developed in early 2013. Once the calculation was defined, the PSM coordinator for the FMO business began calculating and communicating the sites' MOC age scores on a monthly basis. During the first 6 months, the MOC age score was calculated for tracking purposes only. This gave the sites time to communicate expectations regarding timely completion of MOCs and to improve their site scores before the metric was communicated more widely.

Following the first 6 months of use, the MOC age metric was agreed to be useful enough to set a goal for the FY14 CIP. An overall improvement of 9% was achieved in FY14. The MOC age score remains a metric tracked on the CIP through FY16. Figure 3 shows the improvement in overall MOC age scores for the sites.

As with the MOC quality score, the various FMO sites chose to drive improvement using different methods but periodically reviewed their improvement plans with the other FMO sites to enable sharing of good practices. For example, the smallest FMO facility set a goal on their scorecard to reduce the number of open MOCs older than 2 years old for FY14 and then in FY15 set a scorecard goal to reduce the number of open MOCs older than 1 year old. This enabled them to make small, continuous improvements over time. One of the larger sites chose to dedicate a few individuals to address the older MOCs in a targeted window which led to a dramatic score increase over just a few months.

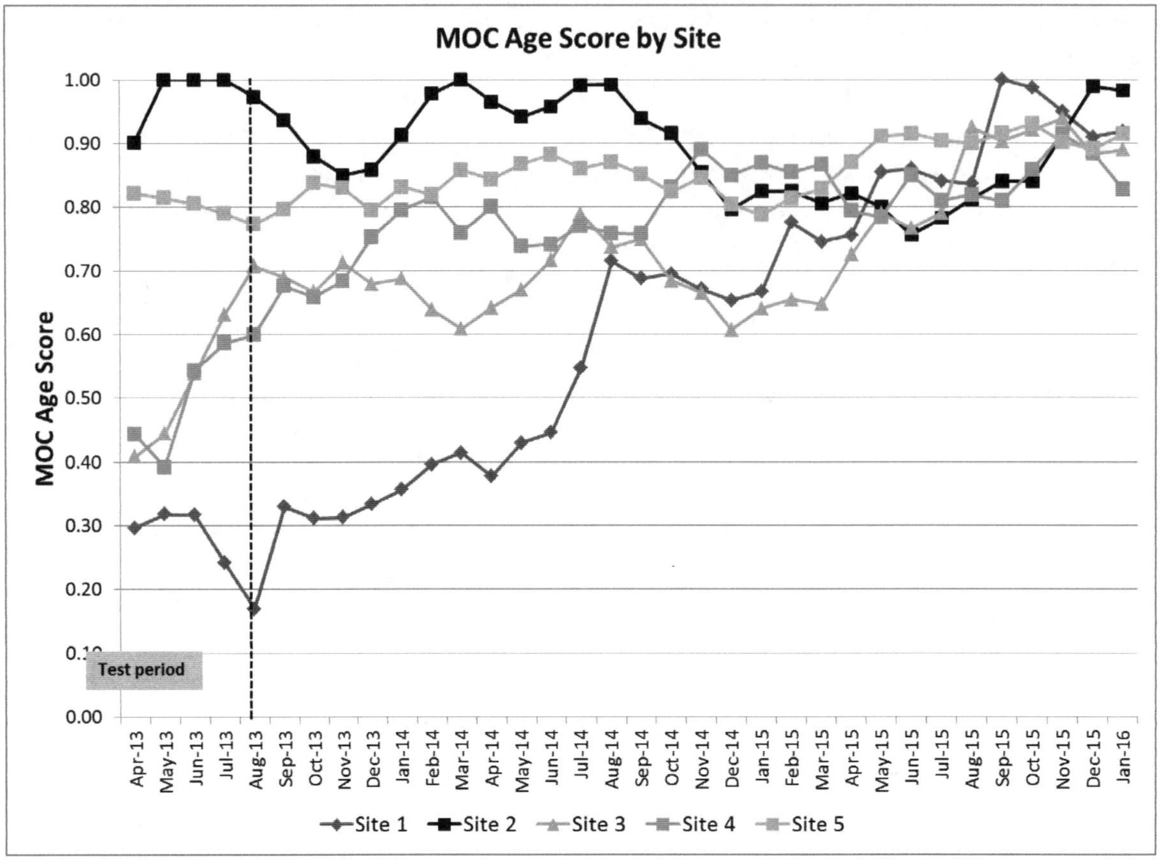

Figure 3. MOC Age Scores for FMO

4 Results and Next Steps

4.1 FMO Results

Due to other concurrent improvement efforts, it is difficult to determine the direct impact of improving the MOC quality and age scores on FMO's overall process safety performance. The FMO business has experienced nearly 40% fewer process safety events in the last fiscal year as compared to previous fiscal year which was the first full year of the MOC metrics being in place. See Figure 4 for comparison of number of process safety events and MOC metrics over the past three fiscal years. Note that both MOC metrics were put in place part way through FY13 and that FY13 was the test period.

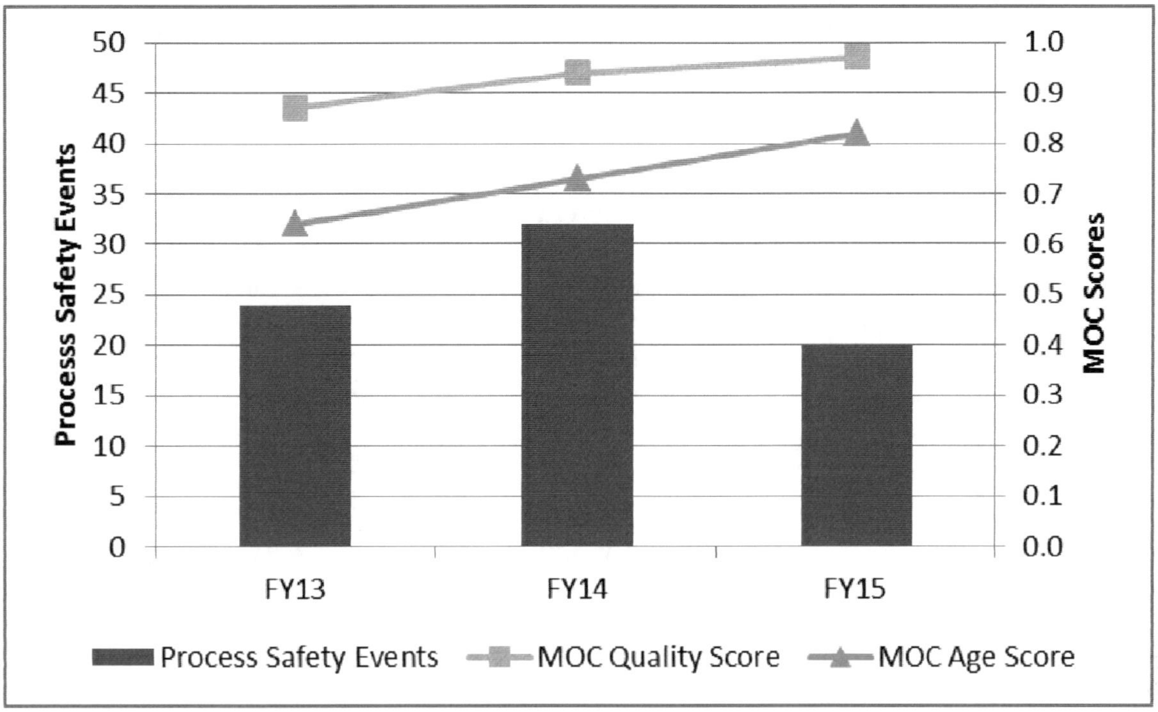

Figure 4. MOC Scores and Number of Process Safety Events for FMO

4.2 Cabot Use

After the FMO business developed the MOC quality and age scores, they were shared with the global Cabot PSM group. Other PSM leaders then began using the scores to evaluate a site's MOC performance before a planned visit to the site.

In FY15, other businesses began to employ the MOC scores. A regional PSM leader leveraged one of the FMO sites to implement the two MOC scores at the other three Cabot sites in the same country. In addition, another regional PSM leader implemented the two MOC scores at the two sites in a nearby country.

4.3 Future at Cabot

Those sites using the MOC scores over the last few years have identified some opportunities that have not yet been tried. Some of these ideas are listed below:

- Cross-site MOC quality checks, e.g. audits of site 1 MOCs by site 2, to validate MOC quality scores are equivalent
- Modifying MOC quality metric to weight elements to give more emphasis to some elements
- More use of MOC quality audit data by person, status, individual elements, MOC type, etc.

- Plotting MOC quality and age scores on SPC (statistical process control) charts

In addition, the existing MOC scores do not address the potential for changes that should be covered by the MOC process but aren't for some reason. The Cabot corporate MOC audit protocol does evaluate whether changes are being managed as required using the MOC process. However, there may be an opportunity to modify or amend the MOC scores to check for uncontrolled changes on a more frequent basis. For example, an additional score could be developed to analyze if there were process safety events where the investigation identified missing MOC or problems with the MOC process. Alternatively (or in addition), the score could review PHA revalidations to identify changes versus the previous PHA that should have covered by the MOC process but were not.

4.4 Keys to Success

Looking back on the implementation of the MOC quality and age scores, several factors were keys to the success of the improvement program. Some of the factors affected just the MOC quality score or the MOC age score, but several of them had an effect on both of the MOC scores. Table 3 lists the various keys to success.

Table 3. Keys to Success for Implementing MOC Quality and Age Scores

<table>
<tr><th>MOC Quality Score</th><th>MOC Age Score</th></tr>
<tr><td>Simple three-level scoring</td><td rowspan="2">Calculation accommodates MOCs open longer for to prepare for annual outages</td></tr>
<tr><td>Audit sample size weighted toward low end</td></tr>
<tr><td>Criteria reflects best practice, not compliance</td><td>Data for calculation easily downloaded from eMOC system</td></tr>
<tr><td>Don't wait until closeout or even after approval to audit</td><td>Older MOCs result in higher penalties</td></tr>
<tr><td colspan="2">Maximum score of 1, rather than 100% as not a true ratio</td></tr>
<tr><td colspan="2">Get input from sites, and test metrics</td></tr>
<tr><td colspan="2">Goals not set right away</td></tr>
<tr><td colspan="2">Keep off scorecard/dashboard initially to allow time to work out any bugs and improve a bit before putting on management's radar screen</td></tr>
<tr><td colspan="2">Share information across sites, but let sites drive their own improvement plans</td></tr>
</table>

MOC Quality Score	MOC Age Score
Before implementing MOC scores, ensure site MOC culture is mature.	
Owner at business/regional level to dump data for sites each month (~15 minutes/month)	

5 Conclusion

The Cabot FMO business has implemented two complimentary MOC metrics: MOC quality score and MOC age score. The two metrics are easy to determine and provide a good overview of the performance of a site's MOC process. Improving these two metrics appears to have contributed to improvement in the overall business PSM performance with a significant reduction in the number of process safety events.

Other companies can certainly use the Cabot version of the MOC quality audit spreadsheet and MOC age score calculation to check a sample of their MOCs to determine if their company would find the MOC scores of value. It is critical though to then modify the Cabot version and tailor the criteria for both scores to the specific business/company. The design of the two scores makes it quite easy to modify the scoring criteria. For example, if a business has a best practice for master PHAs to be updated as MOCs are closed out rather than waiting for the PHA revalidation cycle, then the criteria to achieve a maximum score on the PHA documentation element should include this best practice. As another example, if a company generally has outages every two years instead of annually, the age score calculation may need to be modified to reduce the score only after 20% of MOCs are open longer than two years, or alternatively, to reduce the penalty for having an MOC open longer than two years.

441507 Using Data Analytics to Improve Process Safety

Abstract:

Using Data Analytics to Improve Process Safety

Increasingly, data from multiple sources are compiled into central storage systems for entire facilities and in some cases entire corporations. The low cost of digital memory and the increasing process speed of modern microprocessors and computers continue to make it easier to analyze, manipulate and merge very large databases. Big Data and Data Analytics have been touted as allowing staff at all levels to make real time, data-driven business decisions. These concepts can be used to improve process safety at facilities by proactively analyzing sources of data to identify opportunities for improvement in operating practices, identify and track near misses, and in real time detect incipient instrument and equipment failures that could lead to catastrophic incidents. The aviation industry is one example that is voluntarily implementing similar analytics. Over the last decade the aviation industry has implemented flight operations quality assurance (FOQA) or flight data monitoring (FDM) programs which collect, store, and analyze recorded flight data with the goal of increasing overall safety.

In this presentation we show examples of analytics Exponent performed on data from facilities. These analyses were used to analyze alarm frequencies during specific types of operations, identify near misses, and proactively detect failures of instrumentation and equipment in order to prevent or mitigate catastrophic incidents.

442485 Safety Culture and Leadership- Independent of Language

Abstract:

The existence of safety in the workplace is mandatory for a sustainable business. Strong safety cultures are characterized by human factors, workforce engagement and fully committed safety leadership. With that being said, increased diversity in work environments and international recruitment can oftentimes complicate the process of implementing a common safety culture.

More specifically, those who come from different types of work environments than their current environment may have trouble interpreting the procedures and guidelines that are enforced at their new place of employment. Differences in languages and dialects may also lead to misinterpretations of intended safety guidelines. Lastly, there may be some instances where the workplace follows a non-flexible and hierarchical structure that is not very open to change.

In order to best develop safety practices in the workforce, the senior leaders need to first develop an environment based on the understanding of the different rationales and expectations of a multi-cultural workforce. Once that has been established, communicating with the workforce on safety expectations of the organization and optimizing interpretation of this message is critical.

This paper will continue the discussion from a previous publication; how safety leadership can consider linguistic differences when implementing a safety plan for a diverse workforce. The discussions will emphasize the learnings of the Bradley Curve, showing how important it is to follow safe practices to reach the proximity of an almost 100% incident free area, and will provide real examples from international work assignments and diverse work teams, highlighting this need and offering potential approaches

439897 Indulging in Invincibility (III)- Twelve Steps to Sobriety

Abstract:

<>8/3/2015 <>Indulgence in Invincibility (III)- Twelve Steps to Sobriety <>by Brian P. Kolodji, PE

Could Bhopal, Macondo, or Texas City have been prevented? Directly related examples are presented where PS cultural indulgences contribute to a false sense of invincibility. Light is shed on where the perennial dilemmas of misplaced loyalty, unachievable goals, and belief in corporate higher power contribute to and enable denial of process safety culture failure. Successes and a program are presented where these failed practices are brought out of the shadows and recovery begins.

12TH GLOBAL CONGRESS ON PROCESS SAFETY

Case Study of Relief Analysis of Compressor Stations

Kumud Chaurasia

Process Engineer
Lloyd's Register Energy Americas, Inc.
1330 Enclave Parkway, Suite 200, Houston, TX 77077, USA
Kumud.chaurasia@lr.org

Aniket Patankar

Senior Process Engineer
Lloyd's Register Energy Americas, Inc.
1330 Enclave Parkway, Suite 200, Houston, TX 77077, USA
Aniket.patankar@lr.org

Prepared for Presentation at
American Institute of Chemical Engineers
2016 Spring Meeting
12th Global Congress on Process Safety
Houston, Texas
April 11-23, 2016

Case Study of Relief Analysis of Compressor Stations

Kumud Chaurasia, Process Engineer

Aniket Patankar, Senior Process Engineer

Lloyd's Register Energy Americas, Inc.

1330 Enclave Parkway, Suite 200, Houston, TX 77077, USA

kumud.chaurasia@lr.org, aniket.patankar@lr.org

Keywords: Relief analysis, Compressor Station, case study, concern identification

Abstract

A compressor station is a facility which helps the transportation process of natural gas from one location to another. Compressor stations include several key component parts, the primary being the actual compressor unit, either centrifugal or reciprocating. They also typically include scrubbers, strainers or filter separators which remove liquids, dirt, particles, and other impurities from the natural gas. This study will give an overview of the relief analysis conducted for multiple compressor stations. The relief analysis was system based as opposed to relief device based so as to identify any unprotected systems that may need overpressure protection. It will discuss the overpressure scenarios identified for these systems. It will provide some general statistics on the total number of applicable overpressure scenarios. It will give a detailed breakdown of the various concerns found including, but not limited to, undersized devices, unprotected systems, installation concerns, data needs, and documentation discrepancies. The study will discuss possible mitigation solutions for these concerns.

1. Introduction

Compressor stations are typically located along a natural gas pipeline to facilitate the compression of the gas to a specified pressure, which allows the gas to flow along the pipeline to a target location. Since these pipelines typically extend for several hundreds of miles, the gas tends to lose pressure frequently. This necessitates the requirement to compress it frequently during the transportation process.

The size of the station and the number of compressors (pumps) varies, based on the diameter of the pipe, and the volume of gas to be moved. Compressor stations include several key components, the primary being the actual compressor unit, either centrifugal or reciprocating, scrubbers, and strainers or filter separators which remove liquids, dirt, particles, and other impurities from the natural gas.

It is critical to ensure that the natural gas transported to its intended target is free from any contaminants. This is accomplished through four mains steps. The first step is contaminant removal, which occurs upstream of a compressor station near the well head. Here, the raw natural gas, at very high temperature and pressure, is passed through separators in order to remove oil, water, sand etc. It is then passed through a slug catcher, typically located within the compressor station, to remove any leftover condensate. The third contaminant removal step occurs downstream of the compressor unit (after the gas has been pressurized) via a contactor unit that predominantly removes liquid water and carbon dioxide using triethyleneglycol (TEG) as the absorbent. The TEG that separates out at the bottom of the scrubber unit, carrying the water and carbon dioxide with it, is typically regenerated in a glycol regenerator unit before it is recirculated back into the scrubber unit. The final step prior to delivery of the gas to its intended target is to remove any leftover TEG in the gas using a scrubber unit. There are also several tanks within the compressor station for storage of the condensates and TEG. Additionally, the compressor stations contain several components that are part of the glycol regenerator unit including filters, flash tank separators, pumps, heat exchangers, etc. A general process flow diagram of the compression station is shown in Figure 1.

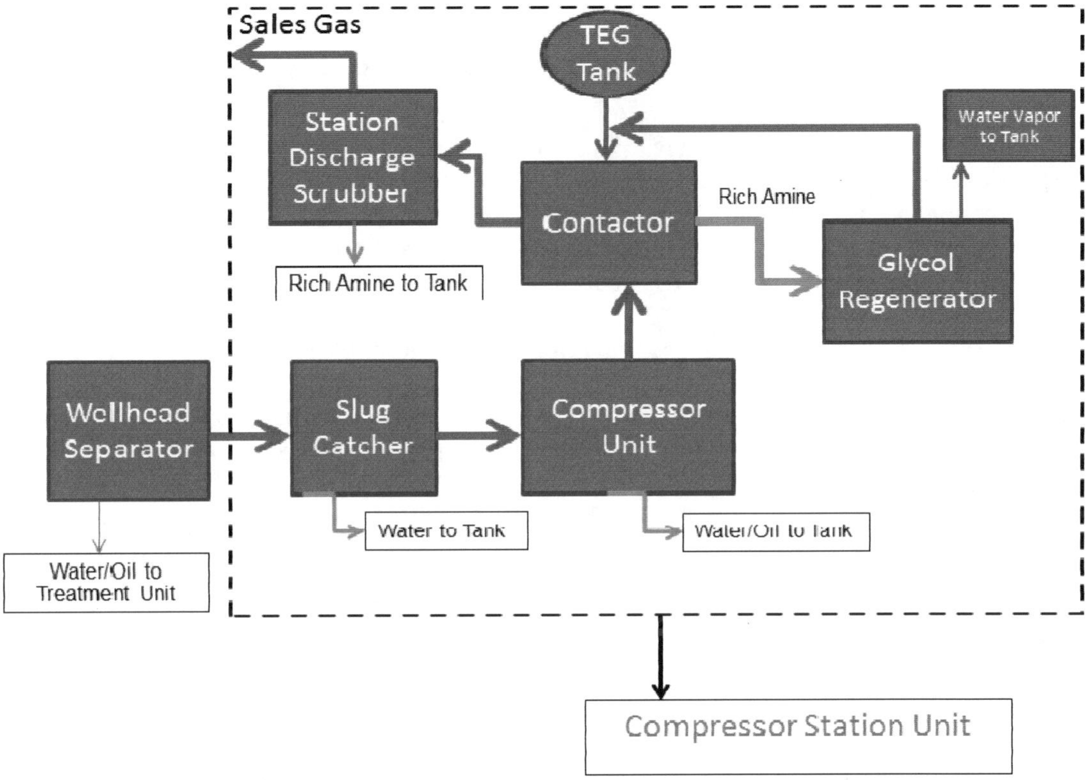

Figure 1: Process flow diagram of Compressor Station

2. Relief Analysis

For this case study, twenty separate compressor stations were analyzed. Specifically, the steps required for performing the analyses, different applicable scenarios, and all the concerns identified with mitigations options are presented. The analysis was conducted using API 520 Part 1 and Part 2 standards, while considering the client's design philosophy as well.

2.1. Steps involved in Relief Analysis

The relief analysis, which is a system based analysis performed for each equipment in the unit, is a methodical four step approach including data and design documentation collection, contingency analysis, pressure relief calculations, concern identification and mitigation. These steps are discussed in detail below.

Data and Design Documentation Collection: Since the client for whom the relief analysis is being performed may not always provide all the information necessary to conduct the analysis, it is imperative that all necessary data and documentation required is collected at the client site prior to commencing the project. Examples of such data collected include isometrics for PSV inlet and outlet piping, PSV and control valve information, equipment and relief device name plate information, U-1A forms, equipment drawings etc. Usually, the piping and instrumentation diagrams (P&IDs) and heat and material (H&M) balance are provided by the client in order to complement the other data collected. All the data collected is then entered into a client project database.

Contingency Analysis: Prior to commencing the contingency analysis, the compressor station is divided into various systems, with each system consisting of several equipment. The interface of systems is defined based on where block valves are situated within the P&ID. Once the systems have been defined, a contingency analysis is performed where all scenarios that could possibly result in an overpressure condition are identified. For the 20 compressor stations that were analyzed in this case study, the following eighteen contingencies were identified: Power Failure, Cooling or Reflux Failure, Steam Failure, Instrument Air Failure, External Fire, Blocked Outlet, Isolated with Heat Input, Equipment Failure, Inadvertent Valve Operations, Failure of Automatic Controls, Gas Breakthrough, Internal Boundary Failures, Reverse Flow, Run Away Reaction, Accidental Mixing, Change in Conditions, Vacuum, Other (applicable scenarios other than listed above).

Calculations: Once the contingency analysis is finished, the required flow rate of the applicable overpressure scenario and the capacity of the relief device (based on the relief device ASME orifice area) are calculated. The inlet and outlet pressure drops, Mach number and sound power level of the piping is also calculated based on the dimensions of the relief device.

Concerns Identification and Mitigation: After the calculations are completed for a system, all the concerns are identified and documented in the database. This is as per the guidelines provided in API 520 Part 1, 2 and the design philosophy provided by client. The concerns are identified based on the calculation results and discrepancies and assumptions made during the analysis. Once all the concerns are identified, recommendations for implementation of mitigation strategies are provided to the client.

3. Results of Case Study

3.1. Relief Device Identification

For the twenty compression stations that were analyzed as part of this case study, the relief devices included all the relief valves, pressure and vacuum vents (PVRVs), goosenecks, and rupture disks. Usually, the relief valves and rupture disks were situated on the pressure vessels while the PVRVs and goosenecks were situated on the low pressure tank. A total of 1099 relief devices were identified for the 20 compressor stations. Depending on the type of relief device, the method of calculation may vary. For example, in the case of PVRVs and goosenecks, it is not necessary to calculate the pressure losses.

The breakdown of the percentage and number of the relief devices, by type, that were analyzed in this study are shown in Table 1 and Figure 2 respectively.

Table 1: Percentage of different relief devices by type

Relief device Type	Percentage
Relief valve	78%
Gooseneck/PVRV's	21%
Rupture Disk	1%

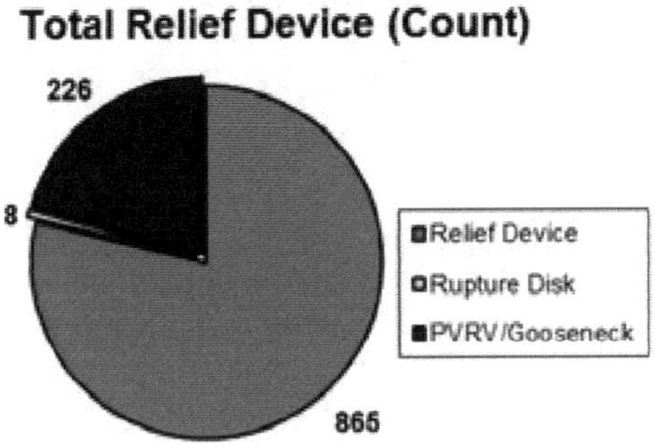

Figure 2: Actual number of different relief devices, by type, for the twenty compressor stations

3.2. Relief Valve Type

There are three types of relief valves that are typically used in the oil and gas industry: conventional, bellows, and pilot valve. Based on the type of valve, the outlet pressure drop limits vary; hence it is important to identify the relief valve type. As shown in Table 2, out of a total of 865 relief valves identified, the majority of the valves in the compression stations analyzed were conventional relief valves,. The breakdown of the number and percentage of the relief valve that were analyzed in this study are shown in Table 2 and Figure 3 respectively.

Table 2: Percentage of different relief valves, by type

Relief Valve Type	Percentage
Conventional	68%
Pilot	23%
Bellows	9%

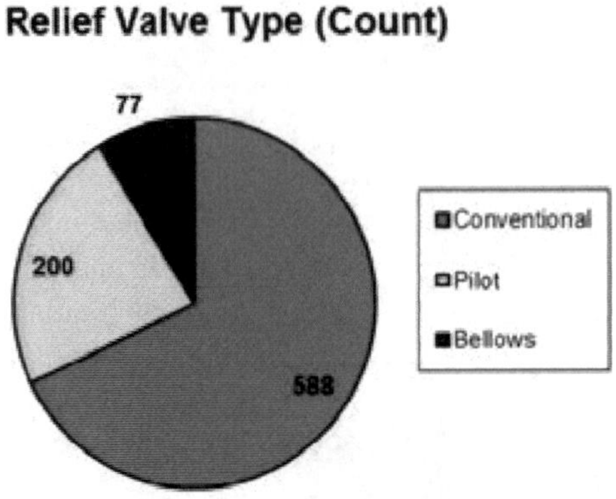

Figure 3: Actual number of relief valves, by type for the twenty compressor stations

3.3. Equipment Type

Based on the equipment type, the overpressure scenario may be applicable or not and the data required to do the analysis will be different. Hence, it is very important to identify the type of the equipment. A total of 2014 equipment were identified in all the compressors stations, of which majority were pressure vessels.

The breakdown of the percentage and number of the equipment type that were analyzed in this study are shown in Table 3 and Figure 4 and respectively. Note that the "Other" type of equipment includes compressors, glycol regenerators etc.

Table 3: Percentage of different equipment, by type

Equipment Type	Percentage
Pressure Vessel	48.5%
Tower	8.1%
Pumps	8.9%
Exchangers	16.6%
Fired Heater	0.1%
Low Pressure Tanks	6.1%
Piping Component	6.4%
Others	5.3%

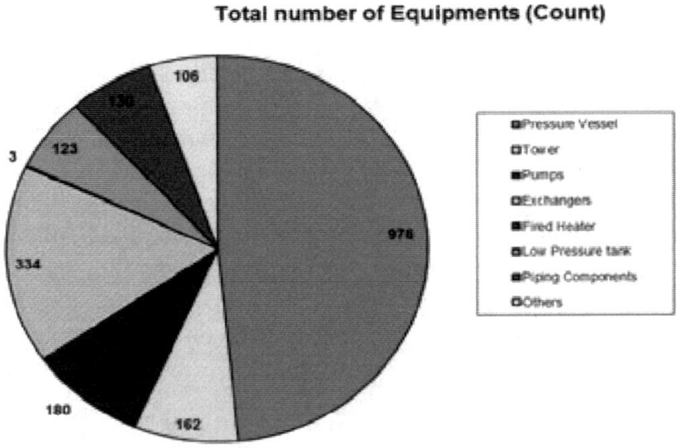

Figure 4: Actual number of equipment, by type, for the twenty compressor stations

3.4. Applicable Overpressure contingencies:

All the applicable scenarios, for each system, were identified after reviewing the P&IDs and other available data. This section shows all the applicable scenarios identified. The breakdown of the percentage and number of the applicable contingencies that were analyzed in this study are shown in Table 4 and Figure 5 respectively. Of a total of 3026 applicable contingencies, the majority of the contingencies in the compression stations analyzed were external or pool fire and blocked outlet.

Table 4: Percentage applicable contingencies by type

Contingency Type	Percentage
External fire	22.8%
Cooling or reflux failure	0.1%
Blocked Outlet	22.6%
Failure of automatic controls	13.3%
Gas breakthrough	8.5%
Inadvertent valve operation	5.6%
Instrument Air failure	2.9%
Internal boundary failure	1.6%
isolated with Heat Input	6.4%
Power failure	0.9%
Reverse flow	3.2%
Vacuum	7.1%
Other	5.0%

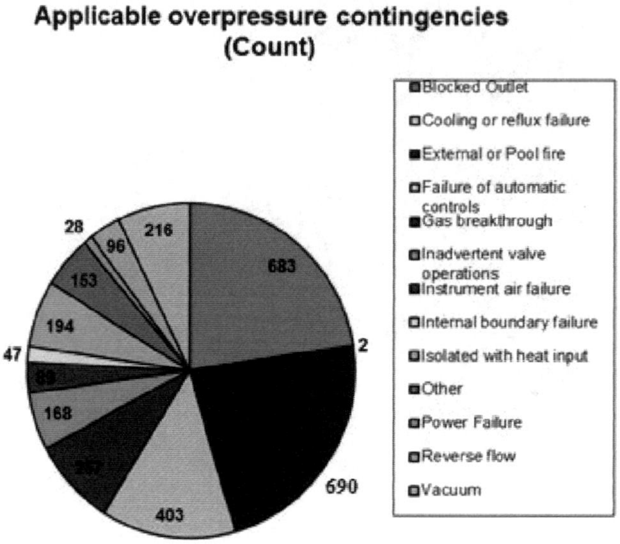

Figure 5: Actual number of applicable overpressure contingencies, by type, for the twenty compressor stations

3.5. Concerns Identification:

This was one of the most important steps in the relief analysis. Based the calculation results, assumptions, discrepancies etc., all the concerns were identified and documented based on the guidelines. The concerns were divided into five major categories: Undersized Concerns, Installation Concerns, Unprotected Systems, Data Requirement, and Documentation Updates. It is important that assumptions made during this stage be documented, verified and approved by the client. The undersized concerns, installation concerns and unprotected system concerns were assigned a higher priority since they could have a direct impact on the operation and failure of the compressor station. Hence, it was recommended that these concerns be mitigated immediately.

The breakdown of the percentage and number of the concerns that were analyzed in this study are shown in Table 5 and Figure 6 respectively. Of a total of 3278 applicable concerns that were identified in this case study, the majority of the concerns in the compression stations were related to Data Requirements.

Table 5: Percentage of concerns identified, by type

Concern Type	Percentage
Undersized	10.4%
Installation	24.6%
Unprotected Systems	9.5%

Data Requirement	50%
Documentation Updates	5.5%

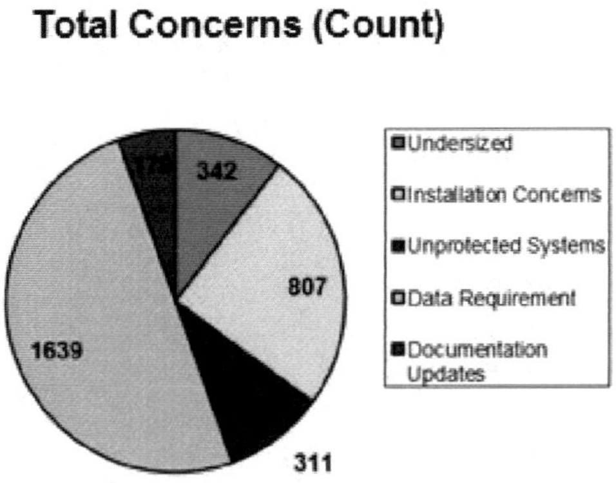

Figure 6: Actual number of different concerns, by type, for the twenty compressor stations

Undersized Concerns:

When the required relief flow rate is more than the capacity of the relief device, the relief device is defined to be undersized. This section shows the overpressure scenarios which were undersized. This includes all the contingencies for which the relief devices were undersized. The most common reasons for relief devices to be undersized were identified as being because of incorrect upstream pressure, incorrect composition of the relief stream, or that the contingency was not identified at all in the previous study.

The breakdown of the number and percentage of the undersized concerns that were analyzed in this study are shown in Table 6 and Figure 7 respectively. Of a total of 342 undersized concerns that were identified in this case study, the majority of the contingencies in the compression stations were Vacuum, Gas Breakthrough, External or Pool Fire and Failure of Automatic Controls.

Table 6: Percentage of type of contingency within undersized concerns

Contingency Type within Undersized Concern	Percentage
Blocked outlet	11.1
External fire	15.2
Failure of Automatic Controls	15.2
Gas breakthrough	15.5
Inadvertent valve operation	1.1
Internal Boundary failure	8.2
Reverse flow	8.5
Vacuum	24

Other	1.2

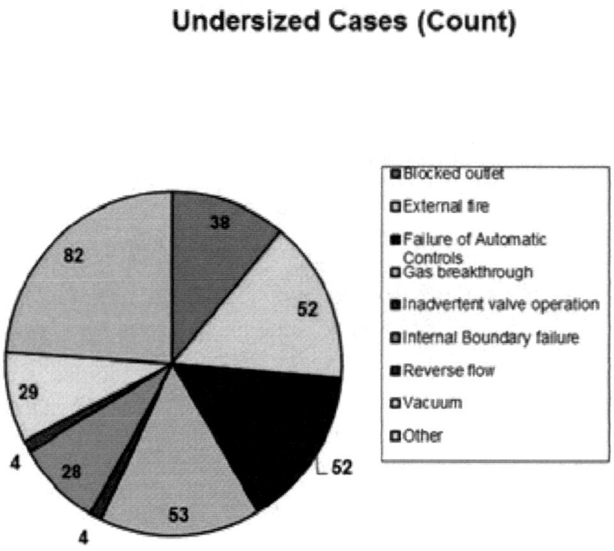

Figure 7: Actual number of different contingency types within the undersized concerns for the twenty compressor stations

<u>Installation Concerns:</u>

The installation concerns (total of 807 were identified) are divided into eleven sub categories as shown below. This is based on the guidelines used to perform the analysis.

Inlet Pressure Drop: The inlet pressure drop is calculated based on the relief device capacity. This includes all the scenarios for which the inlet pressure drop is greater than 3% of set pressure. As per API 520, if the inlet pressure drop exceeds 3%, the potential for capacity reduction and/or mechanical damage may exist for the relief valve. Note that inlet pressure drop concern was not identified for the relief devices which were undersized for one or more overpressure scenarios.

Outlet Pressure Drop: The outlet pressure drop is calculated based on the capacity of the relief device. This includes all the scenarios for which the outlet pressure drop is greater than 10% of set pressure for single conventional valves, or greater than 16% of set pressure for multiple conventional valves except for fire case, greater than 21% of set pressure for conventional valves for external fire, and greater than 50% of set pressure for bellows valves. As per API 520, if the outlet pressure drop exceeds the limit, the potential for capacity reduction and/or mechanical damage may exist for the relief valve. Note that outlet pressure drop concern was not identified for the relief devices which were undersized for one or more overpressure scenarios.

High Exit Velocity: The exit velocity and sound power level of the relief valve piping can affect its stability. All overpressure contingency with Mach number greater than 0.8 and sound power level greater than 155 was documented. Note that the high mach number concern was not identified for the relief devices which were undersized for one or more overpressure scenarios.

High set pressure: The relief device with set pressure greater than maximum allowable working pressure (MAWP) of the equipment was identified since the relief valve can no longer protect the system if this condition exists.

Isolation valve: This includes the block valves that are not chain locked open as required (the most common ones were the block valve on relief device inlet and outlet piping).

Pocketed Piping: In the event of liquid release and if the relief device is pocketed, the liquid can accumulate in the piping and hence the stability of the piping can get affected; hence it is very important to correct this concern.

Restricted line: In the event the inlet segment of the relief device piping is greater than the inlet nozzle as per API 521, these segments needs to be identified and documented.

The breakdown of the number and percentage of the installation concerns that were analyzed in this study are shown in Table 7 and Figure 8 respectively. Of a total of 807 installation concerns that were identified in this case study, the majority was the high outlet pressure drop concern.

Table 7: Percentage of installation concerns by type

Installation Concern	Percentage
High inlet Pressure drop	17.2%
High Outlet pressure drop	24.8%
High Exit Velocity	5.9%
Pocketed Piping	5.3%
Restricted Line	1.7%
Isolation Valves	11.8%
High Set Pressure	0.9%
No Weep Hole	20.8%
Liquid to Atmosphere	1.5%
Valve Stem Rotation	4.0%
Other	6.1%

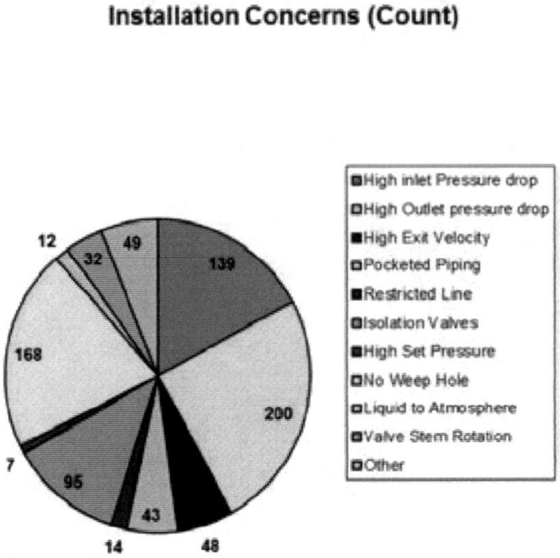

Figure 8: Actual number of installation by type for the twenty compressor stations

Unprotected Systems

These concerns were identified only for those systems that had applicable scenarios without any relief device protecting them. Contingency analysis for all the systems which didn't have any relief device protecting them was performed. As the case study was conducted on system based analysis, it was easier to identify these systems. The systems which had an internal relief device were considered to be unprotected. There were few cases in which there was a block valve present between the system and the relief device; hence these systems were considered unprotected. This list also includes the overpressure contingencies which resulted in over temperature of vessel in the event of external fire. Of a total of 3278 concerns identified, 311 were categorized as unprotected concerns.

Data Requirement

All the assumptions which were made during the analysis were documented under this category. During the analysis, a data need log was created if a proper supporting document was not provided and any follow on assumptions made due to the lack of documentation was approved by the client. Some general assumptions which were made were liquid levels, MAWP of the system (usually was assumed to be same as relief device set pressure), and relief valve information. Of a total of 3278 concerns identified, 1639 was data requirement concerns.

Documentation Updates

During the analysis, some discrepancies were found between different documents such as the U1A-form and the P&ID and these discrepancies were documented, with the correct information needing to be updated on the relevant documents. Documentation updates are usually a low priority concern as they may not impact the operation and safety of plant significantly. Of a total of 3278 concerns identified, 179 were documentation updated.

3.6. Mitigation of Concerns

Once the concerns were identified and documented, different recommendations were provided in order to mitigate the identified concerns. Usually, the clients choose to implement mitigation strategies predominantly for the higher priority concerns such as the unprotected concerns, installation and undersized concerns. An example list (not comprehensive) of the recommended mitigation strategies that were provided to the client for undersized, installation concerns and unprotected systems is detailed below.

Mitigation Strategies for Undersized Concerns

The provided recommendation to resolve undersized concerns depends on the overpressure contingency for which the relief device was undersized.

External Fire: In the event the relief device is undersized for external fire, it was checked if the vessel is insulated and whether credit of insulation could be taken. Usually, in order to overcome this concern, it was recommended that a bigger relief device be installed.

Blocked Outlet: In the event the relief device is undersized for blocked outlet, it was recommended that a bigger relief device be installed.

Isolated with Heat input: In the event the relief device is undersized for "isolated with heat input", it was recommended that a bigger relief device be installed. An alternate recommendation was to see if the undersized concern could be mitigated using administrative controls.

Tank Draining (Vacuum): This was a major concern identified in low pressure tanks. This overpressure scenario was missed in previous studies and hence, a lot of undersized concerns were found. It was recommended that a restriction orifice be installed on the drain line. Another option recommended was to install a bigger PVRV.

Failure of automatic control: For this undersized scenario, it was recommended that a restriction orifice be installed on the line, while still ensuring that it can allow the normal flow through it; however it can regulate the maximum flow such that the relief device is adequately sized. Another option recommended was that a restriction orifice be installed so that the maximum C_v of the control valve can be reduced, which ensures that the relief valve is adequately sized. It was rarely suggested that a bigger relief device be installed for this case.

Inadvertent valve operation: These cases usually included the inadvertent opening of bypass valves around the control valve. For these cases, usually the pipe capacity calculation was reperformed based on the equivalent length of piping from source to destination to check if the concern still existed. If the concern still exists, it was recommended that the bypass valve be chain lock closed or a restriction orifice be installed.

Gas breakthrough: Majority of the time, the failure of automatic controls and inadvertent valve operations resulted in gas breakthrough. For these cases, usually the pipe capacity calculation was reperformed based on the equivalent length of piping from source to destination to ensure if the concern still existed. If the concern still exists, it was recommended that the bypass valve be chain lock closed or a restriction orifice be installed.

Reverse flow: This is one of the most common overpressure contingencies that can be missed while doing the analysis. That is one the reason why the relief device was undersized for this case. In order to overcome this concern, it was recommended that multiple check valves be installed in the direction of flow. An alternate option recommended was the installation of a bigger relief device.

Mitigation Strategies for Installation Concerns

Inlet & Outlet pressure drop: For pressure drop concerns, it was recommended that the concerned piping be modified. The equivalent length of the piping was provided such that it was within the prescribed limit. Alternately, if the relief device was oversized, it was recommended that a smaller relief valve be installed. In this scenario, it is important to still ensure that the relief device is still adequately sized. For inlet pressure drop, it was recommended that a pilot valve with a remote sensing line be installed. For outlet pressure drop, it was recommended that bellows or a pilot valve be installed.

Mitigation Strategies for Unprotected System

If the system is unprotected, the most common recommendation was to install a new relief device. An alternate option recommended was to chain lock open any open block valve to take credit for other relief device. The third option recommended was to mitigate the scenario with administrative controls. This option was common for heat exchangers which have only thermal expansion as the applicable overpressure contingency.

4. Summary

This case study provides an overview of the relief analysis conducted for multiple compressor stations. As stated previously, the relief analysis was system based as opposed to relief device based so as to identify any unprotected systems that may need overpressure protection. It is very important to use the correct operating conditions and fluid composition to perform the relief analysis. The study provided an overview of the various types of overpressure scenarios identified for these systems, with a statistical breakdown of the scenarios by type. It was found that external fire and blocked outlet were the most commonly identified overpressure scenarios. In addition, this case study provided a breakdown of the types of relief devices, a further breakdown of types of relief valves, and the different types of equipment in the compressor stations. A breakdown of the type of concerns identified during the compressor station revalidation was provided along with a general idea of the mitigations that can be implemented to resolve the concerns. When performing relief analysis for compressor stations, it is very important to look out for any contingencies previously not considered, possibly due to changes in facility design or changes in the guidelines.

440451 An Approach to Reducing Risk during Online Isolation, Removal and Maintenance of Relief Devices Using Engineering Assessment

Abstract:

Overpressure protection must be provided for pressurized equipment by means of pressure relief devices, which sometimes require isolation and removal for inspection or repair. It would be advantageous to have a means of isolation without shutting down the process, especially in light of the additional risks associated with shutdown and startup; however, doing so may put the pressurized system at risk of being unprotected in the event of overpressure. Current recognized practices of providing dedicated personnel for monitoring and responding to an overpressure event, or installation of an independent spare, are non-ideal in many instances. An engineering assessment can identify specific ways for providing temporary overpressure protection, including elimination of root causes of overpressure events, provision of alternative pathways to overpressure protection, modification of process operations, and/or establishment of a means of equivalent or better risk reduction for the duration of the isolation. Guidelines and examples are provided to illustrate the analysis, along with recommendations for documentation.

441392 Common "Design Flaws" Identified during Pressure Relief System Analysis

Abstract:

Common "design flaws" identified during a pressure relief system analysis.

It is the goal of a pressure relief system analysis to identify possible gaps in a pressure relief system. In many oil and gas refining and petrochemical plants, it's common to encounter certain "design flaws" which can lead to overpressure scenarios where the system is not adequately protected.

Within these design flaws there are a few which frequently occur, regardless of the process. These can be just simple mistakes or related to consideration not being given to the whole system as part of the design. This paper will focus on five common design flaws:" unusually large bypass valves around a (smaller) control valve"; "liquid head between relief device and equipment being protected"; "how does the suction of the feed pump behaves during the scenario", "tube rupture can affect an entire system and not just the exchanger"; and "common API 650 tank concerns". For each "design flaw" a few common mitigations will also be discussed.

The goal of this presentation is not to discuss the issues in detail, but rather just bring to the industry's attention a few concerns being frequently identified.

Considering Overfilling to be a Non-Applicable Overpressure Scenario

Natalie Doe
Smith & Burgess
190 S. LaSalle St., Ste 550, Chicago, IL 60657
Natalie.doe@smithburgess.com

Dustin Smith, P.E.
Smith & Burgess
7600 W. Tidwell, Ste 600, Houston, TX 77040
Dustin.smith@smithburgess.com

Prepared for Presentation at
American Institute of Chemical Engineers
2016 Spring Meeting
12th Global Congress on Process Safety
Houston, Texas
April 11-23, 2016

Considering Overfilling to be a Non-Applicable Overpressure Scenario

Natalie Doe
Smith & Burgess
190 S. LaSalle St., Ste 550, Chicago, IL 60657
Natalie.doe@smithburgess.com

Dustin Smith, P.E.
Smith & Burgess
7600 W. Tidwell, Ste 600, Houston, TX 77040
Dustin.smith@smithburgess.com

Keywords: PSV, Risk mitigation, liquid overfilling, Safety Instrumented Systems

Abstract

The consequences associated with liquid overfilling are serious for plant safety, as illustrated by many recent incidents (2009, Bayamón, Puerto Rico; 2008, Chesapeake, VA; 2005 Hertfordshire, England; 2005 Texas City, Texas). In many cases, providing overpressure protection solely with pressure relief devices may not protect the facility. Preventing the uncontrolled release of liquid may be the only viable safe option. This paper will present an abbreviated analysis of several incidents that could have been avoided with proper preventive measures. Engineering methods will be presented to eliminate the need to release liquids from a closed system. For instance, the system can be designed such that upstream pressure sources are limited below the allowable downstream pressure. A designer can provide a safety instrumented system with an appropriate Safety Integrity Level (SIL) rating to reduce or eliminate the credibility of overfilling. Alternatively, consideration may be taken for alarms and operator response, with consideration for RAGAGEP, to reduce the credibility of overfilling to an acceptable risk. Systems can be redesigned to be inherently safer by minimizing contained liquids. This paper then gives a few examples of the use of system design, shutdown systems, alarms/operator response combination and inventory controls to eliminate or reduce the potential for liquid release scenarios.

1 Introduction

The consequences of an overfilling overpressure scenario can result in increased risk to the facility when mitigated with pressure relief devices. This could occur due to many different root causes. In order for liquid overfilling to require overpressure protection from a relief device, the following must all be true:

1. The source pressure must be capable of exceeding the system MAWP. Since the relief device is supposed to be set at or below the system MAWP, if the relief device is set lower than the MAWP, a liquid release may occur at a lower pressure.
2. The flow of liquid through the system can be stopped by closing a valve on the system outlet or the loss of a discharge pump (etc). In general, system designers ensure that liquid flows can be stopped, so this is usually a given.
3. There are no acceptable administrative or instrumented controls in place to prevent the scenario.

When reviewing the consequences of an overpressure scenario, the designer must check to ensure that the overfilling event may also not occur during a larger plant upset (power failure, instrument air failure, or other global utility disruption). This would add a simultaneous load to the disposal system, which should be included in the disposal system design.

Overpressure protection

The system designer should consider the root cause of the overfilling scenario to determine how to protect the system or eliminate the scenario. Relief devices are often used to prevent excessive vessel overpressure due to liquid overfilling. For many systems, overfilling is not expected to be the controlling case for relief device sizing; however, a review of incident history shows that many different systems have been overpressured by overfilling with liquid, resulting in relief. Therefore, when relief devices are used to provide protection for overfilling, it is important that proper consideration is given to the effluent handling system.

1. For a system with several pieces of protected equipment, consideration must be taken for the static head throughout the column to ensure equipment at grade is adequately protected. This is particularly an issue for low pressure tanks.
2. When systems overfill, there is the potential for a large amount of liquid to be released to the effluent handling systems. Unless liquid holding drums are sized for the potential liquid rate, there could potentially be an unplanned liquid release with significant consequences.

This paper will outline recent incidents and their consequences due to overfilling, and will overview methods to protect the vessels from overpressure and the system from the consequences of large liquid releases.

2 Overview of recent incidents

The following incidents from the past decade illustrate the consequences of overpressure due to overfilling and the handling of liquid releases.

2009, Bayamón, Puerto Rico;

Per the CSB report, "Tank 409 overflowed with gasoline, resulting in a vapor cloud that encompassed 107 acres of the CAPECO tank farm." [1] The report also states that "Tank 409 overflowed for an estimated 26 minutes before the vapor cloud ignited. The CSB determined nearly 200,000 gallons of gasoline… rushed out of six vents in the tank." These six vents responded to the upset conditions to relieve the excess pressure in the tank; thus, the overpressure protection provided was sufficient to prevent a catastrophic vessel failure. The action of the "engineered" pressure relief devices for this scenario worked and then became the initiating event for this incident. The final result was that the company went bankrupt and 17 of 48 tanks were burned.

2008, Chesapeake, VA;

Per the CSB report, "At about 2:20 pm, as the tank reached a level of 26.72 feet, a vertical split started midway up the shell and rapidly extended to the floor and roof of the tank on the side opposite the workers." [2] In this case overpressure protection to prevent catastrophic failure of the tank would not have worked because the tank was not "full". This incident could not have been protected for by pressure relief devices - only instrumentation, procedures or the like.

2005 Hertfordshire, England;

Per the aforementioned CSB report, "Similar to the CAPECO incident, the vapor cloud explosion and multiple tank fires occurred after a tank was overfilled with gasoline." [1] Thus another incident of overpressure protection for the overfilling scenario working; as such, it operated as designed. The "engineered" overpressure mitigating system was the initiating event for an incident that caused more than a billion dollars' worth of damage to the surrounding commercial and residential property and injured 43 people.

2005 Texas City, Texas

Per the CSB report, "The incident occurred during the startup of an isomerization (ISOM) unit when a raffinate splitter tower was overfilled; pressure relief devices opened, resulting in a flammable liquid geyser from a blowdown stack that was not equipped with a flare. The release of flammables led to an explosion and fire." [3] The pressure relief devices opened, and prevented excessive overpressure of the ISOM tower. As such, the engineered pressure relief devices in this case also worked, preventing a catastrophic overpressure-related loss of containment, but the design for the disposal of the fluid created an additional hazard with significant consequences.

Many of these incidents involved tanks, but that is inconsequential as in each and every case either a relief device was insufficient to prevent overpressure, or the engineered design of the pressure relief device and disposal was the root cause for these scenarios. Not only do the pressure relief devices have to be designed to prevent loss of containment from overpressure, the disposal system has to be designed to ensure the effluent has a safe location to be stored or is adequately destroyed. This can be particularly difficult for liquid releases.

3 Means of overpressure protection

Relief Devices and Disposal

While relief devices are an acceptable method to protect from overpressure due to liquid overfilling, complete risk mitigation requires proper consideration of the ability of the disposal system to handle the liquid effluent. As mentioned previously, additional pressure due to static head should be considered when protecting multiple pieces of equipment. For column systems with relief devices located on the top of the column, this can have significant impact on associated equipment located at grade. It is common practice to perform overfilling relief device calculations at 10% overpressure rather than 16% overpressure when multiple devices are installed, in an attempt to account for the static head. For tall column systems, or for systems with a heavy relief fluid, this may not be sufficient to ensure the allowable accumulation on the equipment located at grade is not exceeded. The relief device set pressure may need to be lowered below the system limiting MAWP to ensure adequate protection for all equipment.

Additionally, the disposal system must be capable of handling the liquid flow from the relief devices. The disposal piping should be evaluated to determine if it can handle the weight of the liquid. A knockout drum must be incorporated as part of the disposal system to ensure liquid is not sent to the flare or vent stack. For large liquid releases, it may not be practical to size one knockout drum to obtain the required particle separation. Multiple knockout drums in parallel can be used to ensure adequate particle separation for these cases. Failure to size the knockout drum(s) correctly can lead to liquid being sent to a system which cannot handle it, which can have serious consequences. The CSB report which reviewed the Texas City incident [3] unduly criticized atmospheric blowdown drums; many of their preferred disposal systems, i.e. flare systems, have had similar incidents [4] with the authors aware of countless near misses from flare systems.

Overpressure protection by system design

The need to design a disposal system for the liquid effluent from relief devices that protect a system of equipment from overpressure by overfilling can be eliminated by designing the system such that relief devices are not needed. The acceptable way to do this is to ensure that the overfilling scenario is not credible. For vessels designed to ASME VIII, UG-140 (formerly known as code case 2211) states "A pressure vessel does not require a pressure relief device if the pressure is self-limiting... and this pressure is less than or equal to the MAWP of the vessel at the coincident temperature..." [5]. This can be achieved in many ways.

Equipment Design

A designer can utilize equipment design to ensure that the disposal system does not need to accommodate effluent from relief devices installed for liquid overpressure protection.

<u>Example 1 – Pump Selection</u>

Instead of feeding a system with a positive displacement pump, a centrifugal pump may be used that is not capable of exceeding the downstream system MAWPs. This is illustrated in Figure 1 below, as P-1 cannot exceed the MAWP of V-1. The defining quality of centrifugal pumps is that their flow will decrease with increased head. Therefore, as the pressure in the downstream system increases, the capacity of a centrifugal pump will decrease. Thus, pump selection can be an effective method of limiting the overfilling rate.

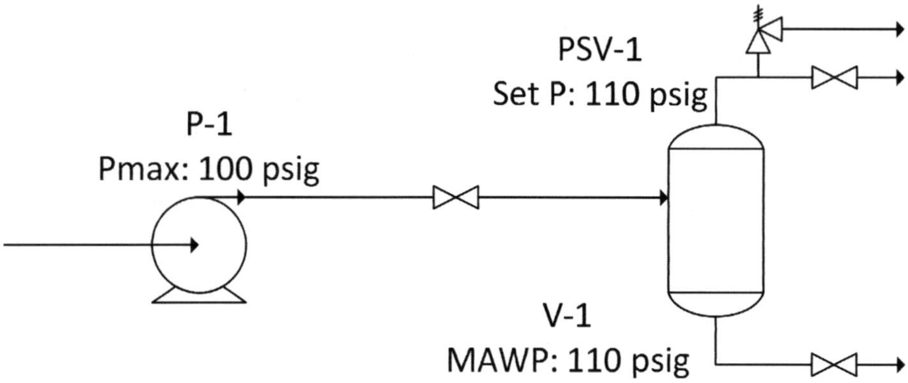

Figure 1: Upstream sources incapable of overpressuring the system

In contrast, positive displacement pumps are constant volume machines that are capable of maintaining high volumetric flow even at increased pressures. Therefore, alternative methods can be applied to protect from overpressure due to positive displacement pumps.

<u>Example 2 – PRV Discharge to Process</u>

If a relief device is installed on the pump instead of the vessel, the discharge piping can be routed to the pump suction. This allows for the liquid relief to be sent to process, rather than to a disposal system. However, the designer is cautioned that this will stop the forward flow of liquid so equipment upstream must be checked for this upset condition. An example is shown in Figure 2 below, as PSV-2A reduces the maximum feed pressure and eliminates the overfilling scenario on V-2.

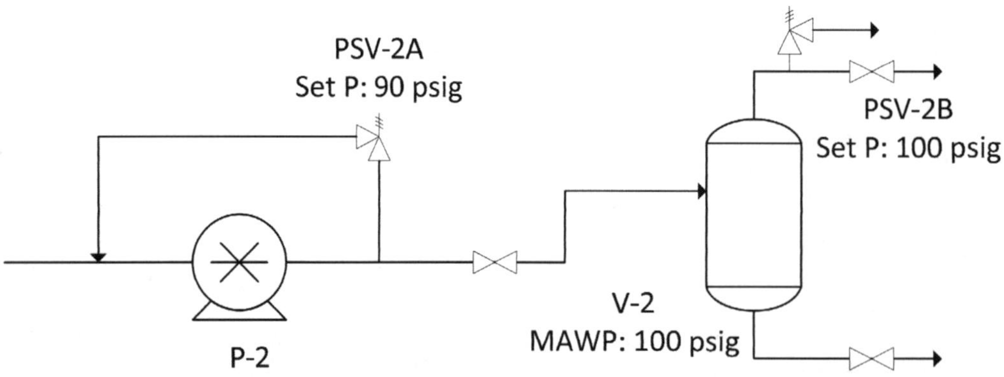

Figure 2: Protecting upstream pressure sources

In other cases, protecting for overfilling by pump selection or upstream protection may not be possible. Inventory controls may be utilized in these situations to ensure overfilling is not credible.

Example 3 - Inventory Controls

If possible, a designer can minimize the liquid inventory in a system. This is also a tactic in inherently safer process design. If the total liquid inventory of a system is insufficient to fill the vessel(s) vapor space then overfilling cannot occur. In this case overpressure does not occur irrespective of the maximum upstream pressure. An example of this is illustrated in Figure 3. The positive displacement pump (P-3) cannot overpressure the vessel (V-3) because the liquid inventory in the upstream tank (T-3) is less than the vessels vapor space.

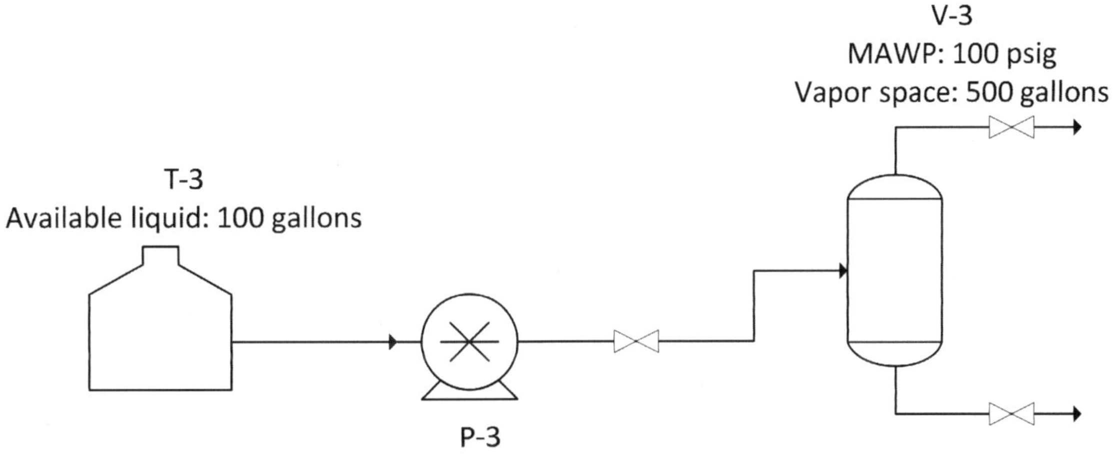

Figure 3: Inventory Controls

Instrumented Solutions

For installations where it is not practical to remove the credibility of the overfilling scenario by equipment design, safety instrumented systems (SIS) can be implemented. There are two ways to protect a facility from the effects of liquid effluent: remove the relief device entirely, or mitigate the consequences of the liquid effluent escaping from the disposal system.

Example 4 – Replace Relief Devices with instrumentation

The system can be designed such that instrumentation is used instead of overpressure protection. A designer can find guidance in API STD 521 which states in Annex E that "High Integrity Protection Systems (HIPS) are designed to avoid overpressure incidents by removing the source of overpressure or by reducing the probability of an overpressure contingency to such a low level that it is no longer considered to be a credible case."[6] ASME VIII UG-140 allows for the use of these systems, provided that detailed description of the safety system is documented, and includes "the identification of all truly independent redundancies and a reliability evaluation... of the overall safety system" [5] The ASME requirements for these systems are further explored elsewhere [7]. Figure 4 below shows an example of a system protected by instrumentation. Three separate level taps are used to determine the liquid level in V-4. Two independent emergency shutdown devices (ESD), each with an independent controller, are used to shut off the feed on high liquid level. The redundancies expressed in this example would be integral to determining the safety integrity level (SIL) of this system.

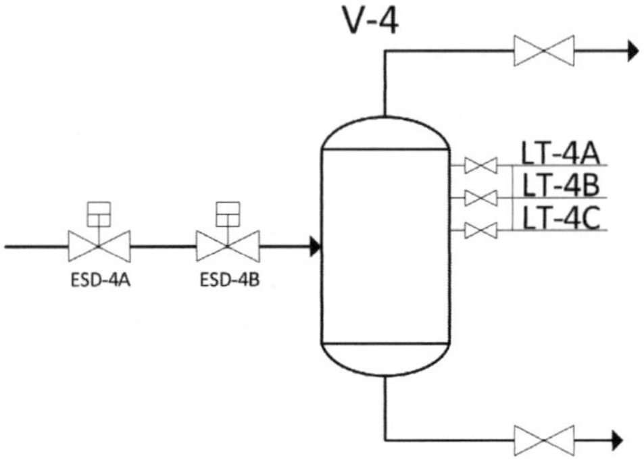

Figure 4: Protection through instrumentation

Example 5 – Add a layer of Protection

Alternatively, the design of a SIS does not require that overfilling is mitigated to a point at which it cannot occur; rather, that the probability is low enough that the total system risk is at an acceptable level. Depending on the existing system, the most cost effective solution may be to mitigate the effluent system release with a SIS but rely on a the relief device to protect the equipment. Due to the uniqueness of each system a risk assessment (such as a LOPA) will be required to determine the reliability of the SIS. The designer may choose to put in a less reliable safety instrumented system in combination with an existing relief device, or the safety instrumented system may replace the relief device. As mentioned above, these redundancies are used to determine the safety integrity level (SIL) of the system. For most cases, a SIL-3 System is needed to remove the credibility of the overfilling scenario. However, a SIL-1 or SIL-2 system may be acceptable when reviewed in a risk analysis [such as Layers of Protection Analysis (LOPA)] to mitigate an incident caused by a release of liquid to the disposal system. Either way, a detailed risk assessment should be performed to determine the SIL rating required to mitigate the hazard [8][1]. An example of a SIS that could provide overfill protection would be level indicators independent of the process control instrumentation from a separate vessel tap feeding a dedicated PLC that triggers a shutdown valve to shut off the feed to the system on a high level alarm.

Operator Response to mitigate overfilling

For many facilities, a combination of instrumentation with an operator response is a reasonable mitigation of risk from overpressure. Research has shown the following factors tend to determine the reliability of operators to take the intended action [9][10]:

1. Trigger – there is a clear indication to the operator that something is wrong triggering the need to take action. It is unreasonable to assume that the operators would have mitigated the overfill scenario at Bayamón PR as they had no indication of level or procedures to know when the tank overfilled [1].
2. Time – Per the work performed by the NRC [9], an individual in a stressful situation with limited time to perform a moderately complex task has a 100 times likelihood of failing compared to the exact same task with sufficient extra time allotted for execution. Others, [10] have put the reliability of an operator completing a clearly understood task on which he/she is trained, with adequate time under stressful conditions, is equivalent to a SIL-III system.
3. Consequences – if the mitigation action causes (or is perceived to cause) a chain reaction that is worse than the failure being mitigated there may be reluctance to perform the action. Additionally, if the operator assumes (or is told) that they will be punished for stopping production, then the likelihood of following the procedure decreases greatly.
4. Training – without the knowledge and training to act, the probability that an action will be taken diminishes greatly, especially if the action is not obvious to the individual operating the facility.

As safety instrumentation systems can be complex and expensive, many facilities choose instead to take credit for operator response to high level alarms. There are several important factors to consider when taking these credits. API 521 4.4.7.3 [6] outlines several factors to consider when relying on level alarms:

a) Whether the level instruments used for safeguards against overfilling are on separate process taps from the process control system
b) Whether level instruments used for safeguards against overfilling are susceptible to the same common mode device failures as those used for the basic process control system.
c) Whether the programmable level transmitter is set to show a low or zero level when the level exceeds the instrument range
d) Whether an open system differential pressure cell is used that can show a false low level due to accumulation of liquid in the upper impulse line if it is overfilled or collects condensed liquid
e) Whether density changes due to temperature effects or composition changes impact readings from differential pressure transmitters
f) Whether the instruments are proven in use for the specific process applications
g) Whether the range of at least one of the level measurement(s) can indicate a valid level reading of the full range between the high critical alarm point and any shutdown or interlock point
h) Whether operating characteristics of the level measurement during off-design, start-up, and shutdown operations are considered in display of level, physical properties, setting alarms, trip points, operator training, and operating procedures
i) Maintenance and testing frequency required for the instrumentation

Even if all of these factors are considered, and the reliability of the operator is evaluated, it is important to analyze the risk associated with human error, and the operator not responding in the allotted time. As with instrumentation systems, relying on level alarms and operator response is not a guarantee overfilling will not occur; rather, it lowers the risk to an acceptable level.

Overfilling of low pressure tanks present a unique challenge as compared to pressure vessels. API has created a specific document to provide further guidance to protect for the overfilling of low pressure storage tanks [11].

Example 6 – Alarms with subsequent operator response

As with Example 4 above, Figure 5 shows a vessel (V-6) with three independent level taps. These level taps are connected to two independent high level alarms (HLA). The designer performed a LOPA, and determined the operator can reliably address the alarm in 30 minutes. This is less than the fill time of V-6 from the time when the alarm sounds (1 hour), so credit may be taken for the operator actuating HV-6A and HV-6B to shut off the feed and prevent overfilling.

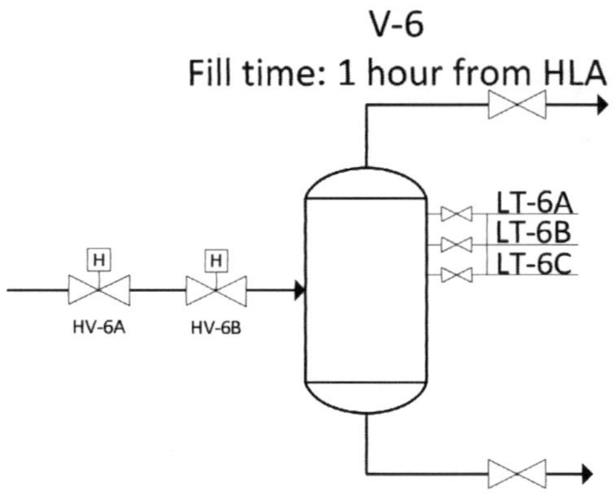

Figure 5: Operator Response

4 Conclusion

Liquid overfill can have serious safety and financial consequences, as seen in the recent incidents outlined in this paper. It is important for the designer to consider both the means to prevent overfilling, as well as methods to respond to an overfilling event. Appropriately designed relief devices and disposal systems are an acceptable method to react to an overfilling event. However, if this is to be the sole means of protection, careful consideration must be given to the large liquid effluent rate sent to the disposal system. Alternatively, the event may be prevented through the proper application of equipment design, instrumented solutions, alarms and operator response, or some combination thereof.

There are a large number of incidents involving overfilling that occurred despite the fact that the installed overfill protection reacted as designed. The protecting relief devices opened and were adequate to prevent overpressure of the overfilled vessel, but the disposal system was not designed for that case. The incidents mentioned in this paper resulted in billions of dollars' worth of damage, and even the bankruptcy of one company. When protecting for overfilling, it is imperative to consider the ramifications of allowing large liquid releases to occur. Overfilling protection design that encompasses consideration of all of the options presented above can prevent an incident with catastrophic consequences.

5 References

[1] U.S. Chemical Safety and Hazard Investigation Board. "Caribbean Petroleum Refining Tank Explosion and Fire" Available at http://www.csb.gov/caribbean-petroleum-refining-tank-explosion-and-fire/. Accessed on 2/5/2016.

[2] U.S. Chemical Safety and Hazard Investigation Board. "Allied Terminals Fertilizer Tank Collapse" Available at http://www.csb.gov/allied-terminals-fertilizer-tank-collapse/. Accessed on 2/5/2016.

[3] U.S. Chemical Safety and Hazard Investigation Board. "BP America Refinery Explosion" Available at http://www.csb.gov/bp-america-refinery-explosion/. Accessed on 2/5/2016.

[4] HSE (Health and Safety Executive) (Eds.), 1997. The explosion and fires at the Texaco Refinery, Milford Haven, 24 July 1994. A report of the investigation by the Health and Safety Executive into the explosion and fires on the Pembroke Cracking Company Plant at the Texaco Refinery, Milford Haven on 24 July 1994, HSE Books, London.

[5] American Society of Mechanical Engineers, 2013 ASME Boiler & Pressure Vessel Code: Section VIII, Division I. s.l.

[6] API Standard 521, "Pressure-Relieving and Depressuring Systems," American Petroleum Institute, 6th ed., 2014.

[7] Smith, Dustin J. "Safety instrumented systems in lieu of pressure relief valves." *Process Safety Progress* 33.4 (2014): 345-349.

[8] A.E. Summers. "Overfill Protective Systems – Complex Problem, Simple Solution," 12th Annual Symposium, Mary Kay O'Connor Process Safety Center, October 2009.

[9] Blackman, Harold S., David I. Gertman, and Ronald L. Boring. "Human error quantification using performance shaping factors in the SPAR-H method." *Proceedings of the Human Factors and Ergonomics Society Annual Meeting*. Vol. 52. No. 21. SAGE Publications, 2008.

[10] Mannan, Sam, ed. Lees' Loss Prevention in the Process Industries: Hazard Identification, Assessment and Control. Vol. 1. Butterworth-Heinemann, 2004.

[11] API Standard 2350, "Overfill Protection for Storage Tanks in Petroleum Facilities," American Petroleum Institute, 4th ed., 2012.

12TH GLOBAL CONGRESS ON PROCESS SAFETY

How Reliable Are Your Process Safety Barriers?

Mike Neill
Petrotechnics USA Inc.
10497 Town & Country Way, Suite 700, Houston, TX 77024
mike.neill@petrotechnics.com

Prepared for Presentation at
American Institute of Chemical Engineers
2016 Spring Meeting
12th Global Congress on Process Safety
Houston, Texas
April 11-23, 2016

How Reliable Are Your Process Safety Barriers?

Mike Neill
Petrotechnics USA Inc.
10497 Town & Country Way, Suite 700, Houston, TX 77024
mike.neill@petrotechnics.com

Keywords: Process Safety Management; Process Safety Barriers; Operational Risk Management

Abstract

Using Process Safety Management techniques, we design and implement multiple layers of protection for our hazardous plants. For these barriers to remain effective over the life of the asset is a challenge, particularly with aging assets. We know barriers deteriorate over time without proper, preventative maintenance and inspection. Often a lack of focus on critical barriers can result in activities being deferred or cancelled without recognizing the potential consequence.

There is generally a lack of understanding in our organizations of process safety techniques and approaches. While the criticality of these barriers maybe well-understood in the Process Safety department, this is unlikely true through other organizations, including Operations and Maintenance.

With so much equipment being tagged "safety critical" on our facilities; it is often difficult to ensure safety barriers receive the focus they need to ensure their effectiveness at protecting against high consequences events when needed. Operators need to manage the health of their barriers in near real-time and mitigate the impact arising from any impairments.

By understanding dynamic aspects of cumulative risk, all members of an organization can easily identify barrier impairments. This level of visibility helps make better operational activity management decisions on account of operational activities' potential impact on Major Accident Hazard barriers.

This paper will examine the traditional approach to barrier design and assumptions on reliability, the approach to maintenance, and the reality of operational dynamics. It will introduce the concept of cumulative risk impacts on safety barriers and examine operational dynamics which impact barrier performance.

1 Introduction

Confucius, the Chinese Philosopher from around 500BC is quoted as saying that "a man who lacks reliability is utterly useless." Well the same can be said for our Process Safety barriers, whose sole purpose is to prevent major accidents or mitigate the consequences of such events. These barriers enable us to operate our hazardous facilities, exposing people to potential danger, since we satisfy ourselves that the calculated risk of incident is acceptable, assuming reliable performance. Unfortunately, when we investigate incidents, we see again and again examples where multiple barriers fail to perform. Occasionally we see evidence where the design was at fault and the assumptions made were incorrect, and we see plenty of evidence where Management Systems designed to prevent barrier failure, fail themselves.

All too often the barriers do not work as intended for a variety of reasons ranging from corroded pipe, seized valves, inoperable switches, blocked flowlines to a simple case of responsible people failing to follow operating procedures.

We focus a lot of attention on making sure we design our barriers properly in the first place but do we do enough during the life of the facility to make sure the barriers continually perform per design? Are our organizations set up to optimally manage these barriers and are our frontline workers engaged sufficiently and armed with the right knowledge to play their role in identifying barrier impairments and to make appropriate operating decisions? These are questions this paper will address and will conclude with examining new solutions which are becoming available to help the industry with this challenge.

2 Evidence of Barrier Failure

The failure of barriers designed to protect or mitigate the consequences of Major Accident Hazards (MAH) often results in an incident such as a significant gas release or an over pressure situation. The failure of multiple barriers can escalate into a major accident such as an ignition barrier failure coincident with a gas release which results in an explosion or fire. The term "barrier", refers not just to equipment hardware such as piping, vessels, instruments, actuated valves, etc., but also includes procedural systems, where plant staff follow agreed practices to operate the plant and respond to emergencies.

The latest Marsh and McLennan report (ref1) provides a current update to property damage losses in the hydrocarbon industry. This report, in its 23rd edition, outlines the 100 largest losses across our industry over the past 40 years. A direct quote from that report

> "*Accidents that result in major losses, such as those reported in this document, generally occur because of the failure of a number of barriers within the process-safety management system — all occurring at the same time; typically, none of these losses are the result of the failure of a single barrier or protection measure.*"

Furthermore, in the same report, other quotes reflecting on the incidents,.

"None of the losses listed in this document should be considered 'black swan' events."

("Black Swan" events are those regarded as inconceivable and impossible to consider as a credible threat.)

The implication of this statement is that the scenarios and failure modes were anticipated so either the design of the barriers was inadequate to meet conditions or the design of the barrier was adequate, but its performance did not match expectations. It could also be a combination of both.

In another report, (ref2) Ron Jarvis from Swiss Re looked at a sample of 45 incidents in the Oil, Gas, and Petrochemical fields over the prior 15 years. Despite this being a random sample and not necessarily a statistical analysis, it is worthy of consideration, not least because its aim was to prioritize those risk control elements whose failure resulted in a majority of incidents.

The study concluded 52% of primary incident cause was failure due to corrosion, 27% failure of operating practices and procedures, and 14% was attributed to unsafe maintenance, equipment isolation and permit to work management. Considering secondary and tertiary causes, poor design and/or inadequate hazard identification is another cause which becomes more prominent. The author goes on to identify inadequate piping and equipment inspection programs resulted in 68% of corrosion failures, the remainder being due to incorrect material selection or defective materials when first installed. Regarding the failure of operating practices and procedures, Jarvis highlights poor management of deliberate by-passing of Safety Critical Equipment (SCE) and the management of abnormal situation on the plant. 57% of all incidents occurred during abnormal conditions on the plant, including start-up, shutdown, or maintenance.

The evidence from these two studies is conclusive that Major Accidents occur as the result of multiple barrier failures. You may ask why it is important to recognize this seemingly obvious fact! This Author believes it is having attended many Process Safety Conferences over the years where the emphasis is on barrier design and static risk scenarios with little focus on what happens to barriers over the life of the asset. This, and working in an industry where there seems only to be a tacit recognition of a dynamic component to MAH risk in which a 5 year review of safety management systems seems adequate. The fact is, barriers fail and over time most, if not all, will fail without periodic inspection, testing, and preventative interventions.

The challenge we face is the data on Major Hazard Barriers, which can indicate an impaired status or diminished integrity, is not easily shared with the people who need to know and who routinely make decisions on daily operations based on assumptions that everything is in good health.

3 Barrier Design Versus Barrier Performance

3.1 Design Principles

When designing Oil, Gas, and Petrochemical plants, many of the principles used are determined by regulatory requirements in addition to generally accepted engineering design standards. Wall thicknesses and material specifications for pipes and vessels will be based on the expected fluid properties and operating conditions as will the specifications for valves and rotating equipment selection. HAZOP and HAZID studies look at potential hazard scenarios and failure mechanisms and identify methods for minimizing risk and/or mitigating the outcome. Risk-based studies look at ways to reduce MAH risk to a level which meets the criteria of acceptance. When the risk levels are excessive, additional barriers are designed.

Unfortunately, barriers themselves are not 100% reliable. This can be the result of a material defect in piping or vessel walls, or corrosion of those same walls, a stuck relief valve, a deluge system plugged with scale, or a standard operating procedure which is not followed. A single safety barrier cannot eliminate risk but it can reduce the likelihood of an event happening or mitigate the magnitude of the consequences if it does occur. As such, the traditional design approach is to apply multiple barriers, some preventative and some mitigative. Together they reduce the overall consequential risk.

Modern techniques of Inherently Safe Design, pioneered by Trevor Kletz, seek to build plants where hazards are removed rather than controlled and where volumes of hazardous fluids are minimized to reduce the magnitude of consequences. This is paving a way towards safer plants. At present, the fundamental needs of Oil, Gas, and Petrochemical plants continue to demand processing flammable, explosive, and toxic fluids often at high pressures and high temperatures and in large volumes. As such, the approach of multiple barriers to reduce and mitigate risk is still essential for plants - old and new.

When a hazardous scenario has been identified, the risk of the consequences occurring is calculated using statistical and Quantitative Risk Analysis (QRA) techniques. Risk acceptance criteria is often prescribed or advised by the regulator or set by internal corporate policy for an Operator. For example, the threshold may be a broadly acceptable risk of death of 1×10^{-6}. If a scenario results in the calculated risk of this outcome being higher than the threshold, additional barriers need to be designed to reduce the risk level to one which is acceptable.

Each barrier is credited with a level of risk reduction. This is calculated using performance characteristics of the barrier which, for hardware, is provided by the manufacturer based on empirical data such as "mean time between failure," etc. The assurance from the manufacturers of this performance is founded on a manufacturing quality system designed to provide reproducible results.

For vessels and piping the wall thickness and material specification are the principle drivers of strength and the ability to contain the process fluids. Once again, manufacturers' quality assurances assume manufacturing standards are followed. Mill certificates demonstrate material specification compliance. Hydrostatic testing of vessels and fabricated flowlines / pipelines, at pressures below the material yield points, are part of commissioning new plant and verifying a high level of integrity prior to start-up.

This process of risk reduction through designing additional barriers where necessary is carried out across the facility until the overall level of risk is reduced to an acceptable level.

Unlike hardware barriers, Procedural Barriers typically are less reliable since they require human interventions to be consistent and repetitive and with a discipline to follow standard procedures. Since most Operators are not robotic by nature, reliability is less than for engineered systems. On balance, this is a good thing since, unlike robots, Operators are free-thinking, unconfined by a narrow scope, and are able to identify a wide spectrum of developing situations and respond appropriately to regain control.

Reliability for procedural barriers is improved if the procedures themselves are well-written and workable so have been repeatedly tested and improved over time. The use of checklists provides a simple go/no-go approach, removing ambiguity in the step-by-step procedure. Training, and of course, experience are the other strong determinants of the reliability of procedural barriers.

3.2 Barrier Categories

The diagram below shows an example of fundamental barrier groupings. These are categories of barriers based on their function. While shown as the familiar "Swiss Cheese" model, the layout is similar to a "bow-tie" diagram where the left hand barrier groups are "preventative" and the right hand are "mitigative." It is useful to understand barriers in the context of multiple layers of protection and to appreciate that the term "multiple layers" does not necessarily indicate multiple redundancy.

As an example, if there is a gas leak on a plant than the failure in itself is a serious incident and has the potential for fire and explosion. Another group of barriers would be ignition control, preventing the escaping gas from igniting. Gas detection is another barrier and fire suppression systems a further barrier categorized under Protection. So each of the multiple layers maybe performing different roles to either prevent an event happening or to mitigate the resulting consequences or preventing escalation.

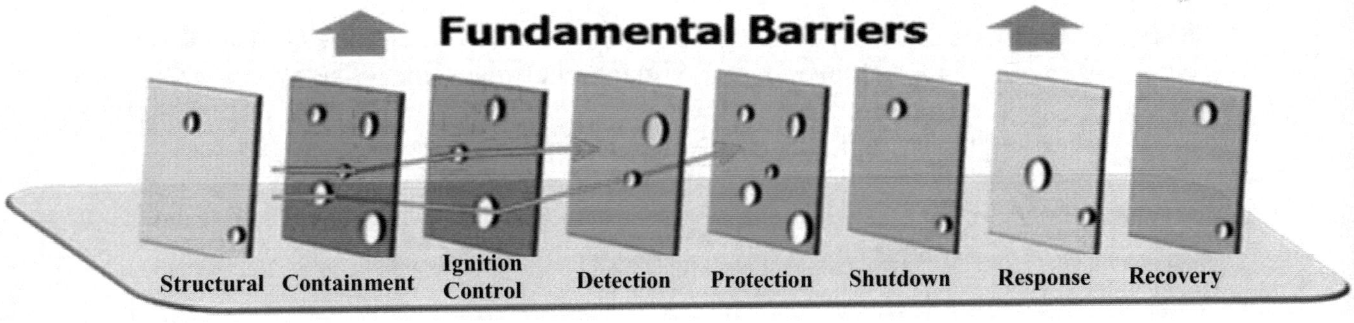

Figure 1. Fundamental Process Safety Barriers

Another example of categorizing is shown in the table below was proposed by Blacklaw, et al. (ref 3), detailing some of the specific barriers for offshore installations:

Table 1. Cumulative Risk Assessment Barrier Model

Barrier System	Example Components
Structural Integrity - (offshore structure)	Jacket, Topsides, Cranes
Process Containment	Vessels, Pipelines, Rotating Equipment, Pressure Systems, Relief Valves
Ignition Control	Ignition Prevention Systems, Flare Systems, Ventilation, Positive Pressure Systems, Gas Seals, etc.
Detection Systems	Fire & Gas Detection and Alarm Systems
Protection Systems	Fire Water Pumps, Deluge & Sprinkler Systems, Safe Refuge, Foam Systems, Blast Walls, etc.
Shutdown Systems	Pressure Trips and Shutdown Valves, ESD Systems, etc.
Emergency Response	Escape Routes, Emergency Lighting, Fire Water Mains & Portable Equipment, etc.
Life-Saving	Life Rafts, Personal Survival Equipment, etc.

3.3 Barrier Failure and Maintaining Barrier Performance

The reliability of engineered barriers will deteriorate over time, eventually to a level where they will just not work. This is simply a factor of material changes due to environmental actions such as corrosion or the rigors of service, erosion, wear, etc. To prevent or slow this deterioration, regular inspection, testing, and preventative maintenance is necessary.

As was explained in section 3.1, the reliability of engineered barrier performance determines the level of risk reduction for which credit is taken resulting in a residual risk which meets the acceptance criteria.

The data used for barrier reliability is based on manufacturer's information but assumes an ongoing schedule of inspection, maintenance, calibration and function testing to maintain that performance. What this means in practice is to take risk reduction credit for a shutdown valve, all the components of the system, from the pressure switch sending the signal to the controller, the actuator, valve stem, valve body and valve seat all have to be periodically inspected, tested, and undergo preventative maintenance. If these activities fail to occur within a specified timeframe, the_only consequence is the risk of major accident increases. Without these sustaining activities, dependability of the system declines, reducing or eliminating the ability to claim a risk reduction credit- thus the consequential risk has to increase.

This simple logic is important because there are often temptations to defer inspection and maintenance of such systems due to other operational priorities. Such deferrals are justified by arguments that it is preventative work and that there have not been reports of any failures.....so there is no reason to assume they will not still work!

The single most important set of barriers on a plant are those classed as "primary containment." Primary Containment includes pipework and vessels, plus a myriad of valves and flanges which connect them all together. It is not only the main flowlines but also includes by-pass lines, impulse lines for pressure switches, temperature gauges, sight glasses, etc. plus small bore pipework for instrument connections, bleed lines, chemical injection, etc.

The performance of these barriers is essential to keep the process fluids contained and under control. They are simpler than instrumented systems, not relying on moveable parts but instead on wall thickness and material specification plus the devices used to couple such as flanges, gaskets, bolt tensions, etc. Corrosion, both internal and external, are the most common failure mechanisms. Erosion, scale deposits, chemical damage and accidental damage will play their part. Inspection programs monitoring for losses in wall thickness are typically in place across the facility. These are supplemented with preventative programs such as protection coatings, corrosion inhibitors, anodic and cathodic protection, etc. Periodic hydraulic testing and recertification of vessels and pipework is also a standard requirement from most regulatory authorities and / or insurance companies.

3.4 Management Systems

Whether specified by the regulator or by internal policy, Management Systems are typically in place to manage the processes described in the sections above. As part of an organization's governance, the suitability of these Management Systems will be periodically reviewed, and their effectiveness tested by audit to ensure they are properly applied. Performance indicators provide a measure of the effectiveness of such systems with API RP 754 (ref. 5) being a standard most commonly used in the O&G & Petrochemical industry for process safety KPI's. In this standard, Tier 1 and 2 measures are known as "lagging indicators" based on historical events such as loss of containment. Tiers 3 and 4 are referred to as "leading indicators" since they are typically based on auditing of management systems in place to prevent or mitigate process safety events. In the UK, the Health & Safety Executive issued guidance, HSG 254, on developing process safety indicators (ref 6) a few years before API published their standard. Here they promoted the identification of MAH scenarios, barrier identification and identifying the "Risk Control System" or RCS which is managing the performance of that barrier. Leading and lagging indicators, similar to API 754 are also recommended in this guide.

While it is essential to the overall safe management of a facility that barrier / risk management systems are in place, both API 754 and HSG254, generate indicators which inform on the system alone and not on specific pieces of equipment or areas of the plant. Furthermore, what are called "leading indicators" are based on data collected months if not years in advance so may not represent what is current. What is clear is that they do not help day to day decisions on risk levels in specific areas of the plant or working on specific equipment.

4 The Reality of Operations and Dynamic Risk

4.1 Deferred Maintenance of Safety Critical Equipment

Maintenance Management Systems are programmed with time-based schedules for inspection and preventative maintenance of all equipment on a facility. Where this equipment forms one of our major accident hazard barriers, it is flagged as "safety critical." A rolling schedule is managed based on different time horizons. As equipment breaks down, repair work is added in based on priority. At two weeks or seven days, the schedule is typically "frozen" and handed over to Operations.

Executing the plan as designed is often a difficult challenge for a variety of reasons and maintenance backlogs, or maintenance activities beyond their planned completion dates, become common as activities slip on the schedule. This becomes more of a concern when the backlog contains safety critical equipment. In such instances, Operators need to justify why the maintenance is not being carried out and a "deferment" review process involving technical and engineering authorities should be conducted.

As we have seen earlier, when inspection and preventative maintenance of safety critical barriers is delayed, the inevitable outcome is that MAH risk levels increase.

Deferred safety critical maintenance is an unfortunate reality in the O&G & Petrochemical industry, especially with old facilities and infrastructure. The magnitude of this issue is difficult to judge since data on the topic is not easy to access as few Operators want to share data which shows them in a bad light; however, there is one exception.

The UK sector of the North Sea is a region where Operators have, for several years, voluntarily submitted data on the maintenance backlog for Safety Critical Equipment (SCE). Oil and Gas UK, the cross-industry body recently published this data in their annual Health & Safety Report in 2015 (ref 4). The diagram below shows how the growth in SCE backlog over the last 4-5 years. The report mentions the regulator has raised concerns the industry has increasingly deferred maintenance backlogs rather than addressing them.

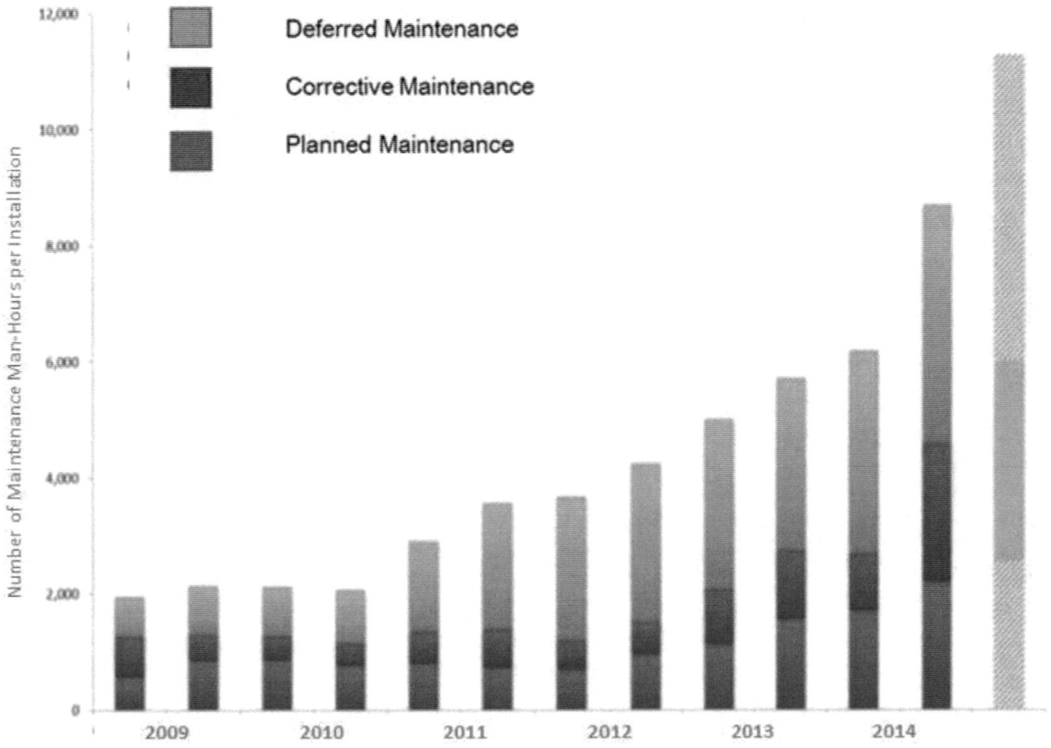

Figure 2. Oil & Gas Health & Safety Report 2015 (ref 4) – Growing Backlog of Safety Critical Maintenance

What this picture says is there are a number of barrier systems on the facility which, with the deferment of inspection and preventative maintenance, are knowingly "impaired. In other words, they are not providing the level of risk reduction that was intended by design. Add to this, the barrier systems which have been inspected but have failed to perform adequately and are therefore also in need of corrective action.

4.2 Deliberate defeat of barrier systems

Another category of compromised barriers must also be considered. Every day, many barriers on a facility are deliberately defeated. This is a necessary on a plant where ongoing maintenance and repair activities have to take place with the plant in a live state or condition. Detection systems, such as gas and fire detectors, have to be deliberately defeated to carry out work in a vicinity which could accidently trip them. Alarms and process trips have to be bypassed for work to be carried out. Ignition sources have to be brought onto the live plant in order for work to be carried out.

The terminology used to describe these situations include: "temporary defeats (of safety system), "instrument override", "interlock by-pass" but they amount to the same thing. A formal, managed process should be followed during this action, often accompanied with a risk assessment. This will often focus on the following critical parameters: **time** – the longer a defeat is in place the greater the risk; **reinstatement** – ensuring the temporary defeat does not get forgotten and becomes permanent; **communication** – ensuring all necessary parties are aware of the status.

Even our most important barrier, primary containment, has to be compromised fairly regularly in order to replace and repair equipment such as valves, pumps, filters, etc. Work involving "breaking containment" is, again, normally managed with a formal process involving energy isolation, draining and purging and gas testing.

All of these situations where safety barriers are deliberately defeated involve an additional level of risk. While the defeats maybe managed with formal process and additional barriers or controls maybe put in place, there will be a residual risk which needs to be considered along with the other sources of risk on the facility.

4.3 Operational Decisions and Corporate Silos

On a daily basis, plant Supervisors make the decision, typically informally and often unconsciously, on whether the plant is safe to operate. They also decide whether it is safe to carry out simultaneously activities - such as maintenance and repair work, project construction work and normal Operational activities. These decisions are often based on incomplete information, specifically with regard to the status of barriers on the plant.

Unfortunately it is quite common within our organizations, that data and information is stored in areas often difficult to access by anyone other than the people who created the data. While this may not be intentional, the result of such "discipline silos" is that updated reports from these discipline groups tend to be too generalized to support specific decision-making.

Given the wide distribution of data, it is not surprising that daily decisions are often made on the basis of exception, i.e. unless told otherwise and in the absence of any visible concerns the assumption is that the barriers are in good health. Visual evidence of barrier impairment cannot be relied upon other than very obvious external corrosion or physical damage to equipment and "local memory" of ongoing status rely on individuals being on shift or very good handover notes and Operations logs. A good supervisor will make great efforts to find the most current information before making decisions but as always when trying to prove a negative….how do you know when to stop looking?

The table below shows some random examples of the types of discrete risk management processes and disconnected data systems which are common in our organizations. This is a short list to demonstrate the principle. The point is that our organizations and discipline groups tend to be very insular and the data they collect is difficult to access from outside the group. This is partly because the work carried out to identify deviations or verify conformance is fragmented and generally covers the whole facility. The output reports tend to be summarized around performance indicators (examples of which are included in the table). These KPI's are useful to judge the effectiveness of the Risk Management Process (RMP) averaged over sizeable time steps (per month , per quarter or per year). They are not necessarily helpful for supporting real time decisions in specific areas of the plant. For that, decision makers need to know the status of specific barriers in particular areas of the plant to identify whether planned activity or plant changes are safe to instigate.

Table 2. Examples of Corporate "Silos" of data and information

Lead Discipline	Risk Management Process (RMP)	Data "silo" system	Performance Indicator
Maintenance	Identified barrier impairment	MMS (Maintenance Management System)	% impaired vs not impaired
Engineering	Corrosion Inspection	Inspection Systems	% systems inspected % failed
Engineering	Safe Operating Limit excursions	DCS / Data Historian	Number of SOL excursions
Instrumentation	Relief Valve testing	Testing Systems	% tests carried out
Instrumentation	Instrument Malfunction	DCS / Data Historian	% malfunctioning instruments
Engineering	Management Of Change	MOC system	% MOC's outstanding
HSE	Incident Reporting	Accident / Incident Database	Tier 1 – 4 reporting statistics
HR	Competency Management	Competency Management / Training database	% workers trained
Process Safety	HAZOP / HAZID	Risk Analysis Database	% Hazops complete
Management	Management Audits	Audit / Action Tracking System	Audit Actions complete

4.4 Cumulative Risk & Dynamic Risk

Two factors of risk management which are gaining some attention in our industry are cumulative risk and dynamic risk. Dynamic risk is perhaps the most simple concept since it is just a recognition that the risk situation is not static in any of our facilities. It changes with a variety of time dependent mechanisms. We have covered extensively the way in which barrier systems fail over time which results in a change in the consequential risk of the considered scenarios. The degree of risk elevation will depend on the length of time the barrier(s) is impaired. Other time dependent components to risk include environmental conditions particularly those at the limit of design such as hurricanes, tornadoes, floods, earthquakes, etc. Ongoing activities on the plant will potentially have an impact on risk either through introduced hazards or introducing potential casualties if an event occurred. Community expansion over time outside the perimeter fence is also a dynamic which changes the risk profile of a facility.

Built into some of our standards and regulatory practice is a 5 year review of hazard and risk scenarios across the facility. This is a suitable time frame for reviewing major equipment items and to ensure that appropriate barriers and risk control systems are in place as well as an opportunity to implement new attitudes towards risk and risk reviews. It does not, however, address the reality of day to day changes in our barriers, temporary situations and the changing hazards of a typical work day and as a result, the changes in overall risk. Instead we rely on informal processes and diligent front line supervisors to recognize when risk levels are reaching unacceptable levels.

Cumulative risk is another component of this dynamic picture which is gaining some recognition in areas, notably, offshore in the North Sea. Oil and Gas UK have recently issued in draft form guidance on the management of Cumulative risk (ref 7). In this document they define Cumulative Risk as follows:

The Cumulative risk on an installation is considered to be the total risk at a particular point in time due to a combination of:

- The risk described in the Safety Case
- Changes to the above caused by one-off operations or other operational changes
- Additional risk from deviations (from a normal state), the interactions between them and the influence of other factors on these risks

The North Sea is a mature basin area with many of the operating facilities having been first installed in the 1970's. Maintaining these facilities is a challenge and, as we saw in section 4.1, a backlog of safety critical maintenance now exists on many facilities. The regulator has a concern on how operators manage the cumulative effect of multiple "deviations", i.e. impaired barriers and how decisions are made.

There is nothing special about the North Sea and it probably represents what Deepwater Gulf of Mexico will look like in 10 to 15 years. Similarly, onshore refineries, chemical plants, storage and distribution facilities suffering the effects of ageing equipment and infrastructure will need to understand and manage the cumulative and dynamic impact of multiple impaired barriers and simultaneous operations. To achieve this successfully, we need to have a better set of tools to help decision makers, in real time, assess overall risk from a number of disparate sources and enable them to more safely and efficiently manage activities and the operation of their plant.

5 A New way to Manage Risk, Barrier Health, and Activity

5.1 New Category of Enterprise Operational Excellence Software

New software platforms are emerging under the category of Enterprise Operational Excellence Management enabling companies to make better decisions in all areas of activity and risk management. These systems are designed to manage activity with the benefit of understanding risk from a number of sources. Barrier impairment, in such systems, is easily tracked and very visible when activities are planned and executed.

These systems help manage some of the following challenges:

- They provide a unified approach to assessing operational risk and governing the status of Process Safety barriers
- They provide opportunities to produce meaningful real-time and predictive risk indicators, to manage the sequence and execution of activities.
- Real time and predictive cumulative risk modeling in time and space dimensions
- Safe execution of frontline activities across the whole planning-to-execution business process.
- Data exchange with other enterprise platforms such as Maintenance Management Systems (CMMS), Inspection databases, Process Data Historians, etc. integrating across many of the data silos in our organizations
- Visualize risk and activity in space and time
- Prioritize work scheduled based on optimizing future reduction of risk
- Easy identification of barrier impairments and management of deviations in risk control systems
- Ability to manage, in real time, all risk and activity in the one place using easy simple graphical representation of the facility.
- Management of all activities including permitted work, energy isolations (LOTO), task risk assessments / Job Safety Analysis, confined space entries, routine activities, shift handover, Operator Duties, etc.
- Efficient and safe execution of work schedule
- Real time and predictive KPI dashboards shwing risk, productivity, plan attainment, time on tools, audit compliance, etc.

The screen capture below shows an example of how graphical views of the plant can indicate where MAH barriers are impaired. Icons show the different categories of impaired barriers. Other icons show ongoing or planned activities. The view can be real-time or predictive depending on the shift date selected. The future position is based on planned activities and estimates for the duration of impacted barriers. The system predicts the overall level of risk for that time period and displays with a simple traffic signal red, orange, green pattern to indicate high, medium, or low level risks. It can also highlight the specific area on the plant where the highest risks are predicted.

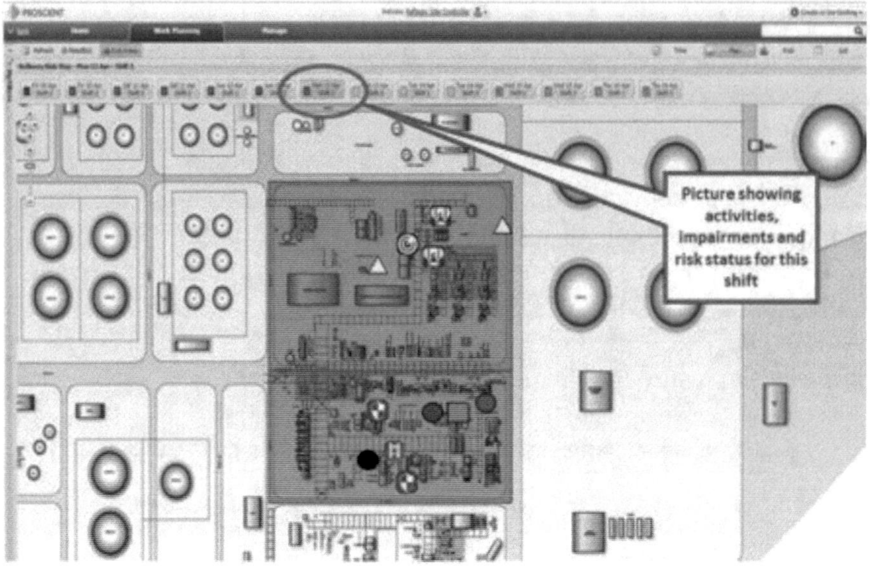

Figure 3. Graphical View of the Plant to Display Operational Risk Impairments and Deviations

Specific risk view screens for each area of the plant allow drill-down into the data to understand which activities or conditions are driving higher risks. This view is also available for current real-time or for future shift periods, with the traffic signals indicating cumulative risk for the facility during that period. This allows an operator to see the current or future state of fundamental safety barriers in each area of the plant.

Figure 4. Risk Views Against Specific Plant Areas for Current or Future Shifts

The impact on risk is driven by specific work activities, plant statuses, or barrier conditions. Each of these is captured as a "line item" and displayed in a Gantt chart style. Much of this data is "sucked in" from the disparate silo stores and is displayed in a single place showing all future work and conditions. This screen automatically shows conflicts depending on configuration rules established. For example, if there was a containment issue in an area and an operator was trying to schedule hot work in the same area, the system would prompt a conflict and shows the activities with a red boundary. Even if it was not a conflict rule, the high risk status for the shift would alert the situation.

Simple drag and drop features allow Frontline Leadership and Planning to simply move conflicts or compounding risk events so that they do not coincide and in doing so, reduce the levels of predicted peak risks.

Digital displays help Operators identify prolonged barrier impairments which are elevating risk levels in specific areas of the plant to unacceptable levels. Simple simulate/fix functionality empowers operators to see the risk reduction impact of fixing an impaired barrier earlier than scheduled.

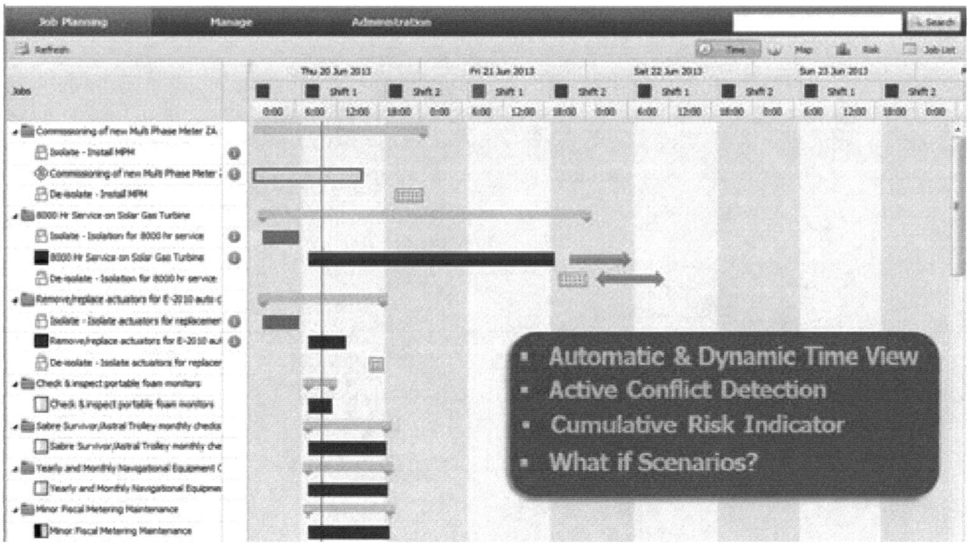

Figure 5. Time View showing Durations of all Planned Activities

A digital risk assessment engine enables Operators to evaluate the risk involved in their planned activities. At the same time, everyone across the plant can see the status of impaired barriers and assess the cumulative risk impact of their activities and those of other departments on all barriers, in all areas, at all times. This means medium-term planning and scheduling (90+ days) can be prioritized using MAH risk criteria. Rather than just blanket labeling of work priorities with the term "safety critical," the true impact on MAH risk can be assessed and activities prioritized accordingly.

With this data being collected centrally and used to assess the overall impact on risks, the quality of analysis and reporting surpasses that which has been achievable to date. The screen below show some of the types of reporting possible such as tracking overdue deviations (i.e. deviations or impairments which are not resolved within an expected timeframe), risk control systems impacts and trends and even trending of MAH risks over time.

These systems are designed for enterprise deployment so reporting can cover a whole organization of assets and identify where barrier impairments and resulting MAH risks are highest.

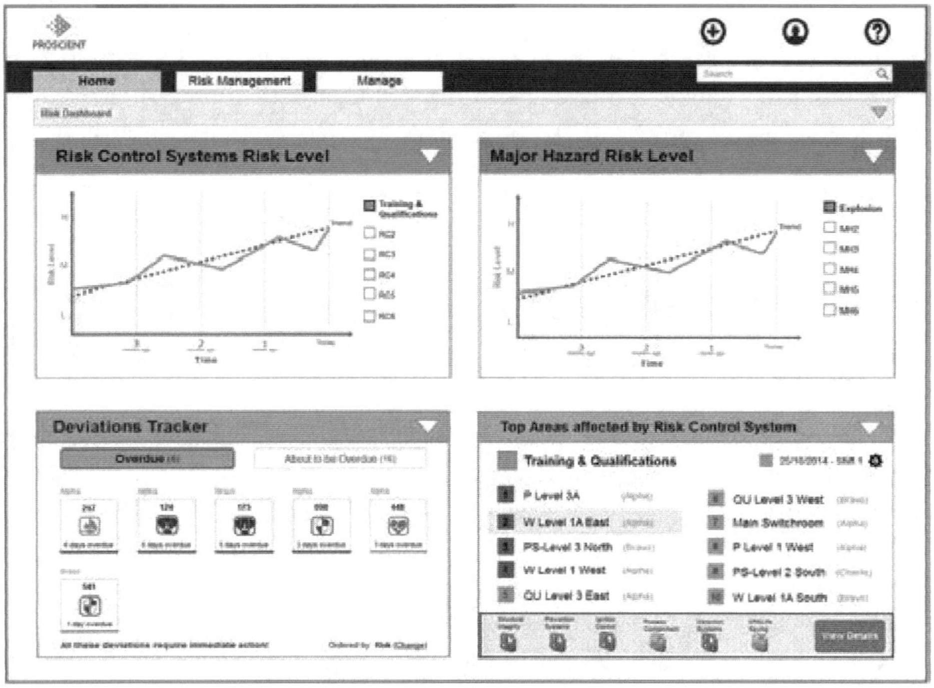

Figure 6. Examples of Dashboard Type Reports

6 Conclusions

We have seen how major accidents occur due to the failure of several of our protective / mitigative barriers. Simply put, the barriers did not perform as designed. Their reliability is dependent on continued inspection and preventative maintenance which, as we have seen, can often be deferred resulting in an increase of MAH risk. This is because the risk reduction credit for the barrier, taken during design, is no longer applicable. Another problem we face is that the data and information regarding these impaired barriers is often not easily accessible, being stored within corporate silos. Frontline Supervisors are making day to day decisions on the basis of assumed barrier status and cannot easily access information which could contradict some of these assumptions. They also need to take into account the dynamic nature of risk which can change from day to day, hour to hour and the cumulative impact of multiple impaired barriers.

Our frontline supervisors and decision makers have the responsibility to manage a safe plant so they need this information in order to make fundamental decisions such as whether the plant is safe to operate and whether the planned program of work is safe to execute. Currently they do not have easy access the relevant data in order to make these decisions.

Fortunately new technology platforms are emerging where the data on impaired barriers, temporary defeats, equipment isolations, work activities, etc. can be sourced through a single place. In this same place, activities can be planned and executed. Planned maintenance schedules can be imported and work carried out safely. Cumulative risk indicators show the aggregate impact of multiple activities and impaired barriers on risk in specific areas.

These new platforms represent an opportunity for a significant change in the way we manage risk on our facilities and should lead to safer and more efficient operations.

7 References.

[1] "The 100 Largest Losses 1974-2013 – Large Property Damage in the Hydrocarbon Industry", 23rd Edition, Marsh Energy Practice 2014.

[2] "An analysis of common causes of losses in the Oil, Gas & Petrochemical Industries" Presented at IChemE Safety and Loss Prevention Special Interest Group, Ron Garvis Swiss Re Risk Engineering Services, July 7th 2015

[3] The Cumulative Risk Assessment Barrier Model, Alan Blacklaw, Al Ward, Kevin Cassity, BG Group, SPE, 146255, 2011

[4] Health & Safety Report 2015 – Oil & Gas UK

[5] API Recommended Practice 754 - Process Safety Performance Indicators for the Refining and Petrochemical Industries, API, April 2010

[6] HSG 254: Developing Process Safety Indicators, UK Health & Safety Directorate, 2006

[7] Cumulative Risk Guidance (Issue Draft for Comment), Oil & Gas UK, February 2016

445034 Donde, Cuando y Que Enseñar y la nueva certification de seguridad de procesos del CCPS

Abstract:

El primer objetivo de este paper es el de tratar de desmistificar que la Seguridad de Procesos debe enseñarse solamente o fundamentalmente en la carrera de Ingenieria Quimica. Basandonos en el concepto de los 20 elementos propuestos por el libro del Center for Chemical Process Safety "Seguridad de Procesos Basada en el Riesgo " (RBPS por sus siglas en ingles) se puede demostrar que ciertos elementos, como integridad mecanica y planificacion de emergencias, cruza transversalmente a otras disciplinas del conocimento academico como Ingenieria Mecanica, Ingenieria Electrica, etc. y ademas, otras no pertinentes al ambito de la ingeniera, como es el caso de divulgacion a la comunidad. Se presentaran ejemplos de ciertas universidades de los Estados Unidos, Canada, y Colombia. Como segundo objetivo, y conscientes del hecho de que no existe en el mundo actual, un modelo unico y probadamente eficaz para implementar la ensenanza de la Seguridad de Procesos, los autores buscan presentar diferentes modelos y llamar a la accion a las universidades de países hispanoparlantes para desacelerar la brecha entre lo que provee la Universidad y lo que necesita la industria.

Ademas, este paper buscara explicar los requerimientos del nuevo proceso de certificación de seguridad de procesos implementado por el CCPS, que es una nueva herramienta para demostrar competencia en seguridad de procesos, que permitirá a los profesionales agregar a sus credenciales "Certified Process Safety Profesional" or CPSP.

440342 PSM Legal Update

Abstract:

Over the last year, there have been many changes to the PSM Standard. There have been two court cases explaining key terms in the PSM Standard such as "change," "process," and "process equipment" and discussing key issues such as the time to implement PHA and audit findings. In addition, OSHA has issued four memoranda expanding the boundaries of PSM by defining RAGAGEP, creating a new mixture rule for Appendix A substances, completely revising the retail exemption and defining "proximate" for the purpose of establishing the boundaries of a process. OSHA issued a Request for Information to gather information for 14 potential changes to the PSM Standard to modernize it. After reviewing all of the responses to the RFI, OSHA has just initiated the review process under the Small Business Regulatory Enforcement Fairness Act (SBREFA) in which a draft rule with the proposed changes to the PSM Standard will be published and small business will have an opportunity to comment on it. The SBRFA process will provide a framework for additional changes being developed for the PSM Standard. This presentation will provide an overview of the key changes in the PSM Standard that OSHA has implemented and a summary of the changes being considered by OSHA.

440951 Application of Computational Fluid Dynamics (CFD) for Liquefied Natural Gas (LNG) Pool Spreading and Vaporization on Water

Abstract:

A marked increase in LNG utilization is forecast with an increase in LNG marine operations. A ship-ship collision of an LNG marine tanker or a rupture of the LNG loading and unloading lines can result in accidental spillage of LNG on water. Upon release, a spreading liquid can form a pool with rapid vaporization, leading to the formation of a flammable vapor cloud. Safety analyses for the protection of the public and property involve the determination of consequences of such accidental releases. The evaluation of consequences resulting from an accidental spill of LNG on water involves the determination of the rate (vaporization rate/source term) at which flammable hydrocarbon vapor is produced and the dynamics of the spreading pool. A specific offshore scenario of rupture of an offloading line is considered resulting in an LNG spill on water, between the space enclosed by the offloading deck and the LNG ship. This scenario was studied experimentally by releasing LNG in a narrow concrete trench filled with water. The experimental study involves the determination of the vaporization rate for a continuous release of LNG in a trench. Along with the vaporization rate, the spreading rate, pool area and pool height were measured. The computational work involves development of a methodology to model the pool spreading and vaporization behavior in Computational Fluid Dynamics (CFD) for the defined scenario. The overall aim of this research is to improve the understanding of pool spreading and vaporization of LNG releases on water.

Modeling an Unconfined Liquid Spill on a Sloped Surface

Robert Deshotels
Retired
21492 Pedroso, Mission Viejo, California, USA
R.L.Deshotels@cox.net

Prepared for Presentation at
American Institute of Chemical Engineers
2016 Spring Meeting
12th Global Congress on Process Safety
Houston, Texas
April 11-23, 2016

Modeling an Unconfined Liquid Spill on a Sloped Surface

Robert Deshotels, PE
Retired
21492 Pedroso, Mission Viejo, California, USA
R.L.Deshotels@cox.net

Keywords: liquid spreading, hydraulic radius, pool evaporation, friction factor, liquid spill modeling, simulation of liquid spreading

Abstract

When determining the consequences of a vapor cloud or fire from a liquid spill, the area of the liquid pool is a key variable. If the pool is contained by walls or berms, dispersion or flame parameters can be evaluated directly from the pool area using correlations that are well known to consequence analysts. For a spill onto an unconfined paved area or sloping deck, there is no simple formula for the pool area. For an unconfined pool, area must be calculated from flow modeling. The progressive movement and size of a liquid spill onto a sloped surface can be accurately modeled with well-known fluid equations using an empirical friction factor. The fluid flow formulas for piping can be applied to a sloping flat spill surface by including a correction for the equivalent hydraulic radius, the ratio of flow area to wetted area of the frictional surface. For uncomplicated flow surfaces, spreadsheet calculation is sufficient. Example calculations point out the need for improved knowledge of friction factors in terms of quantitative surface characteristics, and an improved formula for heat transfer during film boiling of cryogenic liquids.

1 Overview

The surface area of a spill of a toxic or flammable liquid spill is crucial to the analysis of hazardous consequences. The evaporation rate from a pool of volatile liquid, and hence the dispersion rate, depends on liquid properties, meteorological conditions, and the surface area of the pool. In the case of a pool fire, the heat flux radiating to nearby structures or people depends on the height and width of the flame, which depend on the size and shape of the pool. Usually, a liquid spill is confined by berms or curbs, so that the worst-case spill area is known from confinement dimensions. For an unconfined spill, there isn't any obvious maximum pool area, even if the source of the spill has a limited volume, because the pool will tend to continue to spread as time progresses. Therefore, unconfined spills require careful analysis to determine whether the spill can affect sensitive areas or lead to accident escalation. As well as environmentally sensitive areas, sensitive areas include escape paths, structures used for the shelter of employees or the public, and areas where fire or explosion could lead to escalation of the consequences.

Equations for the rate of spread of a liquid pool have been published in standard references on hazard evaluation [Ref. 1, 2], but friction loss was not considered in these reference guides. Friction loss is a major factor in the speed of liquid flow, especially for a thin pool depth. For a fixed-volume release, frictional loss can be the most important factor in whether a pool spreads into sensitive regions.

Much of the recent research on liquid spreading across a surface involves the effects of capillary action and surface tension on the motion of a thin film [Ref. 3 - 5]. There is much to be learned about the interaction between spilled liquid and the surface below in terms of capillary action and the hydrophilic/hydrophobic nature of the surface. However, the complexity surrounding surface tension applies only for very thin films of liquid. It will be shown that very thin films are not usually a significant part of industrial-sized releases.

For liquid spills that present a design challenge to industrial-size facilities, equations that are commonly used for design of piping and open channels, such as the Darcy equation, can be adapted to the purpose of modeling the time-dependent flow of an unconfined spill [Ref. 6 - 8].

For volatile liquids, the open channel flow equations give relatively accurate estimates for dispersion consequences. However, the concern may be that the spill will flow into a sensitive area. For non-volatile pools, the following question will be answered: "At what point is the spill thickness too low to produce a significant hazard?"

With a methodology for evaluating the minimum significant pool thickness in place, the basic steps for modeling an unconfined spill are:

1. Determine an appropriate release scenario.
2. Evaluate flow parameters for a generalized frictional flow formula.
3. Estimate the endpoint minimum significant pool thickness.
4. For sequential time increments, calculate pool dimensions, taking evaporation into account.

At each stage of modeling, uncertainties in the models and parameters should be considered to ensure that conservative choices are made. Uncertainty is especially significant when postulating a release scenario, estimating temperature for the liquid and surroundings, and appropriate modeling for heat transfer and evaporation. The largest pool area and evaporation rate produce the worst toxic or flammable release to the atmosphere. When the concern is distance of pool travel, minimum evaporation will create a worse case. There may be several potential worst case scenarios that must be quantified to determine which one is actually the worst.

Two example cases of unconfined liquid spills are provided in this article:
Example A. A leaking tanker truck located on a large paved area;
Example B. A piping leak onto the sloped deck of a floating oil production platform.

2 Release scenario

Generally, the source of the liquid spill will be a vessel or piping system with many potential leak points and spill trajectories. One possible way to address the multitude of possible release scenarios is to choose a worst credible case. If the worst case results in tolerable risk and acceptable mitigation expense, there is no reason to evaluate any less conservative cases. Sometimes, a release scenario may be so conservative as to be speculative. For example, if a release is conjectured to be the total contents of a tank due to accidental impact, there may not be any good basis for determining the speed at which the tank breaks, the size of the break, or the splashing that would occur upon impingement against nearby objects. In this case, simple assumptions should be made, and there should be a discussion of the sensitivity to uncertainties.

Any individual spill case may require a different approach to conservatism, but in general, the most conservative release scenario involves the greatest volume and flowrate. Concerning scenario parameters, pool spread distance is worse when the slope is high, and evaporation is low. Higher liquid temperature decreases viscosity and increases evaporation. When the spill poses multiple hazards or affects multiple possible targets, several potential worst-cases may have to be run to find the actual worst. When risk acceptability or mitigation cost are sensitive to the choice of release scenario, quantitative risk assessment tools such as event tree analysis should be used to pick the appropriate set of release cases. Data from offshore platform leaks [Ref. 9] can be used to estimate the probability of specific leak sizes from common process piping and equipment items.

Standard fluid dynamics methods [Ref. 7, 10] can be used to calculate the flow for a release of a given opening size and specified liquid conditions. The volumetric flow rate through an orifice, q, can be expressed in terms of the following parameters (in order of mention) flow coefficient, flow area of the orifice, gravitational acceleration, and differential pressure across the orifice, expressed in height of liquid:

$$q = C\,A\,(2g\,h)^{1/2}$$

The value of the flow coefficient, C, is a function of the Reynolds number (discussed in the section "Formulas for Flow of a Liquid Pool") and the ratio of the orifice diameter to the diameter of the pipe or vessel that is leaking. Plots of C are readily available [Ref. 7].

For liquids that flash upon release to atmospheric pressure, classical thermodynamics theory [Ref. 10] will accurately predict the vapor-liquid split and liquid properties after flashing. The mass fraction of the leaking fluid that vaporizes after release, f_v, can be expressed in terms of the exponential of the following parameters (in order of mention) liquid specific heat capacity, initial temperature of the liquid, atmospheric boiling point, and latent heat of vaporization. The heat capacity and latent heat should be the mean values over the temperature range T_i to T_b.

$$f_v = 1 - \exp\{-C_p(T_i - T_b)/\Delta H_v\}$$

Depending on release conditions and liquid physical properties, the liquid release may reach the pool surface as a steady stream or a spray of droplets. If the release is in the form of an undispersed stream, the trajectory can be evaluated using classical Newtonian physics, including the effect of wind pressure on the stream. In some cases, the release will immediately begin breaking into droplets, and in all cases a liquid stream will disperse into droplets if it falls enough distance.

Empirical formulas have been developed to predict the distance that a liquid release will travel before breaking-up into droplets. Droplet formation is characterized by the dimensionless Weber number, which is the ratio of shear force on the liquid surface, $\frac{1}{2}\rho_g v_d^2$, to surface tension pressure on the drop, σ/d. The Weber number is defined in terms of the following parameters (in order of mention) gas density, velocity of the drop relative to the air, drop diameter, and surface tension :

$$W_e = (\tfrac{1}{2}\rho_g v_d^2\, d)\, /\sigma$$

For subcooled liquids, the length between jet release and breakup into droplets, L_b, can be expressed in terms of the following parameters (in order of mention) jet diameter at the narrowest point, Weber number, and Reynolds number [Ref. 11]. This formula applies to leaks large enough to cause a serious hazard, but not to a 2 mm diameter hole or less. The units of measure must be consistent with the dimensionless Weber number.

$$L_b = 540\, D\, W_e^{1/2}\, R_e^{-5/8}$$

For superheated liquids [Ref. 11],

$$L_b = 2.75 \times 10^{10}\, D\, W_e^{1/2}\, R_e^{-2}$$

Based on experimental results for liquids without superheat, the typical droplet size, d, can be expressed in terms of the following parameters (in order of mention) surface tension, gas density, and jet velocity relative to the wind. The drop size depends on the liquid, and could be double the typical value [Ref. 1].

$$d = 24\, \sigma /(\rho_g\, v_d^2)$$

For superheated liquids, the droplet size depends on the difference between the liquid temperature and the atmospheric boiling point [Ref. 1],

$$d = 3 \times 10^4 /(10 + 4.0\, \Delta T_{sh})$$

The terminal velocity of a droplet in air is a complicated function of the liquid density and drop diameter, which can be read from a plot in a standard reference [Ref. 12].

Comparison of the terminal velocity to the wind speed will show whether droplets will rain out onto the pool or drift away like a cloud. The direction of release and Newtonian physics determine the trajectory that the droplets will follow from the liquid breakup point to the point where the pool begins growing. Different wind conditions may have to be considered. The probability of the combination of accident scenario, wind and temperature should be consistent with a tolerable risk for the accident.

3 Liquid pool growth and shrinkage

The pool volume will increase as long as the liquid release rate into the pool exceeds the evaporation rate. The release rate changes with time, depending on source inventory, pressure and effectiveness of process shutdown controls. The evaporation rate depends on pool area and changing temperatures.

It is possible to determine a worst-case maximum pool area using the initial leak rate and assuming a minimum evaporation rate. In this case, the maximum pool area equals the maximum pool volume created over the interval of concern (e.g., maximum escape and rescue time) divided by the minimum significant pool thickness. If the hazards evaluated from this bounding case give a tolerable risk, there may be no need to actually model the growth and configuration of the pool. However, even if the worst case results in tolerable risk, dynamic modeling of the pool may provide insight that would be valuable for planning and training.

Dynamic modeling consists of stepwise calculation of pool characteristics at small increments of time. At each increment, the following variables are computed:

1. Based on process conditions of pressure, temperature and composition, determine the flow properties and vapor pressure of the liquid just upstream of the release point. Using classical thermodynamics, determine the fraction of vapor flash after release into the atmosphere.

2. For the release scenario chosen, calculate the liquid flow rate into the pool using the orifice flow equation that is appropriate to the scenario. Calculate the liquid trajectory and dispersion from the jet breakup formula that applies to the orifice size, Weber number, and flashing superheat. It can be assumed that the liquid volume will be in a conical shape, with the target area being circular on a flat substrate, or elliptical on a sloped substrate. While a liquid release is in the momentum-driven region of flow (about 6 to 100 orifice diameters from the release point), a liquid jet will diverge at an angle of 10° from the centerline [Ref. 6]. From the flowrate, trajectory and divergence angle of the release, the target area of the liquid added to the pool is determined, as well as the volume increase in the target area during the time interval.

3. From a predicted Reynolds number, or the Reynolds number from the previous time increment, the applicable flow equation for pool spread is chosen. The volume flow and velocity at the edge of the pool can then be calculated, from which pool edge growth can be calculated. The equations for flow of the pool are given in the section "Formulas for Flow of a Liquid Pool".

4. A conservative calculation of pool evaporation is done, based on the new pool area calculated in the previous step. Depending on the purpose of the modeling, the evaporation calculation should be either a low estimate or a high estimate. An exact calculation is generally not possible because of uncertainty about rates of diffusion and heat transfer. An approach to providing a conservative estimate of evaporation is provided in the section "Evaporation". Evaporation may result in a cooling effect on the pool. The evaporation calculation is expressed as a uniform decrease in pool thickness, which will result in an adjustment in the pool surface area.

5. Knowing the inventory volume and process conditions, new process conditions are calculated from the liquid loss through leakage flowrate calculated in step 2 above. As the process pressure decreases, any evaporative cooling inside the process containment should be considered in the revised process conditions and composition.

Considering the accuracy of the postulated release scenario and air dispersion models, a simple model for pool thickness should suffice. It may be reasonable to model the pool volume as a cone with an elliptical base (because of the slope of the substrate), with the peak of the cone located at the target point of the liquid release trajectory.

4 Minimum significant pool thickness

The driving force for pool flow is the difference in height between different parts of the pool. For a pool moving down a sloped surface, the uphill part will become thinner as the pool spreads downhill. Furthermore, it is common for the leak rate from the process to stop or diminish after a while, causing the rate of evaporation rate from the pool to exceed the fill rate. At some point, part or all of the pool may become too thin to pose a significant hazard. To see whether this is possible in a specific case, it helps to categorize the different hazards posed by a pool, as follows.

Pool liquid that has not evaporated can pose a hazard of toxic exposure by direct contact, or a fire hazard.

For liquids that are not flammable, but are combustible upon ignition by flame, the pool may be too thin to maintain combustion. This happens when heat transfer to the substrate will quench any fire that starts by cooling the liquid below the flash point. This thickness could be calculated, but instead of solving unsteady-state heat transfer calculations using uncertain heat transfer coefficients and uncertain burning rates, the minimum fire hazard thickness for a non-volatile liquid should be determined by a field test. For one rough example, in a backyard test in a one-foot square confined pool of kerosene on concrete sidewalk at 20°C in breezy air, it was found that the minimum liquid thickness that would maintain a flame was between 2 and 3mm.

Where a liquid is highly toxic to people or the environment (e.g., aqueous ammonia) OSHA Standard 1910.119 has established a threshold quantity. A pool below this quantity no longer poses a catastrophic level of hazard, but mitigation is still imperative. Consequently, there is no insignificant pool quantity or thickness in the case of toxic liquids.

Vapor that is evaporating from a pool can create a toxic or flammable cloud leading to toxic exposure, flash fire, or explosion. The severity of exposure of a person to a toxic cloud is measured as the integrated dose of concentration over time. Therefore, for toxics, the quantity of material in the vapor cloud is more important than the peak rate of evaporation. As a result, the thickness of a pool does not affect the severity of a toxic release much, especially for a thin film that will evaporate soon. For a flash fire or explosion, severity is measured as heat flux or blast pressure, which depend on the size and concentration of the flammable cloud. Thus, for a spill of volatile liquid, the pool thickness just prior to complete evaporation does not have much impact on accident consequence.

When there is a concern that a combustible spilled liquid will flow into a sensitive area, the pool can be considered non-hazardous, provided there is no continuous source of ignition, if the entire pool depth is less than the minimum fire hazard thickness, determined from field testing of flame propagation.

5 Formulas for flow of a liquid pool

All practical formulas for the flow of liquids are derived from Bernoulli's equation, an energy balance that includes all of the involved forms of energy: potential energy (elevation difference), pressure, velocity, and frictional loss. The Bernoulli equation for the energy change for a fluid flowing from point 1 to point 2 can be expressed in terms of the following parameters (in order of mention) elevation, pressure, density, velocity, gravitational acceleration, and frictional loss expressed as height of fluid [Ref. 6, 7]. Consistent units of measure must be used. With U.S. customary units of measure, the gravitational acceleration factor of 32.2 ft/s^2 is needed to convert between pounds force and pounds mass.

$$Z_1 + P_1/\rho_1 + v_1^2/2g = Z_2 + P_2/\rho_2 + v_2^2/2g + h_L$$

Because the entire pool is at atmospheric pressure, there is no pressure energy term, P/ρ, to consider for a pool. The change in the elevation term Z is the only driving energy for pool flow, unless the release into the pool has a significant velocity due to the release trajectory at the target point. Further explanation follows for the velocity head term and friction head term.

Flow of a liquid pool on a sloped surface has two components. The first component of pool flow is the downhill motion of the entire pool. The driving force for downhill motion is the gravitational force multiplied by the slope. To determine downhill motion, the downward velocity due to the slope must be added to any velocity head imparted by the trajectory of the release flow into the pool, and the friction head must be subtracted. The downhill momentum increases as time continues. The second component of pool flow is the spreading of the pool outward from the center of mass, which can be modeled from the difference in height between the liquid and the surrounding surface.

Unlike flow through a pipe or in an open channel, liquid spreading on a sloped surface does not have a constant flow depth or cross-sectional flow area. Consequently, velocity varies through the pool. Therefore the value of the velocity used in the velocity head term, $v^2/2g$, and friction head term, h_L, must be considered carefully. For a pool modeled as a cone with a round or elliptical base, spreading begins at the high point as gravity-induced downward flow, then turns outward to be parallel with the base at the outer edge. To determine pool growth, the velocity head due to the elevation difference must be decreased by the friction head to yield the effective elevation change from the high point to the pool edge. From the effective elevation change, a new pool radius can be calculated for a chosen time differential. The new height can be calculated from the mass balance of the pool.

When the substrate is sloped or there is a velocity head imparted by the release stream, the growth in the pool can be evaluated by considering the cone to have four quadrants: uphill, downhill, and two equal sideways quadrants. Each quadrant is broken into differential elements at intervals of radial length along the distance from center of the ellipse outward to a point on the edge in order to evaluate a velocity that applies to each differential element of the quadrant. This allows a calculation of the friction head contribution of each element, from which the radius increase for each quadrant is calculated, and the new pool height determined as discussed in the previous paragraph.

There are different liquid flow formulas (e.g., Darcy, Fanning, Manning) that can be used to estimate the empirical term for frictional loss, h_L.

Regardless of the formula used for friction loss, there are four regimes of flow that can apply to an unconfined liquid spill:

1. Frictionless flow
2. Turbulent flow
3. Laminar flow
4. Thin film flow

5.1 *frictionless flow*

When the flow path has a low elevation gradient and a high pool depth, frictional loss may be insignificant. Also, it is possible for frictional loss to be very low in the case where the pool is heated by film boiling at the substrate interface [Ref. 4]. Otherwise, in actual spill cases that pose a serious risk to a facility, frictional loss is significant. When significant frictional loss is ignored, the simplified formulas for liquid flow rate [Ref. 1, 2] will overestimate the speed of liquid travel. Because friction loss is not difficult to evaluate, there seems to be little incentive to ignore friction.

5.2 *turbulent flow*

Flow turbulence is characterized by the Reynolds number. The Reynolds number is dimensionless, and can be expressed in any consistent units of measure, in terms of the following parameters (in order of mention) pipe diameter, velocity, density, and absolute viscosity.

For pipe, $R_e = D\, v\, \rho / \mu$

Commonly used formulas for liquid flow through a pipe or channel [Ref. 6 - 8] express frictional loss in terms of a friction factor, f, which is determined from experiments. However, formulas for pipe and channels cannot be used directly; the formulas must be modified to account for the different geometry of flow.

In an unconfined liquid spill, the flow often transitions from turbulent flow to laminar flow as the pool spreads thinner. Although the Reynolds number can be calculated exactly, the transition between the turbulent and laminar flow regimes is not consistent from one case to another. The transition is usually gradual, and can depend on flow disturbances and flow history. Turbulent flow is ordinarily exhibited when the Reynolds number is above 4000, with laminar flow exhibited below 2100. Laminar flow has less friction loss than turbulent, resulting in a higher velocity for a given slope. However, laminar flow results in a lower heat transfer rate. In the transition range between flow regimes, the flow should be modeled conservatively as either fully turbulent or laminar. The conservative choice will depend on the situation.

Unconfined liquid flow on a horizontal or sloped surface is analogous to gravity-induced flow in a pipe, except that wall friction only occurs along the bottom of the spill. The concept of hydraulic radius, R_H, [Ref. 6, 7] must be applied.

$$R_H = \text{cross sectional flow area / wetted perimeter}$$

For flow through a round pipe full of fluid, the flow diameter equals $4R_H$. For unconfined pool flow, the hydraulic radius is the pool height, z, at the point where velocity is measured, and the equivalent flow diameter equals $4z$. The Reynolds number for a round pipe full of liquid is expressed in terms of the inside pipe diameter. For full flow through a narrow slot, the Reynolds number is expressed in terms of the height of the slot. For open flow across a flat surface, we typically assume that there is no friction at the top surface of the liquid or the edges, although this is not strictly true due to relative motion between the air and liquid surface. With negligible air friction, the zero-shear level is at the top of the liquid pool, rather than at the centerline as in a round or rectangular pipe. Therefore, the equivalent hydraulic radius used to evaluate the Reynolds number for a round liquid pool, with a flow area (circumference multiplied by height) of $2\pi\ r\ z$ and a perimeter of $2\pi\ r$, should be corrected to twice the usual value, or $2z$. In the Reynolds number and friction head equations, the D used for full-flow round pipe should be replaced by $8z$ for a pool.

$$\text{For a pool, } R_e = 8z\ v\ \rho / \mu$$

Wind friction can affect pool motion, but the effect is expected to be minor. Weather stations report wind speed at a reference height of 10 m above grade. If the spill is at grade, the wind is reduced to about 8% of the reference value in the daytime, or 5.5% at night (the difference is due to air stability) [Ref. 4]. It should be noted that wind wave height cannot exceed the pool thickness; therefore, the surface of the liquid will be smooth, even under the influence of wind. Wind drag can be evaluated from low-speed aerodynamics. The drag pressure caused by air friction is the drag coefficient multiplied by the wind velocity head. For wind blowing parallel to a flat surface the drag coefficient, C_D, is 0.04 [Ref. 13]. To convert into liquid head, the wind head is multiplied by the ratio of gas density to liquid density. If the wind is in the opposite direction of flow, the wind head is added to the friction head. If the wind is in the same direction as flow, the wind head is subtracted from the friction head. The wind head, h_{drag}, can be written in terms of the following parameters (in order of mention) drag coefficient, air density, air velocity, and liquid density.

$$h_{drag} = C_D\ \rho_G\ v_G^2 / 2\rho_L$$

Table 1

Formulas for Turbulent Flow Friction Loss

Darcy (Moody)	$h_L = f\,L\,v^2/2D$
Fanning	$h_L = 2f\,L\,v^2/D$
Manning	$h_L = f\,L\,v^2/2R_H$

Note that the Moody friction factor is four times larger than the Fanning friction factor. This can be confusing because plots of friction factor do not always mention which friction factor is given. For the Darcy and Fanning equations, the friction factor is a function of the Reynolds number and relative roughness of the pipe wall, ϵ/D. For a pool having a height of z, D should be replaced by $4R_H$, or $8z$. Roughness is determined by experimentation, rather than surface measurement. Based on the limited data available and the difficulty in defining roughness, the friction factor is the major cause of uncertainty in flow rate calculations. The Manning equation is an empirical equation developed from data for water flow in deep open channels. For the Manning equation, a roughness parameter n is used to correlate the friction factor to the hydraulic radius [Ref. 6, 8].

Manning $f = 29n^2/R_H^{1/3}$

Table 2

Roughness Factors for Turbulent Flow

Surface	ϵ ft	n $ft^{1/6}$
Clean commercial steel	0.00015	
Galvanized iron	0.0004	0.012
Cast iron	0.00085	0.011
Concrete pipe	0.001 - 0.01	0.015
Riveted steel	0.003 - 0.03	0.014
Earth, straight and uniform		0.019
Brick open channel		0.017
Concrete open channel		0.022
Gravel open channel		0.033

5.3 laminar flow

Unlike turbulent flow, fully laminar flow along an unobstructed path exhibits no eddies in the axial or radial direction. Friction loss is due to viscous effects alone, and the friction factor is a function of Reynolds number alone, not surface roughness.

Darcy laminar $f = 64/R_e$

Fanning laminar $f = 16/R_e$

The value of f for laminar flow is substituted into the Darcy or Fanning equation for friction head. However, the laminar friction factor was developed for round pipe containing full flow, and will be somewhat in error for open flow on a flat surface, due to the parabolic velocity profile that exists between the pipe wall and zero-shear point in laminar flow [Ref. 6]. Although this paper proposes a correction in the hydraulic radius for the location of the zero-shear point, there is no published data to verify that this proposed correction is accurate or conservative.

Like turbulent flow, as long as there is a height difference between the top of the liquid pool and the surface of the substrate, the Bernoulli equation and derived Darcy formula tell us that the pool will continue to spread without limit, although at an ever decreasing rate. For turbulent or laminar flow, the Darcy formula does not consider the effects of surface tension, capillary action, or hydrophilic/hydrophobic nature of the surface under the spill.

5.4 thin film flow

Recently, investigators have been able to explain why liquid pools do not actually continue to spread endlessly [Ref. 5]. When the liquid height is less than around 100 nm (0.01 cm), long-range intermolecular forces (like van der Waals forces) begin affecting the ability of the liquid to wet the surface of the substrate. There is a minimum thickness of liquid film, beyond which the film no longer wets the substrate; instead, the film breaks into individual droplets. In the case of a level substrate, flow stops. The minimum film thickness is affected by substrate surface properties such as roughness, cleanliness, porosity, and surface tension between the solid and liquid. There are many cases for which the proposed model has not been tested (including a sloped surface), but two conclusions can be drawn:

1. For thin pools, the laminar flow model gives a higher spreading rate than any model that considers the effect of surface tension.

The thin film flow model does not apply for liquid pools thicker than about 0.01 cm.

6 Evaporation

Evaporation will occur if the liquid has a significant vapor pressure at the pool temperature, and the atmosphere above the pool is not saturated with the liquid vapor. The temperature of the pool can change because of heat transfer with the substrate or air, which would affect the vapor pressure.

Evaporation rate from the pool is described in terms of classical thermodynamics [Ref. 10]. The driving force for evaporation is the difference between the vapor pressure of the liquid and the partial pressure of the liquid in the air. The mass evaporation rate, W, can be written in terms of the following parameters (in order of mention) molecular weight, mass transfer coefficient, flow area, vapor pressure of the pure liquid, partial pressure of the vapor in the bulk stagnant air above the liquid, gas law constant, and temperature of the liquid. Pressure and temperature are in absolute units.

$$W = M\,K\,A\,(P_{sat} - P_v)\,/(R\,T)$$

At atmospheric pressure and practical operating temperatures, ideal gas can be assumed. Except in the case of a boiling liquid, the rate of evaporation is restricted by the rate of diffusion from the surface of the liquid. Diffusion coefficients have been measured or estimated for common chemicals [Ref. 14]. The resistance to mass transfer from a homogenous liquid to air is mainly due to the resistance of the air. The resistance of the air depends on the transport properties of the air (expressed as the Schmidt number) and flow regime (expressed as the Reynolds number). There are many theoretical and empirical correlations for evaporation, depending on the geometry of the surroundings, flow regime, concentration of the evaporating component in each phase. A great deal of judgement is required in choosing an appropriate correlation. Key issues involve friction modeling for a boiling pool, wind effects, and heat transfer from the substrate [Ref. 5]. It seems that nearly all of the correlations were developed for equipment design and were tested in laboratories, not outdoors in industrial-sized facilities where nearby objects can obstruct or funnel the wind. Therefore, correlation accuracies claimed in literature may not apply to evaporation from a real pool [Ref. 5]. An empirical correlation, rather than theoretical-based, should be applied when the flow is parallel to the interface through which mass transfer occurs (such as pool evaporation) [Ref. 14]. For a calculation that overestimates evaporation, one can assume that there is no resistance to mass flow within the liquid or across the liquid-air interface, which results in an air concentration just above the interface that is equal to the partial pressure of the liquid relative to atmospheric. When the resistance to diffusion is essentially due to air convection alone, the diffusivity is in the range of 10^{-4} to 10^{-6} m^2/s. A simple but crude estimate of diffusivity, K, based on molecular weight, M, may be sufficient [Ref. 10], by assuming the ratio between the diffusivity and cube root of molecular weight is constant for the liquid of interest and some reference liquid with known diffusivity, K_0.

$$K = K_0\,(M_0/M)^{1/3}$$

An evaporation correlation specifically for a pool of volatile liquid has been developed, accounting for the wind speed and stability class of the air [Ref. 2]. This formula does not consider evaporative cooling of the pool. The evaporation rate, W, involves two parameters, λ and n, which depend on the Pasquill-Gifford air stability classification. The equation is written in SI units in terms of the following parameters (in order of mention) λ, wind velocity, vapor density, vapor pressure, ambient pressure, wind velocity squared, pool diameter, and n.

$$W = \lambda\,v_w\,\rho_v\,P_v\,/P\,\{2/(v_w^2\,D\}^n$$

Table 3

Pool Evaporation Parameters

Pasquill Stability Class	λ	n
A	1.0×10^{-3}	0.078
B	1.2×10^{-3}	0.091
C	1.5×10^{-3}	0.111
D	1.7×10^{-3}	0.130
E	1.8×10^{-3}	0.149
F	1.8×10^{-3}	0.180

To calculate a high estimate of evaporation for non-cryogenic spills, assume the spill occurs on a windy, hot summer day; for a low estimate assume a still, cold winter night.

A separate formula applies to a spill of a cryogenic liquid on land, explicit as a function of time [Ref. 2]. This equation is written in SI units in terms of the following parameters (in order of mention) a surface correction factor, thermal conductivity of the substrate, heat of vaporization, thermal diffusivity of the substrate, substrate temperature, liquid temperature, and time. The surface correction factor, *X*, is assigned a value of 1 for a smooth substrate, or 3 for a rough substrate such as gravel. Note that this equation crudely assumes that heat transfer is proportional to the temperature difference between the substrate and liquid; this assumption does not apply at all beyond the point of maximum nucleate boiling, when partial or full film boiling occurs.

$$W_{cryo} = 2X\, k / \Delta H_v\, (\pi\, \alpha)^{-1/2}\, (T_S - T_L)\, t^{1/2}$$

If the pool is boiling, the evaporation rate is usually limited by heat input. If the liquid flashes upon release from the process, the liquid and vapor reach equilibrium soon after release, usually before the liquid touches the ground. The liquid will not continue to boil unless heat is added from the air or substrate. The main uncertainty involves heat transfer from the substrate under the pool. Different modes of heat transfer are possible, nucleate boiling, film boiling or transition [Ref. 15, 16]. The transition from nucleate boiling to film boiling occurs at a critical temperature difference between the substrate and liquid, which depends on liquid properties. Heat transfer rates for boiling LNG and NPG have been documented from field experiments with liquefied gas inside level confinement [Ref. 16]. In the case of film boiling, which often occurs in cryogens (especially LNG), the friction factor at the substrate interface is markedly decreased. Equations for the friction factor during film boiling of LNG spilled over a water surface have been published [Ref. 4]. If the pool is supported by the ground, transfer through the substrate depends on several factors including moisture content. One bounding option that overestimates evaporation is to assume that the substrate temperature remains constant. A bounding assumption that underestimates evaporation is that for fully developed film boiling, the heat flux equals the minimum for film boiling for that liquid.

Several references show an "S" curve for typical boiling liquid heat flux as a function of temperature difference between the substrate and liquid. The typical curve shows a low flux at low temperature difference, rising like the front end of a sine wave to reach a maximum flux at the point of fully developed nucleate boiling. When the temperature difference increases beyond this critical point, the heat flux decreases like the back end of a sine wave, reaching a minimum flux point caused by a layer of vapor that separates the substrate and liquid. Beyond the minimum point, the "typical" flux curve curves up again at a slightly lower slope than the initial curve for nucleate boiling. This increase is attributed to the increase in radiative heat transfer to the liquid from the hotter substrate. The data that supports this "S" curve is mainly drawn from data from experiments with silicone oil, which boils at a high temperature. In the typical experiment, the substrate is heated beyond temperatures where radiant heat transfer is significant. This data may apply to film boiling of cryogenic liquids inside pressure vessels, but an atmospheric spill of LNG is different. If the increase in film boiling heat transfer is due to radiative transfer, it should not be expressed as a linear relationship with temperature. Radiative heat transfer is a function of absolute temperature to the fourth power. The equation for transfer between parallel plates, or a flat substrate to a liquid suspended by a flat vapor layer is [Ref. 15]:

$$Q/A = \epsilon \, \sigma \, T_1^4 - \alpha \, \sigma \, T_2^4$$

where Q/A is heat flux, ϵ is emissivity, σ is the Stefan-Boltzmann constant, T_1 is the absolute temperature of the substrate, α is the absorptivity of the liquid, and T_2 is the absolute temperature of the boiling liquid. Units of measure must be consistent with the Stefan-Boltzmann constant, which is 1.713×10^{-9} Btu/(ft^2 h °R^4) or 5.67×10^{-8} W/(m^2 °K^4). The emissivity and absorptivity of common materials is given in standard references [Ref. 15]. It would seem that the radiative effect on film boiling heat flux could be calculated, but there has been no experimental data to validate this conjecture.

7 Conclusions

Use of the Bernoulli equation for liquid flow provides a method for calculating the spread of an unconfined liquid pool on a sloping surface. The Bernoulli equation is an expression of the fluid flow energy balance, so there is no need for the user to characterize the flow as gravity-driven or momentum-driven in order to choose an appropriate model. Friction loss can be calculated the same as for piping or open channel flow after flow geometry is accounted for. Although flow can be calculated exactly if all parameters are known, the surface roughness parameter can be uncertain by an order of magnitude. The most important variable, release flow rate into the pool is generally determined by accident conditions that are difficult to predict. Furthermore, the evaporation rate from the pool can be affected by heat transfer rates and mass diffusion coefficients that are somewhat difficult to predict outside of a laboratory. Therefore, care should be taken to assure conservative assumptions, methods and data.

The calculation accuracy of liquid flow across a sloping surface is limited by roughness data and heat transfer knowledge. Within this current limitation, one-dimensional spreadsheet calculations can provide useful bounding results. However, when accurate data has been developed, computational fluid dynamics will provide more accurate results.

The non-skid checkered deck plating used on the decks of land-based processing facilities and floating production platforms has well known surface dimensions, but no documented friction factor for flow calculations. Rather than relying on field testing of qualitatively defined surfaces, isn't it time that someone develop a method for estimating friction factors based on dimensions and some measurable index of roughness?

Spills of cryogenic fuels such as LNG and LPG pose a high risk of catastrophic accident worldwide. The heat flux during film boiling as a function of differential temperature could be more accurately calculated as function of the difference of the fourth power of the absolute temperatures, like the Stefan-Boltzmann equation, rather than a first-order difference. Experiments with cryogenic liquids such as liquid nitrogen could test this possibility and be used to develop a better correlation for film boiling, especially for liquids with low boiling points. Furthermore, experiments with cryogenic liquids flowing on sloped surfaces would be useful for accident analysis.

8 References

1. J.L. Woodward, *Estimating the Flammable Mass of a Vapor Cloud*, "Discharge Characteristics", p 49-59, Center for Chemical Process Safety of the American Institute of Chemical Engineers, 1999.
2. A. Cox, F. Lees, and M. Ang, "Classification of Hazardous Locations," pp 46-47, Inter-Institutional Group Classification of Hazardous Locations of the Institution of Chemical Engineers, Rugby, UK, 1990.
3. M. Apel-Paz and A. Marmur, "Spreading of liquids on rough surfaces," J Hazard Mater, 146, pp 273-279, 1999.
4. S. Brambilla and D. Manca, "Accidents involving liquids: A step ahead in modeling pool spreading, evaporation and burning," J Hazard Mater, 161, pp 1265-1280, 2009.
5. A. A. Pahlavan, et al., "Thin Films in Partial Wetting: Internal Selection of Contact-Line Dynamics," Phys. Rev. Lett. 115, 034502, July, 2015.
6. R. H. Perry and D. W. Green, *Perry's Chemical Engineers' Handbook,* 7th ed., "Incompressible Flow in Pipes and Channels," pp 6-9 through 6-16, 6-20, 6-21 McGraw-Hill, New York, 1997.
7. Crane Co., *Flow of Fluids, Technical Paper 410*, Ver. 15, pp 1-1, 1-4 through 1-7, www.flowoffluids.com, 2009.
8. S.D. Lin, *Water and Wastewater Calculations Manual,* pp 242-252, McGraw-Hill, 2001.
9. UK Health and Safety Executive, "Hydrocarbon Releases Database," https://www.hse.gov.uk/hcr3/, 2005.

10. D.A. Crowl and J.F. Louvar, *Chemical Process Safety Fundamentals with Applications,* 3rd ed., pp 122-140, Prentice Hall, 2011.

11. J.H. Lienhard and J.B. Day, "The Breakup of Superheated Liquid Jets," J Basic Eng. Trans., ASME, pp 515-521, 1970

12. R. H. Perry and D. W. Green, *Perry's Chemical Engineers' Handbook,* 7th ed., "Particles Dynamics," pp. 6-50 through 6-54, McGraw-Hill, 1997.

13. B. Jones, *Elements of Practical Aerodynamics,* 2nd ed., p 17, Wiley & Sons, 1939.

14. R. H. Perry and D. W. Green, *Perry's Chemical Engineers' Handbook,* 7th ed., "Mass Transfer," pp 5-42 through 5-79, McGraw-Hill, 1997.

15. R. H. Perry and D. W. Green, *Perry's Chemical Engineers' Handbook,* 7th ed., "Heat Transfer," pp 5-8 through 5-29, McGraw-Hill, 1997.

16. J.L. Woodward and R.M. Pitblado, *LNG Risk Based Safety: Modeling and Consequence Analysis,* pp 142-152, Wiley, 2010.

17. Western Regional Climate Center, "Cooperative Climatological Data Summaries," http://www.wrcc.dri.edu/climatedata/climsum/ Feb. 2016.

18. ASHRAE, *Fundamentals Handbook,* ASHRAE, Ch. 22, 1977.

Example A

Leaking LNG tanker truck located on a large paved area

A-1 Accident scenario

The concern is that a tanker truck at a LNG terminal could spill LNG into a storm sewer, with the potential that the natural gas could ignite in a sewer line in a public area. A calculation must be performed to see whether the contents of the truck would evaporate before reaching the nearest storm sewer drain.

At a terminal used to fill city vehicles powered by compressed natural gas, the cryogenic storage tanks are filled by a 10,000 g tanker truck. The tanker is double-walled for insulation and impact protection. The tanker contains LNG at about 70 psig and -210°F. In this example, the LNG is assumed to consist of 100% methane. At an actual facility, the LNG composition would be known. At atmospheric pressure, methane boils at -259°F. The tanker is unloaded through a nominal 3" flexible hose.

The terminal equipment and storage tanks are located inside a protective building. The tanker unloading hose is connected to a double-walled pipe that goes through the building wall. To reach the unloading station, the truck must drive along a concrete driveway that is 59 ft wide. The driveway slopes at a 1% grade toward a storm drain that is located a little more than 400 ft down the driveway.

A-2 Release modeling

The intent is to model a bounding case, to see if the worst-case spill of LNG can reach the storm drain. Leaks from tanker trucks seldom occur, and when they do, it is usually due to accidental impact or failure of the unloading hose or coupling. Failures involve some degree of human error, and therefore are difficult to assign a probability. The failure chosen for modeling is a sudden loss of the total contents of the tanker, which may not be credible, and certainly is a bounding case.

The LNG spill quickly flashes upon release to the atmosphere. The flash can be assumed to be adiabatic. After the initial flash, the LNG continues boiling during the life of the pool. The longest spill travel occurs when there is minimum LNG evaporation or lowest surface friction. Evaporation is minimized by cool air, cool pavement, and poor heat transfer. The lowest evaporation would occur on a cold winter night. However, friction is minimized during film boiling, which is favored by a hot summer day. Calculations that follow show that the summer temperature case is worse for pool travel. At the terminal site, the lowest annual temperature is 39°F (4°C), and the highest annual temperature is 96°F (36°) [Ref. 17]. The maximum temperature typically occurs on a clear, breezy afternoon, consistent with a pavement surface temperature of about 180°F (82°) due to solar gain. Frictionless flow is assumed when the temperature difference between the substrate and liquid indicates the possibility of film boiling. The summer case is average for about 60 days a year, or 16% of the year. Therefore, the summer temperatures could be exceeded 16% of the time. It turns out that the ability of the LNG pool to travel 400 ft is not sensitive to surrounding temperature within the range that occurs at the site.

The modeling covers two phases of pool spreading: unconfined and partially confined. For simplicity, the first phase begins with the volume of the tanker in the shape of an upright cylinder, the same height as the center of the tanker, located at the tanker unloading station. After the cylinder expands to the width of the driveway, which takes about 3 seconds, the pool shape changes due to partial confinement imposed by the curbs on each side of the driveway.

A-3 Pool spread methodology

The boiling regime and heat transfer rate depend on the temperature of the pavement, which will vary with time due to contact with the LNG. A preliminary step of analysis is to correlate pavement surface temperature as a function of contact time with boiling LNG. The heat transfer from the pavement to the LNG depends on the temperature profile of the pavement and sub-pavement as a function of depth. The starting point profile was created from typical day and night temperatures as a function of time, using thermal properties of concrete and aggregate and a plot of air-concrete heat transfer coefficient as a function of wind speed [Ref. 18]. The summer daytime case requires data on solar radiation, which varies with location [Ref. 17]. The profile of temperature versus pavement depth depends on the recent history of air temperature and solar radiation. For the winter night case, there is radiative heat loss on a clear night from the pavement to the sky which is at a temperature of nearly absolute zero. However, the winter night radiation effect is insignificant. For the summer day case, there is no historical data of pavement surface temperatures, so a pavement profile was calculated by solving the one-dimensional differential equations for conductive heat transfer, starting with the temperature profile at sunrise at an average daily minimum summer temperature, allowing for night solar radiation. The air temperature and solar gain varied until six hours later, when air temperature and solar gain reached the average summer maximum. Several trial calculations showed that the daily temperature cycles did not have a significant effect on the profile below a depth of about 12". The summer temperature at 12" depth is assumed to be the average of the summer minimum and summer maximum air temperatures for the month. The equations for one-dimensional heat transfer in to a thin section of substrate at the surface of the pavement are:

$$Q_{radiation}/A = F_R \, \Delta t$$
$$Q_{convection}/A = h \, (T_{air} - T_1) \, \Delta t$$
$$Q_{conduction}/A = -k \, x \, (T_1 - T_2) \, \Delta t$$
$$\Delta T = (Q/A)/m \, C_P$$
$$m = \rho \, A \, x$$

where Q is the heat transfer rate
A is the transverse heat transfer area
F_R is the solar heat flux
Δt is the differential time step
T_1 is the temperature in the 1st section of pavement
k is the thermal conductivity of the pavement
x is the thickness of the section of pavement
ΔT is the temperature change in a section during time dt
m is the mass of the section
C_P is the specific heat capacity of the pavement

ρ is the density of the pavement

By taking a small enough time step, the heat transfer equations can be solved accurately enough by using the temperatures from the previous step. The exception occurs at the first section, where the temperature changes rapidly with time. This section was solved by obtaining the difference between the first and second sections by trial and error. Trial and error is easily done by manually adjusting the temperature difference in the spreadsheet for each time step. This procedure was used to calculate the temperature profiles for the summer case influenced by solar gain, and for the profile when pavement is exposed to LNG.

A summary of the summer temperature profile at different pavement depths is tabulated below. The bottom row shows the profile at the hottest part of a typical summer day:

Table A-1
Pavement Temperature Profile before an LNG Spill
Pavement sections are 1" thick. Section numbering begins with the section on top.
The profile for the hottest hour (hour 6) is used for spill calculations.

Time h	T_{air} °F	Solar Btu/h-ft^2	Section 1	Section 2	Section 3	Section 4	Section 5	Section 6	Section 7
0	60.4	0	59.1	61.1	63.1	65.1	67.1	69.1	71.1
1	66.4	50	69.9	68.2	68.1	68.5	69.3	70.3	71.4
2	72.5	100	85.4	80.2	77.6	75.9	74.9	74.4	74.4
3	78.5	150	105.1	96.4	91.1	87.0	83.9	81.7	80.1
4	84.5	200	128.6	116.3	108.0	101.2	95.8	91.4	88.0
5	90.6	250	155.3	139.3	127.9	118.3	110.2	103.4	97.7
6	96.6	250	178.6	161.2	147.7	135.9	125.5	116.4	108.4

When LNG is spilled onto pavement in the summer or winter, the pool stays in the film boiling regime until the surface of the pavement cools to within 20°C (36°F) of the LNG boiling point. In the film boiling regime, there is virtually no friction loss. As the tanker spill spreads to fill the area between the curbs of the driveway, the liquid is modeled as an upright cylinder in film boiling, non-frictional flow. For lack of better information, it is assumed that the vapor film under the liquid is of uniform thickness. Energy loss due to internal turbulence is ignored, which overestimates velocity and distance travelled. In the case of frictionless, gravity-induced open flow, the change in elevation equals the resulting velocity head: horizontal velocity squared divided by twice the acceleration of gravity, where the velocity is the average value across the cross-sectional area of flow.

$$z = v^2/2g$$
$$v = (2g\,z)^{½}$$

The velocity head equation applies to both components of pool motion – downhill motion of the entire pool, and radial spreading. In the case of downhill motion, z_d is the horizontal distance travelled, Δy, (including the downhill radius of the cylinder) multiplied by the slope of the surface.

$$v\,\Delta t = \Delta y = (2g\, z_d)^{1/2}\,\Delta t$$

In the case of radial spreading, z_r is the height of the pool at the center

$$\Delta r = (2g\, z_r)^{1/2}\,\Delta t$$

Neglecting viscous forces, the initial radial increase equals $(2g\, z)^{1/2}\,\Delta t$, where Δt is the time interval. Prior to considering evaporation loss, the cylinder volume is the same as before, but the cylinder is shorter by Δz. Therefore, the new pool height is the volume of the pool divided by the area calculated from the new radius.

In the beginning, the pool is deep, and the time required to reach the curbs is short, so it is not yet important to model the effect of the driveway slope. However, when the elevation difference between the uphill and downhill portions of the pool becomes significant relative to the pool thickness, the slope should be considered. The cylinder can be partitioned into four quadrants: uphill, downhill, and two sides. Without surface friction, the top of the cylinder is level, so each point on the circumference has a different height above the pavement. The sloped z values for the uphill and downhill quadrants are used to calculate new radii of the uphill and downhill quadrants. The net effect of the longer downhill radius is that the cylinder will spread farther in the downhill direction and become slightly elliptical.

The evaporation loss is the combination of evaporation from the surface of the liquid plus the vapor generated from boiling. The boiling heat transfer rate for LNG can be conservatively calculated by using the minimum heat flux value of 20 kW/m^2 when the temperature difference between the substrate and LNG exceeds 20°C [Ref. 16]. Evaporation from the surface is due to the combined effects of heat input from solar heat and convective heat transfer with the air. Evaporation is assumed to be limited by heat flux, not diffusion into the air. The heat flux terms for solar heat and air convection are the same as used for calculating the pavement temperature profile for a summer day.

As the pool spreads thinner, there is a time when Δz exceeds the pool thickness on the uphill edge of the pool. When this happens, the uphill edge is moved so that the thickness becomes zero at the uphill edge.

A-4 Results

Upon release to atmospheric pressure, 17.3% of the LNG flashes, based on an adiabatic flash, taking data from the methane pressure-enthalpy diagram.

The one-dimensional, time-dependent heat transfer calculations showed that on an example hot summer day, concrete that is in contact with boiling LNG can take almost 10 minutes before the liquid is in frictional contact with the pavement. This is because a film of natural gas vapor insulates the concrete, maintaining a high temperature difference between the boiling LNG and the pavement surface. Film boiling occurs for about 8.7 minutes in the summer example. For the next minute, boiling becomes more nucleate, finally reaching the point of maximum heat flux. The same spill on a cold winter night showed that the duration of film boiling was about half as long at this site, but still greatly exceeding the time required for the pool to evaporate. Other factors, such as difference in evaporation rate between summer and winter are not important. Therefore, because flow will be essentially frictionless during film boiling, the hot summer case is the worse case for distance travelled by the moving pool of LNG.

Example B

Piping Leak onto the Sloped Deck of a Floating Oil Production Platform

B-1 Accident scenario

There is a potential for a large fire on a lower deck of a floating oil production platform. To assess the potential for this fire to cause escalating damage to the platform and possibly interfere with escape, the size of the pool must be estimated. The area of the oil spill is the major factor affecting the height and heat flux of the potential pool fire. This particular spill case is one possible event of an event tree performed for a quantitative risk assessment. The length of the spill path is known, but the width of the pool as it flows down the deck must be estimated by flow calculation.

The spill source is a manifold of 12-inch pipe on a deck made of metal checker plate of the kind typically used in industrial applications. The piping is supported every few meters, so that the piping does not interfere with the flow of oil across the deck. The deck plates are galvanized metal with tread ridges to provide a non-skid surface. The deck has a slope of 2.5% from the keel line down to the edge of the deck, 31.5 m away. The lower deck is shielded from sunlight, and there is little difference in temperature between the air, deck, piping, and contained oil.

The potential leak consists of light crude oil that has been separated from produced water and gas. The crude oil contains enough dissolved gas to be flammable, but the oil loses an insignificant volume of light components when flashed to the atmosphere. The crude oil is contained at a maximum pressure of 8 barg. The transport properties of the oil at flowing temperature are:

Density	889 kg/m^3
Viscosity	11 cp, or 0.00112 $kg\ s/m^2$
Mol Wt	250 kg/mol
Surf tension	27 dyn/cm, or 0.027 N/m

Based on the number of potential leak points and the probability of leakage from each type of potential leakage, the probability of a serious leak is on the order of 10^{-3} per year for a leak equivalent to a 50-mm diameter hole. The most important contributor to serious leakage is gasket failure at a flange or valve. Nearly all flanges are arranged vertically, with the opening aimed downhill. Ordinarily, any leak onto the deck will be quickly detected by flammable gas detectors, resulting in a manual shutdown of the piping system. In the accident case for the fault tree branch of interest, the detector alarm system or operator response fails, which is estimated to happen less than 10% of the time, which lowers the probability to 10^{-4} per year. The probability of ignition is about 5% to 10%, which places the accident scenario in the barely credible range. The average wind speed is 4.3 m/s, directed crosswise to the slope of the deck. Extreme weather conditions at the time of the accident would make the scenario beyond credible, and consequently need not be considered.

B-2 Release modeling

The maximum credible leak, equivalent to a 50-mm hole is likely to occur as a narrow gap rather than a round opening, either as a tear in piping due to impact or a gap between flanges due to improper bolt tightening or gasket failure. Round holes often result from corrosion, which is not a problem with crude oil in this case. The gasket thickness is 2 mm, so a 2 mm gap is assumed. A full blow-out of a 2 mm gasket from a 12-inch line results in an opening with an area of 1920 mm^2, which is about equal to the area of a 50-mm hole.

The flow rate through the leak opening can be calculated from the flow coefficient [Ref. 7], which is a function of the Reynolds number for the pipe. The Reynolds number for the pipe is 17,400 and the ratio of orifice diameter to pipe diameter is less than 0.2, which gives a square-edge flow coefficient, *C*, of 0.60. The standard equation for flow through an orfice gives the leak flow rate. Based on the pressure drop, density, and equivalent orifice size, the leakage flow is 0.0500 m^3/s. The liquid velocity at the release point is the volumetric flow rate divided by the flow area, or 26.0 m/s.

For holes less than 2 mm wide, the drop diameter is about 1.9 times the narrow dimension (2 mm in this case) [Ref. 1, 11]. This correlation gives a drop diameter of 3.8 mm. A second method for calculating drop size is based on the Weber number being between 12 and 22 [Ref. 1]. For a Weber number of 12,

$$d = 24\,\sigma /(\rho_g\, v_d^2).$$

For the site air density of 1.204 kg/m^3, surface tension of 0.027 kg/s^2, and drop velocity of 26.0 m/s, the estimated drop diameter is between 0.80 mm and 1.46 mm. The free fall velocity for a drop between 0.80 mm and 3.8 mm is between 3 m/s and 10 m/s [Ref. 12]. The chosen values are 2.3 mm and 2 m/s.

The release will break up into droplets at a distance of L_b, which is a function of the Weber number and Reynolds number. For a drop diameter of 2.3 mm, and an air density of 1.204 kg/m^3, the Weber number is 34.7. The Reynolds number at the release point is 47,500. The distance from release to break-up into drops is

$$L_b = 540\, D\, W_e^{1/2}\, R_e^{-5/8}$$

In this case, L_b = 0.0874 m.

The trajectory of the release begins as a circular sheet 2 mm thick sprayed from a pair of vertical flanges. Before the sheet contacts the deck below, it breaks into droplets and the velocity slows to 2 m/s, relative to the air. For the portion of the spray directed straight down, that liquid lands directly under the leak point. For the portion of the spray directed up, the droplets will be in the air for about 0.36 s, calculated from the free fall drop velocity. The wind force on a droplet F_d, is a function of the drag coefficient C_d, projected drop area facing the wind A_d, air density ρ_a, and relative velocity v between the drop and air [Ref. 12]:

$$F_d = \tfrac{1}{2}\, C_d\, A_d\, \rho_a\, v^2$$

From Newtonian physics, acceleration of the droplet is force divided by mass:

$$a = F_d/(V_d\,\rho_d)$$

Distance travelled by a droplet is:

$$y = \tfrac{1}{2}\,a\,t^2$$

At the free-fall velocity, the droplet Reynolds number is 12,900. For droplet Reynolds numbers exceeding 1,000, C_d has a value of 0.445. The volume of a spherical droplet is:

$$V_d = \pi\,d^3/6$$

Solving the above, the wind will move liquid droplets downwind around 0.2 m, crosswise to the slope of the deck. At the same time, the trajectory of the leak will move droplets up to 2 m downhill, depending on the position on the vertical flange from which the droplet was released.

B-3 Pool spread methodology

The exact direction of leakage (and splashing against objects) cannot be predicted, but the model should be plausible. The release of crude oil is assumed to be continuous and at a constant flow of 0.0500 m^3/s. The starting point of the pool is directly below the leak point, located 31.5 m from the downhill edge of the deck. The initial pool area is 2 m in radius, based on droplet dispersal radius from the assumed leak due to gasket failure. After the release lands on the deck, the effect of wind is insignificant compared to gravity. Evaporation is not significant. The fluid properties of the oil remain constant. Friction loss is calculated from the Darcy equation for laminar flow. The pool flow is in the laminar regime at the edges of the pool, and is a conservative bound on areas that are in the turbulent or transition flow regime.

The pool is modeled as a conical volume with the peak located directly under the leak point. In the case of gravity-induced open flow, the change in elevation equals the resulting velocity head plus head loss from friction. The velocity head is the horizontal velocity squared divided by twice the acceleration of gravity. The friction loss is a linear function of the area of contact and the Reynolds number.

$$z = v^2/2g + h_L$$
$$v = (2g\,(z - h_L))^{\frac{1}{2}}$$

To include the effect of the sloping deck, the conical pool is partitioned into four quadrants: uphill, downhill, and two sides. The height of the peak is the same for all quadrants, but in each quadrant the vertical distance between the peak and the edge can be different. The radius for the uphill quadrant is adjusted as needed so that the uphill edge is never higher than the peak. As pool spreads faster in the downhill direction, the pool model changes from a right cone with a circular base to an angled cone with an elliptical base. The contact area for a quadrant of a cone is $\frac{1}{4}\,\pi\,r^2$. In this example, the quadrants are calculated as sections of cones of different radii, which do not match at the edges. The result is a pool area that looks like a wind rose. This mismatch results in an error in pool area calculated as an ellipse. It would have been possible to calculate the area more correctly or increase the number of sections from four if it had seemed that this error was important.

Because the friction factor is a linear function of Reynolds number in laminar flow, the average friction factor occurs at the radius associated with half of the area, r divided by the square root of

2, or approximately $\underline{r} = r/1.414$. The Darcy equation for friction loss, with the hydraulic diameter, $8z$, substituted for pipe diameter, is

$$h_L = f\,L\,v^2/(16\underline{z}\,g)$$

where $\underline{z}$ and v are evaluated at $r/1.414$, and the quadrant radius, r, is substituted for L. For laminar flow, the Darcy equation for friction factor is

$$f = 64/Re$$

where

$$Re = 8\,\underline{z}\,v\,\rho/\mu$$

giving the following for laminar flow friction head

$$h_L = \mu\,r\,v/(2\underline{z}^2\,\rho\,g)$$

For turbulent flow ($Re > 4{,}000$), the Churchill equation for the Fanning friction factor can be used [Ref. 6]

$$1/f^{0.5} = -4\,\log\{0.27\epsilon/D + (7/Re)^{0.9}$$

where ϵ is the surface roughness. The roughness for galvanized steel is 0.152 mm. The variable D is replaced with the equivalent value of $8z$. The Fanning f must be multiplied by 4 for use in the Darcy equation.

After the average friction loss is calculated at $\underline{r}$, the velocity at that point can be calculated. Then the outward spread at that point can be calculated by

$$.\ \Delta\underline{r} = v\,\Delta t$$

The outward spread at the radial edge of the pool, Δr, is calculated by assuming that the ratio between $\underline{r}$ and r does not change:

$$\Delta r = 1.414\,\Delta\underline{r}$$

The equation for Δr and even the proposal that pool spreading can be modeled with a conical shape has not been validated by computational fluid dynamics or actual experiments.

The deck material is assumed to be flat after installation, and has typical checker plate ridges. The ridges rise from the base at an angle of 45° to a height of 1.5 mm. The ridges occupy about 25% of the surface area. Each ridge has about 1.4 times the surface area of the flat base, which increases the overall area of the plate by 8%. Therefore, the ridges increase the hydraulic radius by a factor of 8% above that of flat plate. The roughness value, ϵ, is estimated to be the same as for galvanized steel pipe.

B-4 Results

Table B-1

Oil Spill Spread on a Sloping Ship Deck from a Constant Continuous Leak

The radii for the side, downhill, and uphill quadrants are labelled ***a***, ***b***, and **c**, respectively.
The height of the conical peak, **z**, is relative to the deck directly below.
The velocities for the quadrants are calculated at the centroid.

Time s	V m^3	a m	b m	c m	z m	v(a) m/s	v(b) m/s	v(c) m/s
0	0.005	2.00	2.00	2.00	0.001	0.004	0.790	0.088
1	0.055	2.28	3.65	0.33	0.012	0.271	1.136	0.175
2	0.105	2.67	5.80	0.44	0.012	0.269	1.409	0.123
3	0.155	3.03	8.40	0.44	0.011	0.238	1.675	0.000
4	0.205	3.35	11.46	0.44	0.010	0.204	1.944	0.000
5	0.255	3.62	14.98	0.44	0.009	0.175	2.207	0.000
6	0.305	3.85	18.96	0.44	0.008	0.150	2.486	0.000
7	0.355	4.05	23.40	0.44	0.007	0.130	2.759	0.000
8	0.405	4.26	28.30	0.44	0.006	0.141	3.034	0.000
8.6	0.435	4.43	31.46	0.44	0.006	0.152	3.199	0.000

As an overall characterization of a spill on a sloped surface, the spill can be viewed as having a roughly triangular footprint. In this case, with a surface slope of 2.5%, the spill could be viewed as a triangle 31.5 m long and 8.9 m wide at the base. This corresponds to a divergence angle of about 8° from the centerline of flow.

Uncertainties in the results include the question of whether the conical model is representative of the action of an actual liquid spill at an industrial site. There is no published data that could be found containing a friction factor for checker plate decking. The checker plate ridges may have an effect on flow pattern, such as deflecting the flow. In this case, the flow pattern that has the largest area (greatest width before stopping at the edge of the deck) gives the worst consequences. There was no study done for this report to investigate the effect of various conditions affecting the width of the flowing pool. However, results are based on recognized flow equations and should be viewed as useful in the absence of experimental or computational fluid dynamics results.

440778 Apparent Discrepancies in Methods for Predicting the Explosion Energies of Bleves

Abstract:

The authors have used calculation methods for BLEVE blasts as described in the CCPS book *Guidelines for Vapor Cloud Explosion, Pressure Vessel Burst, BLEVE and Flash Fire Hazards*, and most recently similar methods in the PHAST process industry hazard analysis/consequence analysis software tool, developed by DNV-GL. Predictions made using PHAST appeared to be overly conservative, which instigated a comprehensive comparison of PHAST to other methods in the literature. There is not universal agreement concerning the dominant mechanisms driving the air blast pressure wave from a BLEVE, and there are major differences in methods used to predict the explosion energy driving the blast wave. The CCPS guidelines book and PHAST consider the explosion energy driving the blast wave to be from both the vapor and liquid phases present at the time of loss of containment. For each phase the contribution to explosion energy is the difference in internal energy between the thermodynamic states of the liquid and vapor at the time of loss of containment and the final states of the materials when isentropically expanded to atmospheric pressure. Recently, Birk compared various prediction methods to experimental data for propane BLEVEs and concluded that the flash vaporization of the liquid may be too slow to contribute significantly to the blast, therefore the blast energy is primarily from the expansion of the vapor in the vessel at the time of loss of containment. Other authors also base the explosion energy on the gas expansion and others on the liquid superheat. Ogle and coworkers have discussed the available energy of a BLEVE (exergy) and compared this to the BLEVE explosion energy predicted using several methods in the literature. These various approaches can lead to drastically different predictions of explosion energy and the resultant blast pressures and impulses. BLEVE theories and models are discussed in this paper and predictions are compared to limited experimental data. This paper is meant to provide comparisons between the many various approaches, including many more that are not listed above, and summarize their differences.

441089 Automated Optimization of Gas Detector Placement

Abstract:

Chemical process facilities are moving toward the use of performance-based design techniques for fire & gas detector placement. These approaches are displacing methods for layout of gas detection based on uniform "grids" that date back more than 25 years. First generation performance-based design techniques utilized 'gas volume detection' principals that are valid in terms of detecting and mitigating a large explosion event, but can result can result in an excessively large number of detectors and a costly design. This paper presents a superior alternative to the first generation technique of volume detection that uses sophisticated modeling of gas release scenarios with computational fluid dynamics (CFD). The techniques described in this paper align with ISA TR 84.00.07 guidelines for scenario-by-scenario basis. Furthermore, the paper breaks the paradigm that CFD can only be used sparingly on complicated gas dispersion problems. With the proper definition of the process, optimal selection of boundry criteria, and automated scripting of multiple scenarios and data collection from those scenarios, the thousands of CFD runs required to develop results can be efficiently and cost effectively performed. This paper will demonstrate practical application of CFD modeling and probabilistic risk analysis in order to improve on the performance of gas detection systems while at the same time, reducing unnecessary conservatism in the number of detectors. The paper will present methodology for determining scenario-based gas detector coverage and include a worked application example for an offshore gas processing platform.

441414 Quantitative Risk Assessment in the Process Industries: Beyond Numerical Estimates of Individual and Societal Risk

Abstract:

The Chemical Process and Oil and Gas industries (both on and offshore) have largely adopted Quantitative Risk Assessment (QRA) with open arms following its application in the nuclear industry. Currently, the methodology not only forms a core aspect of the risk management programs of numerous organizations who design, build, own, manage or operate high hazard facilities (refineries, LNG terminals etc.). It also occupies a central role in the legislative and regulatory framework in many jurisdictions across the world. It has become a tool of choice for both the regulators and the regulated parties.

As its application has become more widespread, it has been the subject of extensive critical inquiry, much of which has revolved around the primary product of the process, namely numerical estimates of risks to exposed people. Advocates highlight the utility and rationality such numbers offer in the often challenging decision making process associated with risk management. Critics make reference to the wide variability and uncertainty associated with such predictions as well as the implicit value judgments they embody. Less emphasis has been placed on the wealth of other insights (often latent) afforded by a well scoped out, executed and documented QRA.

The aim of the paper is to examine and explore the value of QRA over and beyond the well-known orthodox output of individual or societal risk estimates. It identifies and outlines what these added benefits are and elaborates on their significance - particularly in the context of achieving risk management objectives. It argues that such latent benefits should be considered to be equally or more important to the typical output and that more attention should be paid to these whilst conducting a QRA. A popular adage goes thus "sometimes, it's the journey that teaches you a lot about the destination"; so it is with QRA.

Quantifying the Importance of Accurate Physical Property Data in Dispersion Modeling

Devin Averett
Baker Engineering and Risk Consultants, Inc.
11011 Richmond Avenue, Suite 700
Houston, TX 77042
daverett@BakerRisk.com

Jef Rowley
Baker Engineering and Risk Consultants, Inc.
11011 Richmond Avenue, Suite 700
Houston, TX 77042
jrowley@BakerRisk.com

Prepared for Presentation at
American Institute of Chemical Engineers
2016 Spring Meeting
12th Global Congress on Process Safety
Houston, Texas
April 11-23, 2016

Quantifying the Importance of Accurate Physical Property Data in Dispersion Modeling

Devin Averett
Baker Engineering and Risk Consultants, Inc.
11011 Richmond Avenue, Suite 700
Houston, TX 77042
daverett@BakerRisk.com

Jef Rowley
Baker Engineering and Risk Consultants, Inc.

Keywords: Dispersion Modeling, Physical Properties, DIPPR

Abstract

Modeling discharge and dispersion of materials from processing vessels is an important part of a facility siting study. In this paper, we investigate the dependence of dispersion modeling on physical property data accuracy, including the impact of uncertainty in the lower flammability limit, heat of vaporization, liquid density, and vapor pressure data. Vapor pressure is further analyzed by comparing correlations obtained from the Design Institute for Physical Properties (DIPPR) to the Clausius-Clapeyron equation using the normal boiling point as the reference.

Overall, the predicted dispersion distances are relatively resilient to uncertainty in physical properties. However, errors in dispersion distances can be large under certain conditions. It is important to have accurate data for these cases. To ensure data accuracy, we recommend using a modern, critically reviewed database such as the latest DIPPR database, which contains thermophysical data for a large number of compounds and routinely employs rigorous analysis and consistency checks to assure the reliability of the data.

1 Introduction

Dispersion modeling can be a main driver for many major economic and safety decisions. Modeling of a toxic release may help dictate where and what type of personal protective equipment (PPE) is needed, where to place high-occupancy buildings and work areas, and whether risk to personnel or the surrounding community warrants investment of capital into mitigating the effects or decreasing the likelihood of such a release. Modeling of a flammable vapor cloud may provide similar information regarding safety from the risks posed by flash fires and explosions, but it can also aid in estimating the economic impact from explosion events by determining which buildings and equipment are likely to be damaged. This impact is not just in the cost of replacing equipment, but also in the loss of production during the repair downtime. This information may in turn determine insurance rates and proper amounts of coverage.

With such important decisions hinging on the results of a model, it is vital that the model is reliable and provides as accurately as possible a representation of what will happen during an event, especially when there is already so much inherent uncertainty due in large part to the natural variability in the weather. So selecting a good model that will make the best of a difficult situation is vital. When modeling a dispersion scenario, there are many models available. Typically the most accurate models are based on computational fluid dynamics (CFD), as most other models do not explicitly account for the presence of obstacles; however, use of CFD is computationally expensive and impractical if it is necessary to model numerous release scenarios subject to various weather conditions. Other models have been developed that sacrifice some accuracy in the presence of obstacles to gain computational speed.

Whatever the model, similar inputs are typically required, many of which are scenario-dependent (e.g., atmospheric conditions such as wind direction and speed, weather stability class, etc.). The inputs that remain constant across all different scenarios are the physical properties of the material being released. Unlike the weather, a material's chemical properties do not exhibit large and unpredictable fluctuations. So if there is any input to the model in which one can place a reasonable level of confidence, then surely it is the physical property data.

Of course there is disagreement in the literature pertaining to physical properties. Aside from the more obvious sources of error (human error, bad equipment calibration, etc.), different experimental techniques may yield differing results, and there is even variability in experiments employing the same or similar techniques. The trouble, then, is determining which literature value is likely to be the most correct, which may not be an easy task.

In this paper, we explore whether the effort of finding accurate material properties for consequence modeling is worth the effort and cost. If the physical properties are not exact, does it make a significant difference in dispersion model results? We take a quantitative approach to answer this question by utilizing the dispersion model currently used in SafeSite$_{3G}$®, BakerRisk's in-house software for consequence modeling. By varying a number of different properties, we can determine the impact they have on the final dispersion results. Conclusions can then be drawn on the importance of obtaining the best physical property data in dispersion modeling.

2 Methodology

Dispersion model results can't be summarized in a single number. There are many points of comparison that could have been used in our analysis. In this study, we use for our basis of comparison the horizontal distance at which the centerline of our release reaches a concentration equal to the lower flammability limit of the material.

The dispersion model used is detailed by Rowley [1]. In short, it solves the conservation of mass, momentum, and energy in one dimension, and then concentration profiles in the vertical and cross-wind directions are obtained based on similarity concepts. The model itself actually draws from several other ancillary models for handling phenomena such as droplet trajectory, pool formation, and droplet and pool evaporation. Physical properties for materials used in the model were obtained from the most recent version of the DIPPR® 801 database [2].

Several different physical properties were varied to monitor the effect any variance in the values would have on dispersion distances. These were chosen based on which properties intuitively would seem to have the greatest impact on dispersion results; namely lower flammability limit (LFL), heat of vaporization (HVP), liquid density (LDN), and vapor pressure (VP).

Several chemicals were chosen for the analysis, selected to give a wide range of normal boiling points (NBP) to ensure materials of different volatilities were adequately explored. The chemicals, in order of increasing normal boiling points, are methane, ethylene, propylene, ethylene oxide, methanol, *n*-octane, phenol, 1-decanol, glycerol, and *n*-eicosane. The intent was to select fairly common chemicals that differed in normal boiling point by approximately 50 K.

Dispersions were run at two different weather conditions, both at atmospheric pressure of 1 atm and an ambient temperature of 70 °F. The first weather condition used a wind speed of 1.5 m/s and a Pasquill stability class of F, and the second used a wind speed of 3 m/s with a D stability class. In addition, different dispersion runs were performed for different storage temperatures and pressures. Pressures of 10, 100 and 1000 psig were run at temperatures ranging from -275 to 675 °F in 50 °F intervals. Releases where the material exists as a solid at those storage conditions were ignored.

To establish a baseline set of results, dispersions were run for the different materials with physical properties set equal to the accepted values from DIPPR. Additional dispersion runs were then gathered where one physical property was altered slightly. Each property was altered by 5% in either direction. (Note that if altering the property resulted in a phase change at the storage conditions, these cases were omitted from the analysis, the presumption being that even if a physical property may not be accurately known, it is known in which phase the material is being stored.) We were then able to obtain an estimate as to how much the dispersion distances change in relation to a standard step change in a specific physical property. The vapor pressure analysis was further extended by looking at how dispersion results differed when using the Clausius-Clapeyron equation using the normal boiling point from DIPPR as the reference point. This was done to see how important it is to obtain accurate data across a broad range of temperatures rather than simply at a single point and extrapolating from there numerically with a basis on theory.

3 Results and Discussion

Taking one property at a time, results from the dispersion runs are summarized below, accompanied by some discussion on implications that can be drawn from the data. All results are plotted as:

$$\frac{\%\Delta(Distance\ to\ LFL)}{\%\Delta(Physical\ Property\ of\ Interest)}\quad vs.\quad \frac{Process\ Temperature\ (K)}{NBP(K)}$$

where $\%\Delta$ refers to a percent change in the pertinent value.

3.1 Lower Flammability Limit (LFL)

The flammability limit has the most straightforward effect on dispersion distances. This is because it has no effect on the dispersion model whatsoever. It merely changes the target concentration we are looking for when calculating the extent of the plume to LFL. Figure 1 shows the results for all runs.

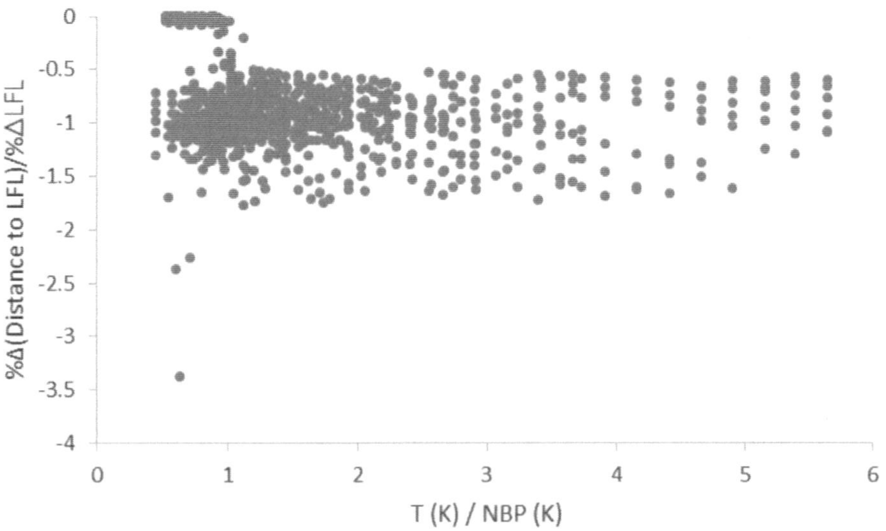

Figure 1. Ratio of change in dispersion distance and change in LFL vs. ratio of process temperature and NBP.

The uppermost cluster showing virtually no change in dispersion distance represents those scenarios that form highly non-volatile pools. Otherwise, all scenarios show a decrease in dispersion distance with an increase in LFL, which intuitively makes sense. The largest deviations occur for materials near saturation (T/NBP = 1). On average, dispersion distances changed only slightly, with an average change of -0.81 % in dispersion distance for a 1 % increase in LFL and a standard deviation of 0.45 %. While this may not appear to be very significant, the trouble with LFL is the inaccuracy in measuring LFL. Literature values may vary widely, and these large discrepancies will result in large differences in dispersion distance. For example, within the DIPPR database, looking at only methane, the accepted value of 5 % vol [3] differs by as much as 12 % from other experimental values (a value of 4.4 % vol [4]). A previous study has shown that large discrepancies can occur simply due to the method of measurement. Data obtained using the US Bureau of Mines tube method can give dispersion distances to LFL that differ by as much as 200 % from data obtained using the EN 1839 tube method [5]. Indeed, the same study discusses the reality that rather than a discrete value, the LFL exists as a range of concentrations that may yield varying degrees of flame propagation, and different methods of measurement may give measurements biased towards one end of this range or the other. Fortunately, in this study we found that small differences in LFL generally didn't have a large impact on dispersion distances, but certainly caution should be taken in cases where the LFL is low or wide ranges of LFL data are available in the literature.

3.2 Heat of Vaporization (HVP)

For heat of vaporization, we examine a much narrower range of conditions. This is because for HVP to play any significant role at all, an appreciable amount of both liquid and vapor phases must be present. Thus high-temperature vapor releases and low-temperature liquid releases would give virtually the same results for large changes in HVP because a negligible amount, if any, of vaporization is occurring. Figure 2 shows the results from the analysis.

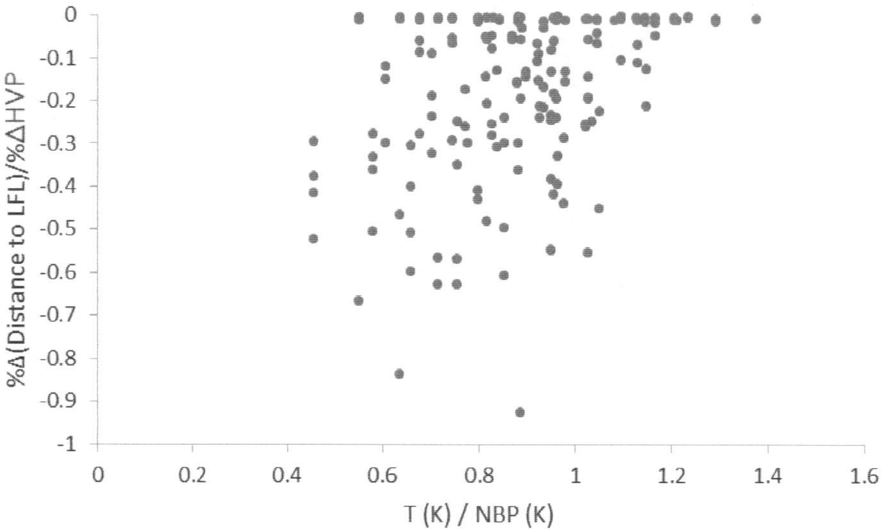

Figure 2. Ratio of change in dispersion distance and change in HVP vs. ratio of process temperature and NBP.

Similar to LFL, we see that small changes in HVP result in approximately the same order of magnitude of change in dispersion distances. For every one percent increase in HVP, the distance to ½ LFL will on average change by -0.17 %, with a standard deviation of 0.19 % and a minimum of -0.93 % occurring in the cases we analyzed. And this, remember, is only for cases where any appreciable phase change is taking place. If at least somewhat reliable HVP data can be obtained, then we can be confident that it will not affect dispersion distances to any significant degree, and especially not if the scenario does not include a phase change.

3.3 Liquid Density (LDN)

The effect of liquid density on dispersion models is much more complex than the effect of LFL or HVP. First of all, liquid density will affect the discharge rate. It would seem logical that a higher density would lead to a material that would discharge at a greater rate from a specific orifice, suggesting greater dispersion distances. However, this would affect other characteristics of the flow as well such as the choke pressure. The change in choke pressure can cause additional issues, particularly in cases where the liquid flashes upon release. Any change in pressure will then have a large effect on the vapor fraction of the discharge, and hence will cause a large swing in the overall density of the initial discharge.

The second issue to consider is of plume buoyancy. For plumes of high liquid fractions, an increase in liquid density will cause a decrease in plume buoyancy, resulting in the liquid raining out at a nearer distance. This would likely result in shorter dispersion distances, which could either supplement or compete with any effects from the change in discharge.

Although there are other issues affected by LDN, the last issue we will mention here is that momentum must be conserved. Given two identical droplet-filled plumes (both plumes start with the same velocity and have the same number of droplets with equal diameter) with the only difference being the density of the material in the droplets, and assuming an equal air entrainment rate between both plumes, conservation of momentum dictates that the denser plume will decelerate at a slower rate than the less dense plume. So with these simplistic plumes, the liquid in the denser plume will rain out at a farther distance, exactly the opposite conclusion that was reached when considering plume buoyancy. We see that with LDN, trying to reason out what effect changing LDN might have on a single scenario without running the model can become very complicated.

We turn to the model, then, for some answers. Figure 3 shows results for LDN dispersion runs. Again, some data was omitted because it was not relevant to the analysis, namely the cases where no liquid was present throughout the scenario. Note that unlike previous properties, both negative and positive deviations in dispersion distances are seen, and the differences become more pronounced the more the temperature is decreased. The point that gives the largest deviation (a surprising 13.1 % increase in dispersion distance per one percent increase in density) is phenol at P = 1000 psig and T = 275 °F for the F-classified weather. Under these storage conditions, phenol exists as a liquid. Looking at this specific case in more detail, we see that the droplets in the denser plume rain out more quickly (less than a second compared to about 11 seconds), but the velocity in the less dense plume is knocked out very quickly, causing rainout to occur at about 28 ft compared to the denser plume at about 60 ft.

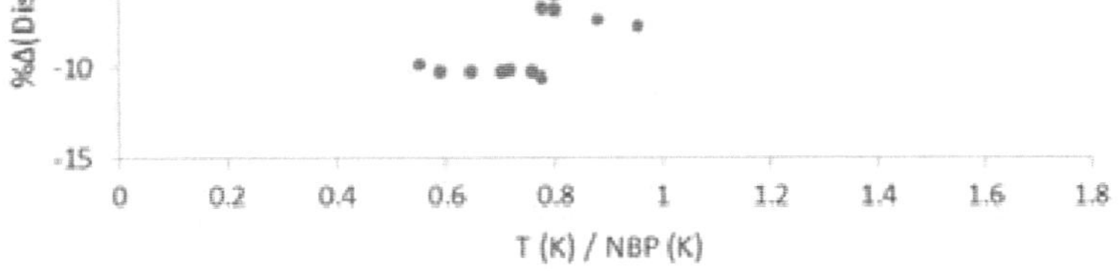

Figure 3. Ratio of change in dispersion distance and change in LDN vs. ratio of process temperature and NBP.

On average, a one percent increase in LDN results in only a 0.08 % increase in dispersion distances, but that has a standard deviation of 2.21 %. Thus, having inaccurate LDN data can cause large differences in dispersion results, and those differences can be difficult to predict. Fortunately LDN is a relatively simple property to measure, but if the scenario deals with a wide range of temperatures it is imperative to have accurate temperature-dependent data for LDN.

3.4 Vapor Pressure (VP)

3.4.1 Deviation in Vapor Pressure

Like HVP, vapor pressure can really only affect releases where two phases are possible. Hence, pure vapor releases are ignored, as well as releases low enough in temperature to form pools with no significant amount of evaporation. Also ignored are cases where the change in VP would result in a phase change at process conditions, the assumption being that the initial phase is known.

Like LDN, VP may have competing effects on dispersion modeling. An increase in VP would cause a flashing liquid upon discharge to have an increased vapor fraction and thus a decreased mass flow, likely suggesting shorter dispersion distances. However, it also means droplets and pools evaporate at higher rates, serving to lengthen dispersion distances.

Figure 4 shows the results from the VP runs. As would be expected, most of the variability occurs when the temperature of the fluid is near the normal boiling point where a two-phase release is much more probable. The most extreme result (an 11 % increase in dispersion distance per 1 % increase in VP) come from *n*-eicosane at P = 100 psig and T = 625 °F for the D-classified weather. In this case, the difference is largely buoyancy driven. For higher VP, the resulting excess vapor is sufficient to increase the buoyancy of the plume and significantly delay rainout, keeping the concentration and mass of material in the plume at higher levels for a longer time.

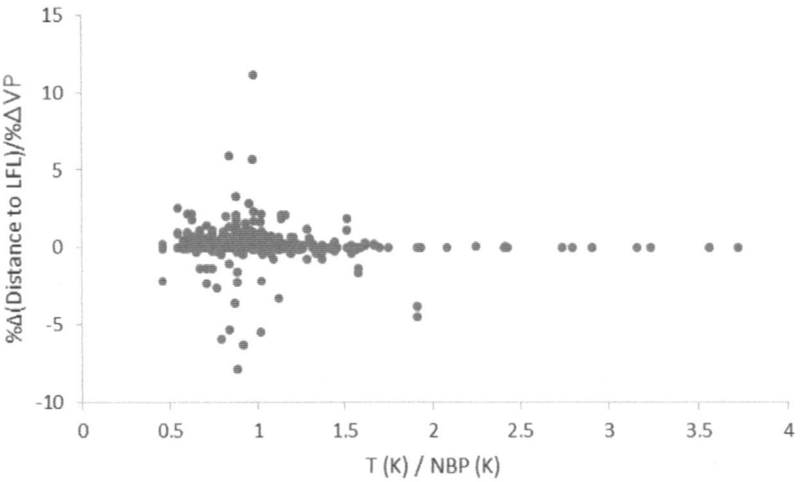

Figure 4. Ratio of change in dispersion distance and change in VP vs. ratio of process temperature and NBP.

The average result is only a 0.18 % increase in dispersion distance per 1 % increase in VP, but with a standard deviation of 1.19 %. While it does not tend to affect dispersion distances to the same degree as LDN, VP does have the potential to significantly alter dispersion results.

3.4.2 Clausius-Clapeyron Equation

It was also of interest to see if one could still get reliable results using the Clausius-Clapeyron equation, or at least the well-known approximation of it, shown here:

$$\ln\left(\frac{P_1}{P_2}\right) = \Delta H_{vap}\left(\frac{1}{T_2} - \frac{1}{T_1}\right) \tag{1}$$

where P_1, T_1 and ΔH_{vap} are the vapor pressure, temperature, and heat of vaporization, respectively, at a reference state, and P_2 and T_2 are the vapor pressure and temperature, respectively, at some new state. Hence, at the reference state, the equation is exact, but errors will develop and worsen as the desired temperature moves farther from the reference temperature. Often, heat of vaporization data is available at the normal boiling point of a material. To establish a basis for this study, then, we select the normal boiling point of the material to be our reference point, using the HVP value obtained from the DIPPR database at this temperature. The DIPPR database base correlations for HVP and VP are based on rigorously evaluated experimental data, so we can be confident that the DIPPR correlation will always result in a more accurate answer than the Clausius-Clapeyron equation, regardless of the temperature. Hence, in the following discussion we assume any deviation from results derived from DIPPR data are unfavorable. Again, pure vapor releases, releases low enough in temperature to form pools with no significant amount of evaporation, and cases where the change in VP would result in a phase change at process conditions are ignored.

Earlier results relating to vapor pressure would suggest that using the Clausius-Clapeyron equation would likely have little effect on dispersion distances since significant deviations seen in Figure 4 are concentrated around the NBP, the area in which the Clausius-Clapeyron (as we have defined it) should be the most accurate. Following the same procedure as the other analyses (testing several materials at several temperatures and pressures), there surprisingly were some significant deviations in dispersion distances. Figure 6 and Figure 6 show the results. Note that because we are not dealing with a strict percentage deviation in vapor pressure, the y-axes in these figures are not presented as ratios, but rather Figure 5 shows straight percent differences in dispersion distance and Figure 6 shows absolute differences in dispersion distances. This makes it difficult to draw a comparison with the results from earlier in this study. Nevertheless, it is clear that particularly at temperatures just below the NBP, significant deviations can occur from the use of Clausius-Clapeyron. Over all runs, there was an average change of 10.0 %, with a standard deviation of 40.4 %.

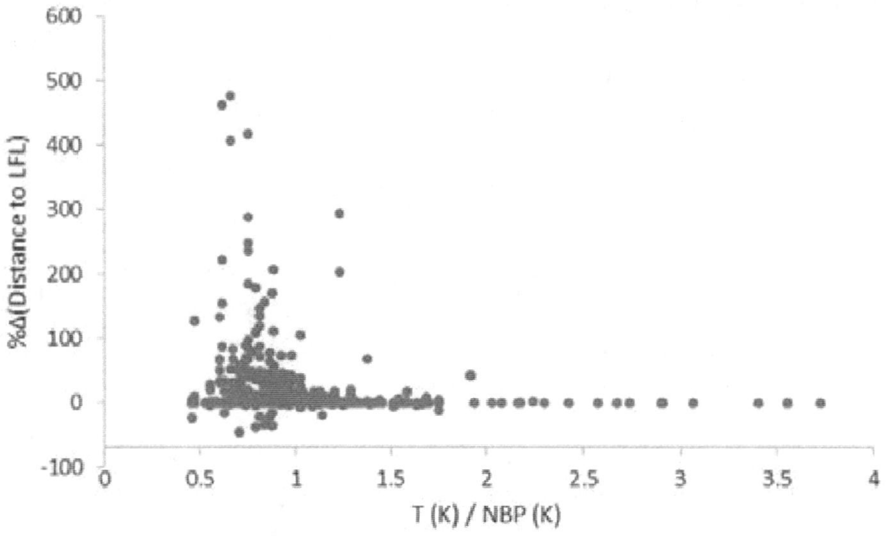

Figure 5. Percent change of dispersion distance vs. ratio of process temperature and NBP.

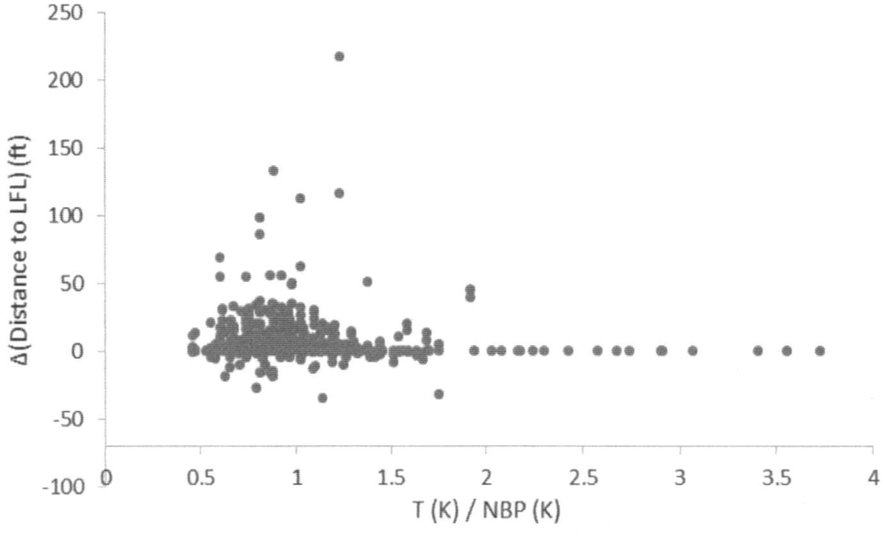

Figure 6. Absolute change in feet of dispersion distance vs. ratio of process temperature and NBP.

Still, the magnitude of these deviations are at times so extreme that it warrants further inspection to determine the reasons behind it. Figure 7 helps make results from Figure 5 and Figure 6 more relatable to results from Figure 4 by showing exactly how much VP as calculated from Clausius Clapeyron differs from VP calculated from the DIPPR correlation, once again plotted against T / NBP. Figure 4 shows deviations per percent change in VP, and Figure 7 shows that using Clausius Clapeyron will frequently result in a deviation much higher. The general trend seen in Figure 7 is that higher-boiling components reach greater deviations at lower T/NBP ratios. (Note that if plotted against T - NBP, the data would look much more uniform across all compounds.) It is the high-boiling compounds that make up the majority of the higher deviations in Figure 5. So not only do we find that Clausius Clapeyron can give very different results from correlations based on temperature-dependent experimental data, but we also see that the magnitude of those differences have a positive correlation with the NBP of the material. For heavier compounds, then, it becomes even more important to use accurate VP data in dispersion modeling.

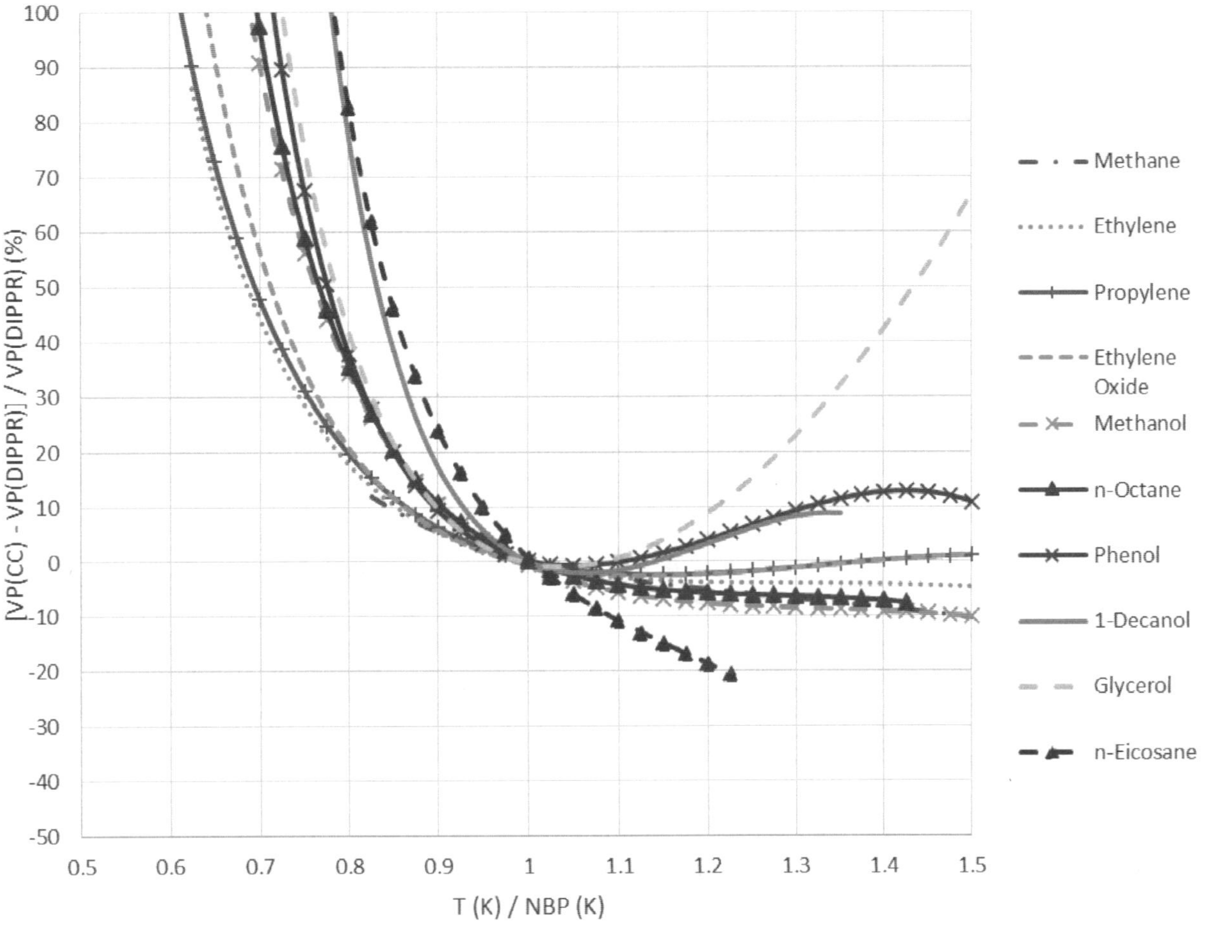

Figure 7. Percent difference between VP as obtained from the DIPPR correlation and the VP as given by Equation 1 plotted against T / NBP.

4 Conclusions

Overall, dispersion distances turned out to be quite resilient to deviations in physical properties. Minor changes in HVP or LFL resulted in only slight deviations in dispersion distances. In many cases, minor deviations in LDN and VP resulted in only minor deviations of dispersion distances. However, under the right conditions (far into the liquid regime for LDN or near the NBP for VP) large deviations in predicted dispersion distances can occur with only minor changes in physical property data. Consequently, great care should be taken as to where one obtains the physical property data used in dispersion modeling.

Furthermore, the Clausius-Clapeyron equation may not work well across the wide range of conditions that may be encountered during even a single dispersion. The processing temperature may be different from the temperature upon discharge, the temperature of the droplets in the plume, or the temperature of any pool formed, and we see in this study that it is vital to have reliable temperature-dependent vapor pressure data rather than rely on an extrapolation from a single point.

Obtaining accurate physical properties may be difficult and time-consuming, and we can often be tempted to settle for any values we can find, regardless of the source. Particularly in the case of LDN and VP, this study has shown that all physical property data should be met with some level of scrutiny when attempting to model dispersions.

We recommend using a critically reviewed database such as the current version of the DIPPR database [2]. The DIPPR database contains thermophysical data for a large number of compounds, and it routinely employs rigorous analysis and consistency checks to assure the reliability of the data. The database is continually updated and reviewed by experts in thermodynamics to ensure the recommended data reflect the latest advancements in thermophysical property measurement and evaluation. For properties where literature values are not available, state-of-the-art prediction methods are utilized to offer a best-guess value for situations where an estimate is needed.

5 References

[1] J. R. Rowley, "A new integral dispersion model based on one-dimensional turbulence theory," in *IChemE's 24th Hazard Symposium*, Edinburgh, UK, 2014.

[2] R. R. Rowley, W. V. Wilding, J. L. Oscarson and N. F. Giles, DIPPR® Data Compilation of Pure Chemical Properties, New York, NY: Design Institute for Physical Properties, AIChE, 2015.

[3] M. G. Zabetakis, "Flammability Characteristics of Combustible Gases and Vapors," in *Bulletin No. 627*, U.S. Bureau of Mines, 1965.

[4] K. Nabert, G. Schön and T. Redeker, Sicherheitstechnische Kenngrößen brennbarer Gase und Dämple, Deutscher Eichverlag, 2004.

[5] J. R. Rowley and J. E. Bruce-Black, Proper application of flammability limit data in consequence studies, Southport, UK: IChemE's 23rd Hazard Symposium, 2012.

[6] G. H. Tryon, Fire Protection Handbook, 12th ed., Boston: National Fire Protection Associations, 1962.

441423 Sustainable Risk Reduction with Dynamic Execution of Procedures for Any Operating Condition

Abstract:

Managing the volume of new information created daily is a challenge. Add to that the complexity of new equipment, employee turnover, and third party contractors working on-site, documenting clearly and concisely is critical to the operation. To truly mitigate risk we need look to operating companies focused on leading indicators of risk management where improved work conditions and profitability of the operation are the by product.

- Reduce human error
- Improve user adoption
- Improve operator efficiency
- Optimize SME time
- Quick resolution incident investigation
- Improve time "on the job"
- Extend equipment life

These objectives, met through the implementation of procedure best-practices to significantly reduce risk and improve the reliability of the operation.

During this presentation we will share the results of our procedure best-practices benchmark report and review the impact of asset level operating procedures for our clients.

As operating companies evolve through the knowledge management maturity model, they not only standardize best-practices but leverage that knowledge to optimize performance with a collaboration hub and dynamically generated procedures to fit any situation.

We will demonstrate how asset level operating procedures, designed for routine operator and facility maintenance, directly relates to improvements in asset life and performance, site-safe incident reduction based on human factors, and improve equipment start-up predictability when returning from planned maintenance.

This is so much more than simply documenting procedures; this is about sustainable adoption to improve operating discipline. Innovatia will introduce a complete procedure management solution to facilitate knowledge transfer in real-time to mitigate risk and improve reliability of the operation.

441155 Process Hazard Analysis of River-Crossing Natural Gas Pipeline

Abstract:

In order to control the process hazard of the river-crossing natural gas pipeline, HAZOP analysis is performed to qualitatively analyze its potential risk. Based on the results of HAZOP analysis, high risk scenarios are identified. So as to calculate the risk caused by pipeline's catastrophic rupture, FLACS software which based on CFD data simulation is used to quantitatively calculate the flame propagation and explosion overpressure of the explosion of natural gas. With the results of HAZOP analysis and quantitative calculation, safety measures are put forward to ensure the long-term safe operation of river-crossing natural gas pipeline.

Safe Depressurization of Large Thickness Vessels Under Pool Fire

Subhash S. Kale and George Maliakkal [(1)], Bechtel Corporation

(1) To whom all correspondence should be addressed

Abstract:

Vessels fail at pressures lower than the Maximum Allowable Working Pressure (MAWP) during fire due to degradation of strength at high temperatures. Analytical solution of the vessel wall temperature profiles as a function of fire exposure time available in API 521 Annex A for thicknesses of 3.2 mm (0.125 in.), 12.7 mm (0.5 in.) and 25.4 mm (1 in.) was extended for vessel thicknesses up to 152.4 mm (6 in.) to include high pressure/large diameter vessels encountered in gas processing. For fire depressuring, plots of ratio of the maximum vessel pressures to avoid failure to MAWP were developed for various wall thicknesses and compared to a plot of the ratio of vessel internal pressure to MAWP, both as a function of time. If the vessel internal pressures throughout fire depressuring were less than the reduced vessel failure pressures, then vessel rupture is avoided. The methodology is illustrated for the case of a high pressure vessel of wall thickness 57.15 mm (2.25 in). in natural gas service.

Introduction:

Large vessels, for e.g. vertical compressor suction KO drums, have a substantial flammable hydrocarbon liquid inventory, which if leaked can form a source of pool fire. The vessels usually have a large diameter 3 - 6 m which requires vessels with thicknesses higher than 25.4 mm (1 in.). API 521 [1] Section 4.4.13.2.4.1 gives information on wall temperatures for plates 3.2 mm (1/8 in.), 12.7 mm (1/2 in.) and 25.4 mm (1 in.). Also, Appendix A of API 521 gives information on maximum pressure as % of Maximum Allowable Working Pressure (MAWP) for plates of thickness up to 25.4 mm (1 in.). No data is available in API 521 for plates of thicknesses greater than 1 in. This work uses the analytical method given in Appendix A of API 521 to extend the calculations for vessels of thickness higher than 25.4 mm (1 in.) - for e.g. 50.8 mm (2 in.) to 152.4 mm (6 in.). thicknesses. The analytical equation given in Appendix A of API 521 is solved to develop wall temperature profiles for various wall thicknesses including those beyond 25.4 mm (1 in.). The temperature versus allowable stress data given in Table A-7 of the Appendix A can then be used to develop temporal profiles of maximum allowable pressure to avoid failure as a function of MAWP. The vessel internal pressure experienced during fire case depressurization is calculated using HYSYS v 7.3 depressurization utility. The internal pressures during depressurization, expressed as a ratio of the internal pressure to MAWP, is overlaid on the failure pressure temporal profiles to ascertain whether the vessel can avoid rupture during a pool fire scenario. The following describes the assumptions used in the calculations, the methodology to compute the temporal profiles of wall temperature and the maximum pressure to avoid failure.

Assumptions:

1. A carbon steel vertical vessel with hemispherical heads is assumed.
2. The surfaces are assumed to fully engulfed in flames hence both convection and radiation terms in equation A-1 of API 521 Appendix A apply.
3. The parameters given in Appendix A are slightly modified to better fit the temperature profiles given in Figure C-9 of Appendix C of API 521.
4. Since the vessels have thicknesses above 25.4 mm (1in.), the depressurization is from the design pressure, which is assumed to be same as MAWP, to half the design pressure in 15 minutes per API 521 guidelines. For these calculations, a design pressure of 80 barg is assumed.
5. The calculations are illustrated for a vessel of controlling thickness 57.2 mm (2.25 in.) for the hemispherical heads. The methodology can be extended to larger thicknesses from 25.4 mm (1 in.) to 152.4 mm (6 in.).

Wall Temperature Calculation by Analytical Method:

The analytical equation given in A-1 of Appendix A of API 521 is stated as:

$$q_{absorbed} = \sigma\ (\ \alpha_{surface} * \varepsilon_{fire} * T_{fire}^{\ 4} - \varepsilon_{surface} * \ T_{surface}^{\ 4}\ ) + h\ (T_{gas} - T_{surface}) \quad (1)$$

Also, by a heat balance,

$$q_{absorbed} = m\ c_p\ (\text{Delta } T_{surface} / \text{ Delta time}) \qquad (2)$$

The following gives a description and the values used for the various parameters in this equation per API 521 Table C-3:

$q_{absorbed}$, Heat absorbed by the wall

σ, Stefan Boltzmann Constant = 5.67 E-8 W/m^2.K^4 (1.713E-09 BTU/H-ft^2-°R$^{4)}$

$\alpha_{surface}$ is the equipment absorptivity = 0.8

ε_{fire} is the fire emissivity = 0.85

$\varepsilon_{surface}$ is equipment emissivity = 0.75

T_{fire} is the fire temperature = 904 °C (1652 °F)

$T_{surface}$ = equipment temperature initially assumed to be 59 °C (122 °F or 582 °R)

The surface temperature increases as the surface heats up due to radiation from the fire

h is the convective heat transfer coefficient 20 W/m^2-K (3.52 BTU/hr-ft^2-R)

Plate area is assumed to be 2.3 m^2 (24 ft^2) (per API Section 4.4.413, Figure 1)

Density of carbon steel is 7850 kg/m^3 (490.1 lb/ft^3)(from internet)

Plate thickness varies from 3.2 mm (1/8 in.), 12.7 mm (1/2 in.) and 25.4 mm (1 in.).

Heat capacity of carbon steel = 502 J/kg°-K (0.12 BTU/lb-°F)

Additionally, larger thicknesses 50.8 mm (2 in) and 76.2 mm (3 in) are also used.

From equation (1), $q_{absorbed}$ is calculated.

The plate volume is calculated for each thickness from the given area.

Using the given carbon steel density, the plate mass is calculated.

From equation (2) Delta $T_{surface}$ is calculated.

Starting with the initial surface temperature, the change in the surface temperature is calculated from equations (1) and (2) for different times in 0.5 second increments. The resulting wall temperatures are plotted and compared to values given in API 521 Appendix C (Figure C-9) as shown in Figure 1.

Figure 1 Calculated wall temperature profiles as a function of time

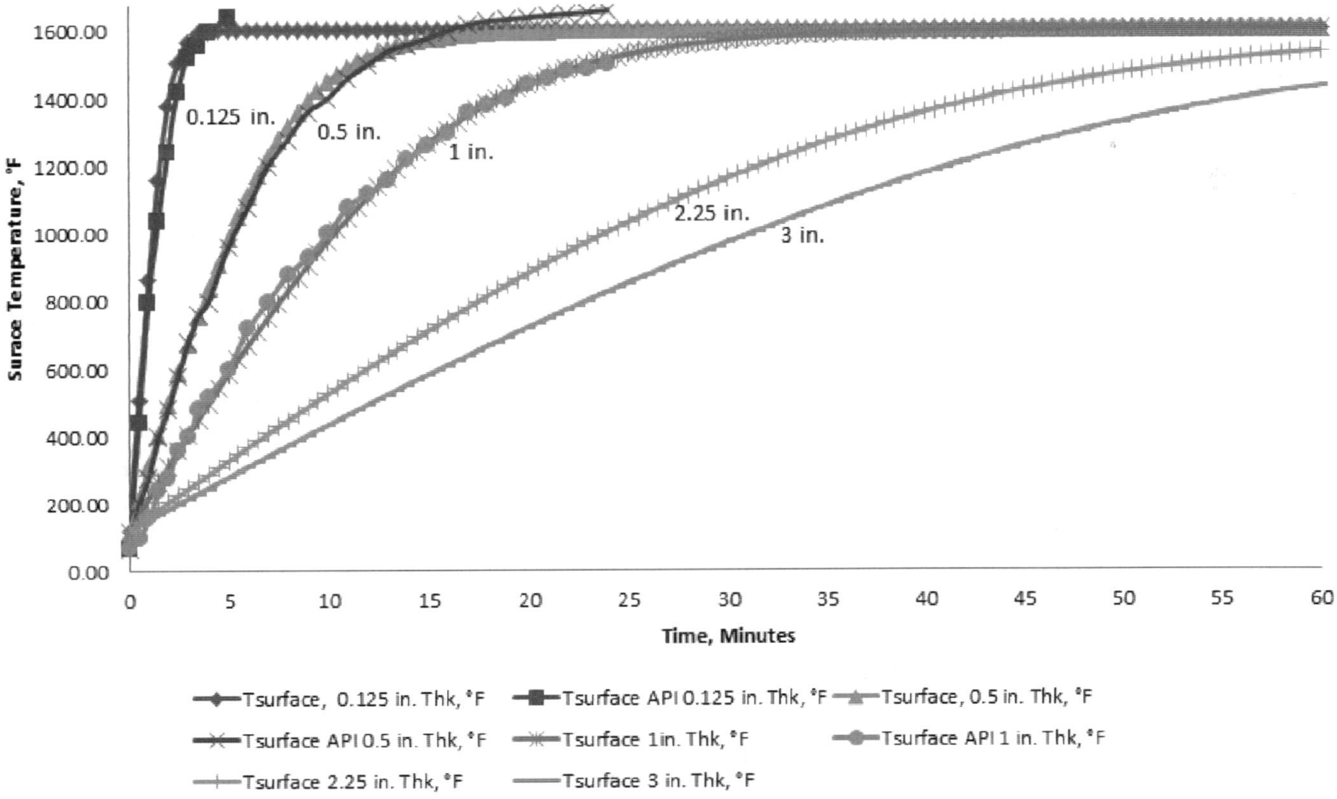

Calculation of the maximum pressure to avoid failure for various wall thicknesses:

As described in Appendix A Section A.3.5.4.6, the wall temperatures can be combined with tensile strength versus temperature data for carbon steel to determine the tensile strength, or the maximum pressure to avoid failure of a vessel, as a function of time in a pool fire scenario.

The tensile strength versus temperature data for carbon steel from Table A-7 in API 521 Appendix A is given in Table 1 for carbon steel.

Table 1 Tensile strength versus temperature for Carbon Steel (SA-515, SA-516)

Temperature (°C)	Temperature (°F)	Tensile Strength (MPa)	Tensile Strength (psi)
482	900	313.7	45,500
538	1000	251.7	36,500
593	1100	187.5	27,200
649	1200	137.9	20,000
704	1300	93.1	13,500
760	1400	62.2	9025

As the wall temperature heats up to 482°C (900°F), the tensile strength or the maximum pressure to avoid failure, decreases. These values are normalized with respect the tensile strength of 313.7 MPa (45,500 psi) and a safety factor of 0.75 is applied to generate Figure A-5 in Appendix A of API 521.

These values are extrapolated to larger thickness greater than 1 in using regression to determine the times to reach failure pressures expressed as a ratio of maximum pressure to avoid failure to MAWP for various thicknesses.

Table 2 Vessel rupture strength as % of MAWP versus tire exposure times

% of MAWP	100%	80%	60%	40%	34%	30%	26%	24%	
Thickness									
in	Time (mins)	Time (mins)	Time (mins)	Time (mins)	Time (mins)	Time (mins)	Time (mins)	Time (mins)	Comment
0.125	1.2	1.5	1.9	2.4	3.1	6.0	12.0	17.5	API 521
0.5	3.6	3.9	4.5	6.1	7.6	10.5	17.0	28.0	API 521
1	12.6	13.1	14.1	17.0	20.0	23.1	31.0	37.0	API 521
2	25.19	25.91	27.58	33.24	38.83	42.15	52.16	59.66	Extrapolated
2.25	28.52	29.28	31.14	37.48	43.74	47.11	57.68	65.18	Extrapolated
3	38.49	39.44	41.81	50.20	58.44	62.00	74.22	81.72	Extrapolated

Figure 2 shows a plot of the maximum allowable pressures to avoid failure for plates of different thicknesses from 25.4 mm (1 in). to 76.2 mm (3 in.)..

Figure 2 Failure pressures for various thicknesses as a function of time

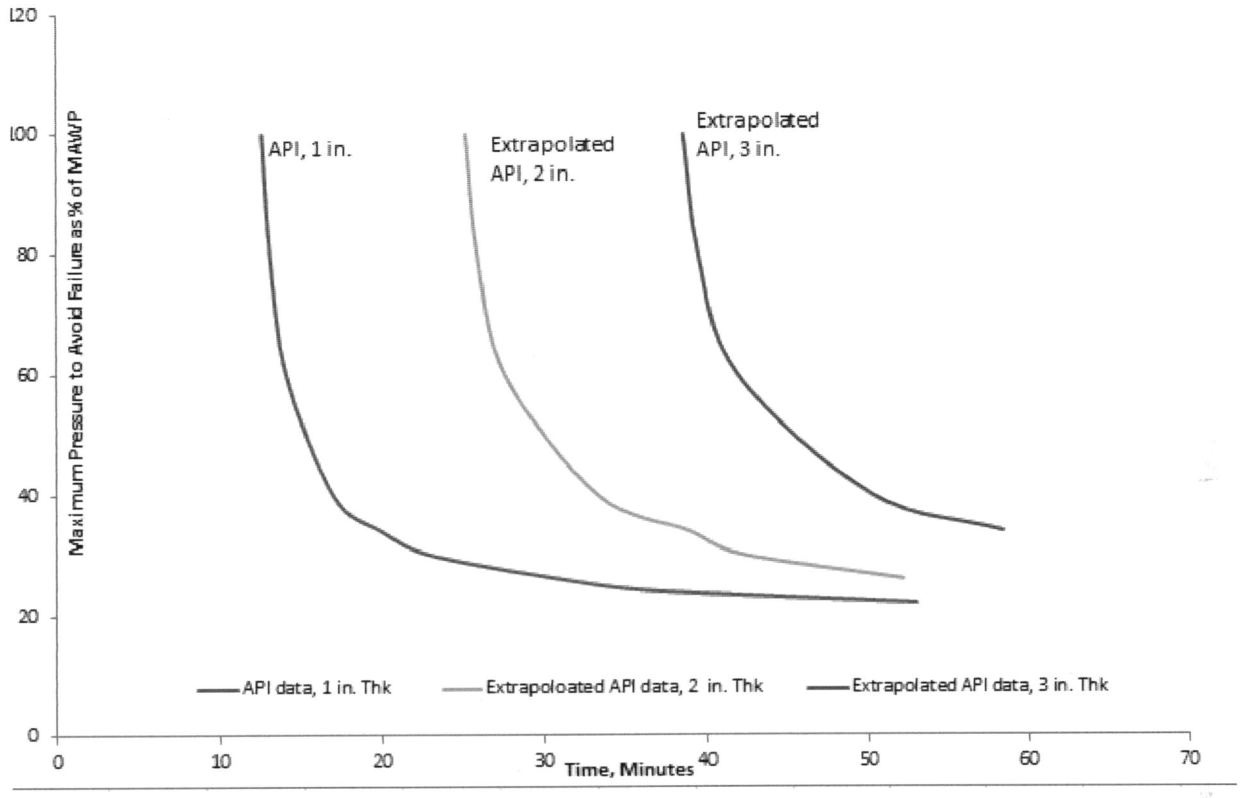

A hypothetical vessel of 4 m diameter and 7 m height with a vessel thickness of 101.6 mm (4 in.) having a hemispherical heads of thickness 57.2 mm (2.25 in.) is depressuerd from a design pressure of 80 barg to a pressure of 40 barg in 15 minutes using HYSYS v 7.3 depressuring utility. The resulting profile of the vessel internal pressure as a function of MAWP is overlaid on the temporal profiles of the maximum allowable pressures to avoid failure for various thicknesses. Data for about 60 minutes is plotted for comparison with the profiles of internal pressures as shown in Figure 3.

Figure 3 Plot of failure pressures for various thicknesses and the vessel internal pressures as a function of time

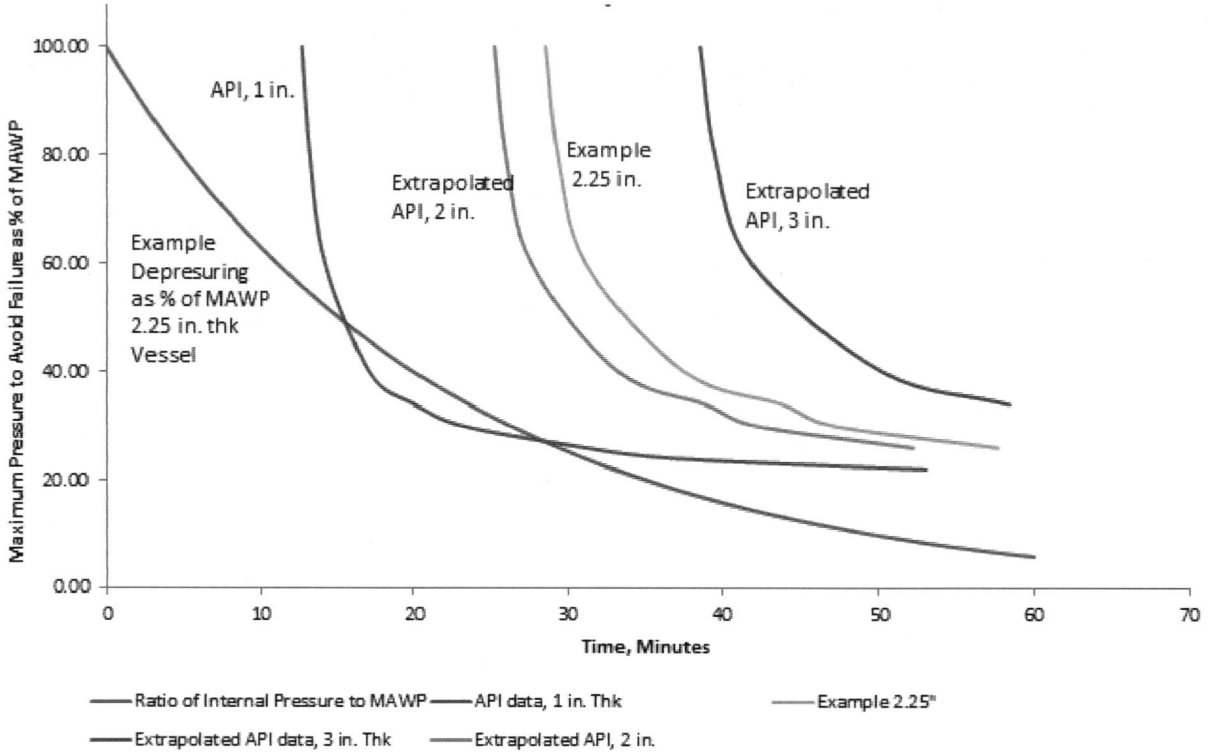

Discussion:

The kinks in the plot are due to round off errors and errors in the statistical regression. However, as seen from Figure 3, the vessel internal pressure for depressuring in a pool fire scenario is less than the maximum allowable pressure to avoid failure for thicknesses greater than 1". Thus, a methodology is developed for calculating wall temperature profiles during pool fire for vessels of various thicknesses and to check if the vessel internal pressures during depressurization exceed the failure pressures for vessel thicknesses greater than 1in.

References:

1. Pressure-relieving and Depressuring Systems, API Standard 521, Sixth Edition, January 2014.

Quantifying the Impacts of Human Factors on Functional Safety

Julia V. Bukowski, Ph.D.
Dept. of Electrical & Computer Engineering
Villanova University, Villanova, PA 19085
Julia.Bukowski@villanova.edu

Loren Stewart, CFSP
exida
64 N. Main St., Sellersville, Pa 18960
lstewart@exida.com

Prepared for Presentation at
American Institute of Chemical Engineers
2016 Spring Meeting
12th Global Congress on Process Safety
Houston, Texas
April 11-23, 2016

Quantifying the Impacts of Human Factors on Functional Safety

Julia V. Bukowski, Ph.D.
Dept. of Electrical & Computer Engineering
Villanova University, Villanova, PA 19085
Julia.Bukowski@villanova.edu

Loren Stewart, CFSP
exida

Keywords: systematic failure rate, product failure rate, site-generated failure rate, FMEDA, PFDavg

Abstract

It is not difficult to think of numerous ways in which human factors might impact functional safety. For example, imperfect repair, improper calibration, faulty installation, etc. can all contribute to decreases in functional safety performance. However, if actions are to be taken to mitigate these impacts then it is essential to have some quantitative measure of the impacts so that the actions' effectiveness can be assessed. This paper describes how the Site Safety Index^{TM} (SSI) is used to adjust safety metrics, computed under the assumptions that human factors play no part in safety system performance, to reflect the effects of human factors on safety system performance on a site by site basis. Examples of the application of the method showing the safety performance changes attributable to human factors are provided.

1. Introduction

A safety instrumented function (SIF) has an inherent product failure rate based on its design and implementation and the assumption that the end user will take all necessary actions (i.e., meet all IEC 61508 [1] and/or IEC 61511 [2] requirements) to insure that the inherent product failure rate will be achieved. In practice, end-user activities and practices often fall short of this ideal and, consequently, the safety performance achieved by the SIF is often less than the ideal [3]. The SSI measures, on a scale of 0 to 4, the extent to which end-user actions support the achievement and retention of ideal SIF safety performance. The SSI is described in detail in [4]. This paper explains how end-user actions as measured by SSI affect various parameters that are used to calculate SIF safety performance and show how different levels of SSI impact various safety metrics.

2. Notation

FFD	field failure data
FMEDA	failure modes, effects and diagnostic analysis

IEC	International Electrotechnical Commission
MTTFS	mean time to fail safe
OREDA	Offshore and Onshore Reliability Data
PFDavg	average probability of dangerous failure on demand
RRF	risk reduction factor
SIF	safety instrumented function
SIL	safety integrity level
SRS	Savannah River Site
SSI	Site Safety Index™
1oo1	1-out-of-1 architecture

3. Using SSI to Measure Impacts of End-User Activities and Practices on Safety Performance of SIFs

3.1 Why a Site Safety Index™?

Studies [5] have shown that the failure rates estimated from field failure data (FFD) for identical modules in the same industrial application exhibit a difference in estimated failure rates for the same product from site to site. Typically, when compared to inherent predicted failure rates for the same products, the FFD estimates are approximately 1.2 to 3 times higher depending on product type. Site audits have attributed these differences to dissimilarities in end-user activities and practices which vary from site to site. This variation in end-user activities and practices can be captured by the SSI [4]. Further, these variations in activities and practices directly impact many variables used in models which assess the safety performance of SIFs.

3.2 The Key Variables Impacting PFDavg and Other Metrics

A key metric for measuring the safety performance of SIFs is average probability of failure on demand (PFDavg). Other closely related metrics are risk reduction factor (RRF), safety integrity level (SIL), and mean time to fail safe (MTTFS). Based on studies of FFD and proof test reports, nine key variables impacting PFDavg have been identified [6]. These nine variables are

1. Failure rates of each device including failure modes and any diagnostic coverage from automatic diagnostics (an attribute of devices chosen and of end-user practices);
2. Proof Test Effectiveness (an attribute of the proof test method and, to the extent that there is more than one possible proof test method, an attribute of end-user practices);
3. Proof Test Intervals (assignable by end-user practices);
4. Mean Time to Restore (an attribute of end-user practices);
5. Probability of Initial Failure (PIF) (an attribute of end-user practices);
6. Mission Time (an attribute of end-user practices);
7. Proof Test Duration (an attribute of end-user practices);
8. Site Safety Index™ (an attribute of end-user practices); and
9. Redundancy of Devices including common cause failures (an attribute of SIF design).

3.3 How Key Variables Are Modified to Quantify the Impacts on PFDavg and Other Metrics

Clearly, redundancy of devices is not an attribute of end-user practices or activities. The SSI itself is not modified in safety performance calculations; rather it is used to modify many of the other variables which impact PFDavg and other metrics. Specifically, whenever SSI is less than 4, the following five key variables are modified:

- failure rates of each device
- proof test effectiveness
- proof test intervals
- mean time to restore
- PIF

Failure rates of each individual device in the SIF are increased by a specific multiplier which is determined by the SSI value and the device itself. It is known that final elements are more likely to be negatively impacted by less than ideal end-user practices than are sensors or logic solvers. By increasing device failure rates on an individual device basis, it is possible to more accurately account for the effects of site practices on safety performance. Proof test effectiveness, i.e., the likelihood that all failures are discovered and completely repaired, is also decreased with decreasing SSI.

Proof test intervals, mean time to restore, and PIF are increased by specific multipliers determined by SSI value. It is common to find that less than ideal end-user practices include delaying proof testing, reducing proof test coverage, requiring additional restoration time, and introducing initial failures through imperfect pre- or post-installation testing. Final elements that may develop failures simply due to lack of motion of the moving parts over an extended time (a few months) are especially vulnerable to initial failures in excess of 1% [7, 8, 9].

The two variables not modified are mission time and proof test duration. Mission time is not modified because its impact is generally small unless the mission time is sufficiently beyond the SIF useful life so as to cause the SIF to see failures due to ageing, i.e., failures no longer modeled by a constant failure rate, without renewal during the mission time. Proof test duration has no impact if proof testing is performed with the process off-line. For proof testing performed on-line, proof test duration generally has only a very small effect unless proof testing frequency is very frequent, i.e., greater than or equal to once per month – an impractical situation for most industrial processes.

3.4 Calibrating the Multipliers

As with the development of the Failure Modes, Effect and Diagnostic Analysis (FMEDA) failure rate predictive technique [10] and its associated validated databases [11, 12, 13], it is necessary to calibrate the database of multipliers for the various modifications and to provide on-going validation against actual FFD. The initial calibration phase is still in progress and uses the extensive FFD of the Offshore and Onshore Reliability Data Project (OREDA) [14, 15] as well as the detailed proof testing and root cause failure records of the Savannah River Site (SRS) [8].

4. Examples of Quantifying the Impacts of SSI on Various Metrics

The same simple 1oo1 SIF was analyzed under five different levels of SSI. Figure 1 shows the impacts on PFDavg and SIL of these variations in SSI. For the same 1oo1 system, Figure 2 shows the impacts on RRF and SIL of variations in SSI. Finally, Figure 3 shows the impact on MTTFS of variations in SSI.

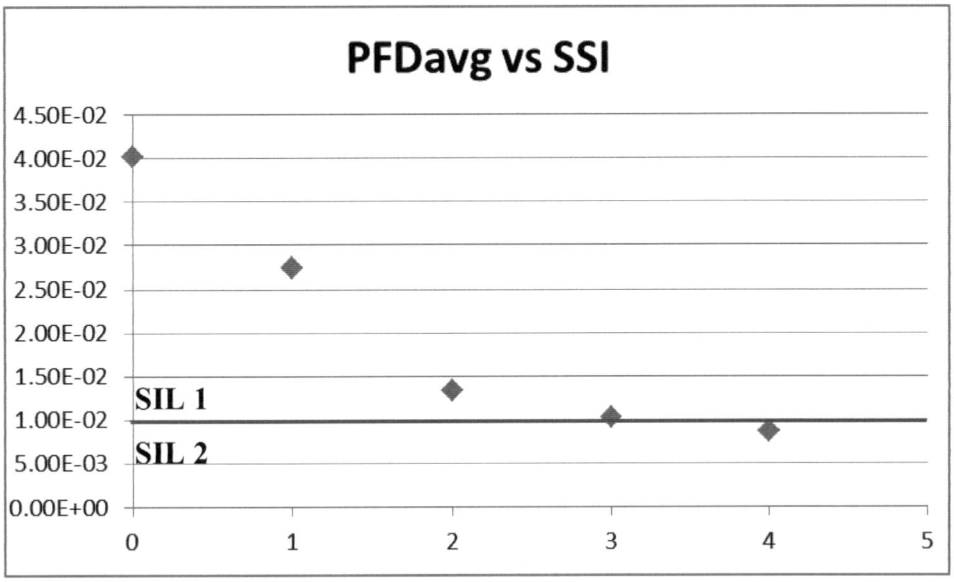

Figure 1. Impacts on PFDavg and SIL of Variations in SSI.

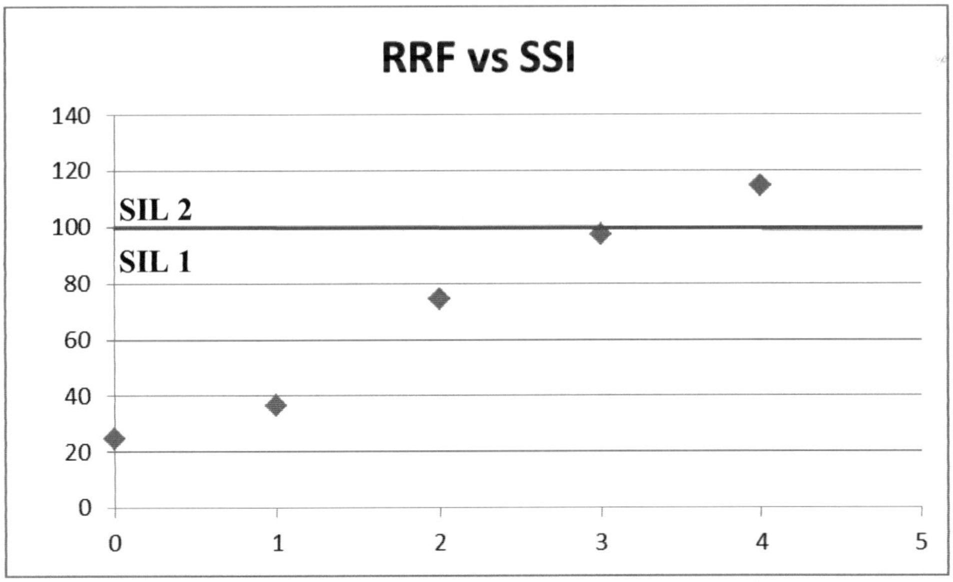

Figure 2. Impacts on RRF of Variations in SSI.

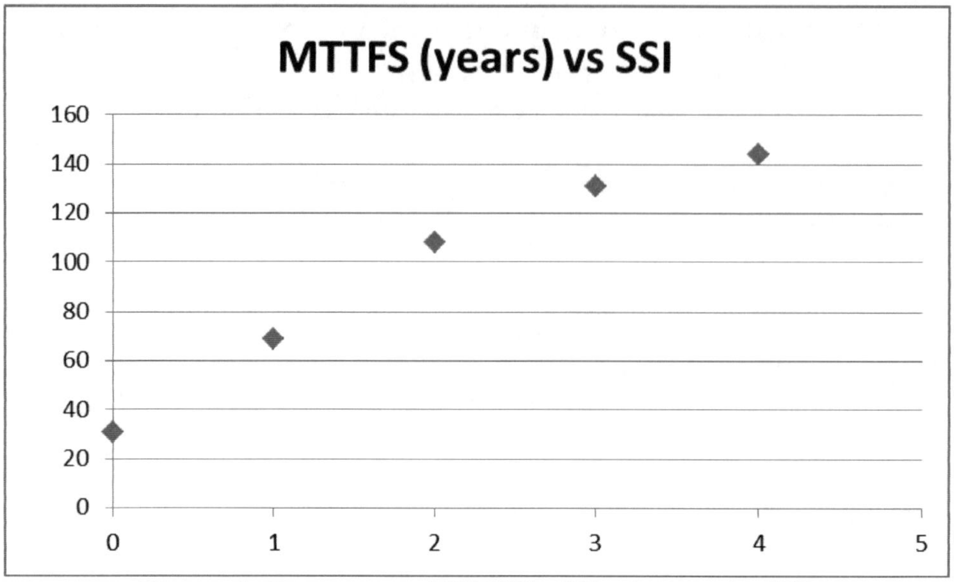

Figure 3. Impacts on MTTFS of Variations in SSI.

5. Summary

This paper clearly shows that the impacts of end-user activities and practices on safety performance of a SIF can be quantified through a simple model like the SSI. It also explains how the SSI modification multipliers are being calibrated and will be updated as increasing amounts of FFD become available. The methodologies presented here and in [4] allow end-users to measure quantitatively their own activities and practices on a site-by-site basis and to measure the impacts of those activities and practices on SIF safety performance.

6. References

[1] IEC 61508, *Functional Safety of electrical / electronic / programmable electronic safety-related systems*, Geneva, Switzerland, 2010.

[2] IEC 61511, *Application of Safety Instrumented Systems for the Process Industries*, Geneva, Switzerland, 2nd edition, 2004.

[3] J. V. Bukowski, W. M. Goble, and I van Beurden, "Product Failure Rates vs Total Failure Rates at Specific Sites: Implications for Safety," *Proceeding AIChE 11th Global Congress on Process Safety*, Austin, TX, April 2015.

[4] J. V. Bukowski and D. Chastain-Knight, "Assessing Safety Culture via the Site Safety Index™," *Proceeding AIChE 12th Global Congress on Process Safety*, Houston, TX, April 2016.

[5] W. M. Goble, "Maintenance capability: Does it matter in the long term?" *Hydrocarbon Processing,* June 1, 2013.

[6] "The Key Variables Needed for PFDavg Calculation," White Paper, exida 2015, http://www.exida.com/Resources/Whitepapers/the-key-variables-needed-for-pfdavg-calculation

[7] J. V. Bukowski, R. E. Gross and W. M. Goble, "Probability of Initial Failure for Spring Operated Relief Valves," *Proceedings of the ASME 2011 Pressure Vessels and Piping Division Conference*, Baltimore, MD, July 2011.

[8] J. V. Bukowski, R. E. Gross and W. M. Goble, "The Adhesion Failure Mode in Stainless Steel Trim Spring Operated Pressure Relief Valves," *ASME Journal of Pressure Vessel Technology*, Vol 135 No 6, December 2013.

[9] L. Stewart, J. V. Bukowski and W. M. Goble, "Improving Reliability & Safety Performance of Solenoid Valves by Stroke Testing," *Proceedings AIChE 9th Annual Global Conf on Process Safety – Process Plant Safety Symposium*, San Antonio, TX, April-May 2013.

[10] W. M. Goble and A. C. Brombacher, "Using a Failure Modes, Effects and Diagnostic Analysis (FMEDA) to Measure Diagnostic Coverage in Programmable Electronic Systems," *Reliability Engineering and System Safety*, Vol. 66, No. 2, Nov 1999, pp. 145-148.

[11] W. M. Goble and J. V. Bukowski, "Development of a Mechanical Component Failure Database," *Proceedings of the 2007 Annual Reliability and Maintainability Symposium*, Orlando, FL, January 2007, pp. 451-455.

[12] J. V. Bukowski and, W. M. Goble, "Validation of a Mechanical Component Constant Failure Rate Database," 2009 *Proceedings Annual Reliability and Maintainability Symposium*, Fort Worth, TX, January 2009, 338-343.

[13] J. V. Bukowski and L. Stewart, "Comparing Failure Rates for Safety Devices: FMEDA Prediction vs OREDA Estimation," exida White Paper available at www.exida.com

[14] SINTEF, *OREDA Offshore and Onshore Reliability Data Handbook, Vol 1. – Topside Equipment and Vol. 2 – Subsea Equipment*, 6th Ed, OREDA Participants, 2015.

[15] SINTEF, *OREDA Offshore Reliability Data Handbook, Vol 1. – Topside Equipment and Vol. 2 – Subsea Equipment*, 5th Ed, OREDA Participants, 2009.

441468 Using HAZOP/LOPA to Create an Effective Mechanical Integrity Program

Abstract:

Many people view the conduct of a HAZOP/LOPA to address regulatory requirements as a chore, and stop there. However, the implementation of a quality HAZOP/LOPA has the potential to provide a framework for addressing numerous safety and operational optimization issues at plants, including the formulation/refinement of the Mechanical Integrity Program. The purpose of this paper is to focus on the mechanical integrity program, illustrate how a quality HAZOP/LOPA can support the effective implementation of some of the new Damage Mechanism Review requirements for California Refineries (e.g., 5189.1(k)), and optimize key elements of an effective Mechanical Integrity Program, e.g.:
- inspection/testing methods
- testing intervals
- maintenance outage periods
- repair prioritization and allowable outage
- identification of low priority equipment

441486 Reactive Chemical Hazards of Azide Compounds: a Case Study and Lessons Learned

Abstract:

While a researcher was servicing a vacuum pump an unintended detonation occurred. This pump had been used in a sulfonyl azide synthesis ($R-SO_2-N_3$). The detonation occurred when he was applying effort to the hose barb from the outlet of the pump to connect it with a rubber hose. Differential Scanning Calorimetry (DSC) and headspace Gas Chromatography/Mass Spectrometry (GC/MS) were employed for the root cause investigation. The DSC showed a large exothermic decomposition (-1388J/g) for the residual sample from the pump fitting. Headspace GC/MS identified the presence of hydrazoic acid in the pump oil. The investigation concluded that either the azide precursor (sodium azide) or hydrolysis of the product (sulfonyl azide) was responsible for generating hydrazoic acid, which escaped from the dry ice trap and appeared in the downstream equipment, and probably reacted with metal parts (brass) to form a highly shock sensitive material (copper azide).

This paper reviews the reactive chemical hazards for azide compounds involved in this case study. Sodium azide is not particularly sensitive to mechanical impact, but its propensity to form hydrazoic acid in water is well-known, even at neutral pH. Likewise, the test data on sulfonyl azide product showed that it is not particularly sensitive to mechanical impact, but it can generate hydrazoic acid upon hydrolysis with water. Metal azides which are formed by the reaction between hydrazoic acid and metal surfaces of equipment parts are highly hazardous due to their extreme impact and friction sensitivity.

Based on lessons learned, this paper also describes further steps for synthesizing organic azides using aqueous sodium azide. These steps include, but are not limited to: (1) Utilizing a dedicated hood and equipments to limit potential HN_3 vapor streams from contacting any reactive metal surfaces. (2) Using a dual caustic trap system designed to scrub the outlet gases swept from the azide reaction step, instead of dry ice trap. (3) Using a 5% aqueous solution of ceric ammonium nitrate as a decontaminating agent for metal surfaces inside the fume hood at the completion of the campaign. (4) Eliminating all brass or copper fittings from equipment used in the synthesis, due to the potential formation of the unstable explosive copper azide that can form from exposure of brass or copper parts to HN_3.

441495 The Application of Frequency Data in Quantitative Risk Assessment

Abstract:

Quantitative Risk Assessment (QRA) is a vital resource for estimating risk in the process industry. There are four primary steps involved in conducting a QRA: hazard identification, frequency analysis, consequence analysis, and risk calculation (the combination of frequency and consequence to determine the risk). Frequency analysis is a key component of the QRA and can have a major impact on the risk profile.

The primary method of estimating event frequency for QRA utilizes historical leak frequency data, of which the Hydrocarbon Release Database (HCRD) is the most widely used leak frequency database. One main issue with applying historical frequency is not having transparency of the causes of the events. Using the historical data approach, a single value is generated for the frequency of an event occurring, with no association to the cause (e.g. corrosion, dropped object, etc.). It has also become widely accepted to apply these offshore databases to onshore processes. There are many concerns in applying these databases to onshore processes, as there are differences in the types of equipment, the hazards the equipment is exposed to, and the safety management systems in place.

When QRAs were first conducted during the 1980s, they were an exhaustive and detailed process. In today's fast track industry, often frequency analysis is conducted without undertaking an evaluation to determine if the frequency being applied is appropriate for the hazard at hand. Leak frequency analysis has become an isolated activity, rather than relying on frequencies established during HAZOP and reliability assessments such as fault tree analysis.

An alternative method to employing historical databases to estimate event frequency is conducting fault tree analysis. Fault tree analysis involves dissecting a hazard into its causes, using a "top down" approach, beginning with hazard identification and working down to a level in which applicable frequency data is available. Fault trees provide a logical relationship between the hazard and its causes.

If historical leak frequency data is employed, which is independent of the cause, the resulting risk is also independent of the cause. Therefore, mitigation measures and the decision making process tend to focus on the back-end consequence of the event, rather than the cause of the event.

Fault tree analysis, although it has had limited application in onshore and offshore risk assessment, provides a major advantage of providing a link between the cause of an event and the associated frequency. This transparency between the event frequency and cause enables us to identify major contributors to the risk and provide risk reduction measures based on what is actually causing the event to occur.

The objective of this paper is to provide an alternative to employing offshore historical frequency data to QRA and demonstrate that the alternative method can provide major advantages in mitigations by linking the estimated risk to the cause and focusing the decision making process on the front-end of the problem rather than solely on the back-end.

441528 Uncertainty in Quantitative Risk Assessment and Decision Making

Abstract:

Including uncertainty in Quantitative Risk Assessment (QRA) is recommended in most published Best Practice Recommendations for QRA, including those published by TNO, CCPS, and Norwegian authorities. Unfortunately typical commercial QRA techniques and models used in Process Risk Analysis do not include the capabilities of uncertainty within the model formulations. Effects on uncertainty may be based on qualitative estimates or ignored, or at most accounted for by sensitivity analysis by "swinging" variables to estimate individual variable sensitivity. By not presenting the uncertainty, the risk assessment results may be misrepresented as possessing an accuracy greater than is warranted.

Failure to account for uncertainty also tends to support centrist thinking and poor quality decision making. The natural tendency of users of risk analysis is to believe in their results, especially if much effort has been put into developing the results. If the result is a single number, that number may be assumed by the analyzer and users as "right". In addition, there is a tendency in people to assume that by picking the most likely values for the inputs into a model, the end result is also the most likely result. This may not be true if the model is complex and non-linear. The most likely results must be determined by including variable distributions and uncertainty modeling in the analysis.

Decision analysis is also deceptively simple when single point criteria or simple curves are used for decision making purposes. A challenge in risk assessment is utilizing results and presenting them in a straight forward way that capture the uncertainty and complexity while allowing clear, high quality decisions by end-users.

This paper presents sources of uncertainty, analysis methods, examples of potential errors introduced by ignoring uncertainty, and illustrates how one can include uncertainties in your quantitative risk assessment (QRA) and benefit from it in your risk assessment decisions.

441529 Analysis of Process Systems during Accidental Fire

Abstract:

In the past 40 years, there have been over 60 major fire accidents reported on offshore platforms. The thermal effects imposed on process equipment and systems (such as pressure vessels, flare scrubbers, etc.) has the likely potential to result in catastrophic consequences such as explosions given that these process systems typically operate at very high pressures. In the event of an accidental fire, blowdown valves are engaged to relieve the excess pressure of the process system to acceptable values in a specific time period. During that time, however, the increased temperature of the blowdown system due to the impinging fire may result in the degradation of its structural capacity. The combination of a degraded vessel shell and a high internal pressure typically leads to sudden ruptures and explosions which could result in the escalation of the initial fire or injuries to personnel. Therefore, it is critical to accurately capture the pressure and temperature of the process system in a fire blowdown event.

A Blowdown analysis is typically carried out using simple thermodynamic software. While these software provide cost-effective approaches to studying process systems, they are unable to accurately compute the local effects of heat transfer and fluid behavior. Moreover, stress concentrations on the pressure vessel due to nozzle effects are not captured correctly using classical approaches and advanced analysis is required. Therefore, to accurately capture the rupture time of pressure vessels, scrubbers, etc., during a fire scenario, advanced approaches such as Computational Fluid Dynamics (CFD) are required.

In this paper, we will demonstrate a new approach in analyzing onshore and offshore process systems in an extreme fire scenario. Heat transfer through the vessel shell as well as the internal fluid will be accounted for by computing the rates of conduction, convection, and radiation. The time history of temperature and pressure as well as the change of phase of the fluid inside the pressure vessel will be captured in the analysis. A degradation study of the structural steel will be performed to predict the time of rupture of the process system to internal pressure using advanced analysis. Moreover, the effect of jet fire momentum on the vessels and their supports will be computed according to industry guidelines.

The methodology is validated by comparison with experimental data. The rupture time and rupture location will be computed and design recommendations will be provided to increase the strength of the process system at rupture locations. The current approach could provide accurate results to compute the response of process systems such as high-pressure vessels, scrubbers, etc., during offshore or onshore fire scenarios.

12TH GLOBAL CONGRESS ON PROCESS SAFETY

Assessing Safety Culture via the Site Safety Index™

Julia V. Bukowski, Ph.D.
Dept. of Electrical & Computer Engineering
Villanova University, Villanova, PA 19085
julia.bukowski@villanova.edu

Denise Chastain-Knight, P.E., CFSE
exida
64 N. Main St., Sellersville, Pa 18960
dchastainknight@exida.com

Prepared for Presentation at
American Institute of Chemical Engineers
2016 Spring Meeting
12th Global Congress on Process Safety
Houston, Texas
April 11-23, 2016

Assessing Safety Culture via the Site Safety Index™

Julia V. Bukowski, Ph.D.
Dept. of Electrical & Computer Engineering
Villanova University, Villanova, PA 19085
Julia.Bukowski@villanova.edu

Denise Chastain-Knight, P.E., CFSE
exida

Keywords: Safety Lifecycle, safety culture, systematic failure rate, site-specific safety performance, PFDavg

Abstract

How can a company establish a baseline measurement of its safety culture against which to gauge improvement? The Site Safety Index™ (SSI) quantifies in part (on a scale of 0 – 4) the degree to which a company's end-user practices support the attainment and retention of an appropriate safety culture for operations and maintenance. Further, the SSI can be used to appropriately adjust parameters that directly impact measures of safety such as probability of failure on demand (PFDavg). Thus, the impacts of safety culture can be further quantified and the effects of changes to safety culture can be assessed.

1. Introduction

Consider the following hypothetical scenario. A company assesses the risk intrinsic to one of its plant processes (given its current level of mitigation) to be greater than the tolerable risk dictated by company policy or regulatory standards. Further mitigation to reduce the process risk to an overall risk at or below tolerable risk level will require a safety instrumented function (SIF) which will result in a risk reduction factor (RRF) of at least 110, i.e., a SIF of safety integrity level (SIL) 2. The company may choose to design and implement the required SIF itself or to deploy an existing SIF design and implementation.

In either case, the company will need to ascertain that the SIF meets its risk reduction requirements. There are several ways to do this. For example, it may rely on proven-in-use data to claim the company has met its risk reduction goal. This proven-in-use data may come from another company or the database of a consortium of companies. Alternately, the company may predict the risk reduction and SIL level by using manufacturers' published data or other available databases perhaps in conjunction with a commercial tool. If by either or both means of risk reduction/SIL verification the company determines the risk reduction goal is met by the SIF, is this alone sufficient for the company to state that, when implemented in its plant, its process risk will be at or below tolerable levels? **No!**

When the company implements the SIF in its process, the actual risk reduction achieved by the SIF will be determined not only by the SIF attributes taken into account by the verification methods but also by the company's safety culture, i.e., by the overall impact of end-user practices that take place within the company in general and especially within the plant/site itself. Further, these practices may differ from site to site even in the same company.

For example, if the risk reduction/SIL was estimated from proven-in-use data, unless "systematic" failures were eliminated, these data inherently contain the impacts of the end-user actions/omissions which took place at the plant or plants during the time period in which the proven-in-use data were collected. When the company implements the SIF, it may actually experience the same, higher or lower risk reduction than that estimated from proven-in-use data depending on whether the company's end-user current practices are the same, better or worse than those practices in the plants *when* and where the proven-in-use data were collected. Especially for consortium-based data, it is not generally known exactly from where the proven-in-use data were gathered much less what the specific end-user practices were at those plants. Thus, estimated risk reduction/SIL based on proven-in-use data is difficult to adjust to more accurately represent the actual risk reduction/SIL the company will experience.

If the risk reduction/SIL was predicted based on the manufacturers' published failure rates or other failure rate data, these rates are likely predicated on the assumptions that the end-user faithfully performs all required actions to sustain the safety integrity of the equipment. Such actions include, but are not limited to, insuring that

- the equipment implementing the SIF is appropriate to the task and environment,
- the equipment is properly installed and calibrated,
- the equipment is correctly functioning when installed,
- all in-service maintenance is correctly and completely performed on schedule,
- the equipment is maintained so that no ageing occurs prior to proof testing,
- proof testing is completely and correctly performed,
- the equipment is replaced when its useful life can no longer be extended by maintenance and refurbishment.

Note that the list extends beyond the usual concepts of maintenance alone. Any required action that is omitted, delayed, or performed incompletely or imperfectly will reduce the risk reduction (and perhaps SIL) achieved in the SIF's actual use in the plant. However, in the case of predicted risk reduction/SIL, which is based on manufacturers' data or Failure Modes, Effects and Diagnostics Analysis (FMEDA) [1], it is possible to adjust the initially calculated risk reduction/SIL to account for a company's end-user practices. This adjustment occurs through the use of a SSI.

This paper explains the way in which the SSI is measured and illustrates how the SSI can impact typical safety parameters such as PFDavg. A second paper [2] explains how the SSI is used to adjust the initially calculated risk reduction/SIL.

2. Notation

FMEDA	failure modes, effects and diagnostic analysis
IEC	International Electrotechnical Commission

PFDavg	average probability of dangerous failure on demand
RRF	risk reduction factor
SIF	safety instrumented function
SIL	safety integrity level
SSI	Site Safety Index™

3. Measuring a Site's Safety Culture

3.1 Why a "Site" Safety Index?

The reader may wonder why the terminology "Site" Safety Index is used as opposed, perhaps, to "Company" Safety Index. It is well documented that a safety product may exhibit a safety performance less than that specified by the manufacturer or by a FMEDA [3, 4] and that even within the same company, failure rate estimates derived from field failure data can vary significantly from site to site [5]. Thus, while company-wide policies and procedures will impact the safety achieved, the possibly different ways company-wide directives are implemented on a site-by-site basis requires that a safety index be measured at the site level.

3.2 How is the Site Safety Index™ Evaluated?

There are currently two ways that SSI can be evaluated: via an external audit, which provides the most accurate measure of SSI, and via a self-administered questionnaire, which relies on less information than the audit and provides an approximate measure of SSI. These two evaluation methods are detailed next.

3.2.1 Via an External Audit

The international standard IEC 61511 [6] lists the requirements that must be met for the entire Safety Lifecycle process. In an on-site audit, it can be verified that the site meets these requirements in combination with additional detailed requisites for the operations and maintenance activities. Additionally, all safety and lifecycle procedures are audited to be sure: a) that the procedures are adequate to support the requirements and b) that they are being correctly followed at the site. The outcome of the audit is a report which includes the SSI score obtained as well as recommendations for changes and improvements. The interpretation of the measured SSI level itself is discussed below.

3.2.2 Via a Self-Administered Questionnaire

A site that is interested in acquiring an approximate measure of its current state of safety culture may choose to complete a questionnaire of about 40 items which represents a subset of the information gathered in a complete audit. These questions cover activities and practices spanning a portion of the lifecycle of the SIF and include installation and commissioning, safety validation, operations and maintenance, SIF modification and decommissioning. Table 1 lists the questions for the operations and maintenance phase of the lifecycle. The complete set of questions is available at [7]. The questions, which are answered Always, Often, Never, or Not

Applicable, are subject to weighted scoring. The total score received determines an approximate SSI level for the site.

Table 1 Questions Regarding Site Activities and Practices during the Operations and Maintenance Phase of the SIF Lifecycle

Do your operation and maintenance procedures include competency requirements for those performing tasks to maintain the "as designed" functional safety of the SIF?
Do your operation and maintenance procedures include the information which needs to be maintained on device failures, proof test results, false trips, demand rates and restore times on the SIF?
Do your operation and maintenance procedures include procedures for fault diagnostics and repair after a failure in the SIF?
Do your operation and maintenance procedures require instruments to be rebuilt or replaced at the end of their useful life?
Do your operation and maintenance procedures require the use of a computer tool like SILStat™ to track demands, false trips, failures and maintenance performance?
Do your operation and maintenance work tasks proceed only in accordance with the relevant procedures?
Are operators that work with the SIF in their area trained to ensure they understand how the SIF functions and what hazards are being protected against?
Do you perform periodic proof-tests using the written procedure?
Is your frequency of proof tests decided using the PFD_{avg} calculation?
Are scheduled proof test performed on schedule?
Do you repair any deficiencies found during the proof testing in a safe and timely manner?
Do you maintain records of proof tests including descriptions of tests and inspections performed, equipment tested, dates, and names of personnel performing work?
Do you maintain records of proof tests including "as found" results of the tests and inspections?

The basis of the validity of using a questionnaire for determining SSI is the accepted practice per IEC61508 of using a questionnaire to determine the β factor that describes the extent of common cause failure to be modelled when evaluating the safety performance of redundant SIF architectures [8].

3.3 Interpreting the SSI Level

The SSI is a number from 0 – 4 which is an indication of the level of site activities and practices that contribute to the safety performance of SIF's on the site. Table 2 details the interpretation of each SSI level. Note that the levels mirror the levels of SIL assignment and that SSI 4 implies that all requirements of IEC 61508 and IEC 61511 are met at the site and therefore there is no degradation in safety performance due to any end-user activities or practices, i.e., that the product inherent safety performance is achieved.

Table 2 Details of the Interpretation of Each SSI Level

SSI Level	Description
SSI 4	Perfect - Repairs are always correctly performed, Testing is always done correctly and on schedule, equipment is always replaced before end of useful life, etc.
SSI 3	Superior - Repairs are usually correctly performed, Testing is usually done correctly and on schedule, equipment is usually replaced before end of useful life, etc.
SSI 2	Typical - Repairs are mostly correctly performed, Testing is mostly done correctly and on schedule, most equipment is replaced before end of useful life, etc.
SSI 1	Good - Repairs are often correctly performed, Testing is often done and often on schedule, some equipment is replaced before end of useful life, etc.
SSI 0	None - Repairs are not performed, Testing is not done, equipment is not replaced until failure, etc.

4. An Example of the Impacts of SSI

The specific ways in which SSI modifies key variables (e.g., failure rate, proof test effectiveness, etc.) which impact PFDavg, SIL and RRF, as well as other safety metrics, is explained in detail in [2] where additional examples are presented. For the purposes of this paper, Figures 1 and 2 provide a simple example of how SSI can impact safety performance. In Figure 1, computations were performed under the assumption that SSI level is 4 and thus end-user activities and practices do not degrade the safety performance of the SIF. Note that in this case the SIF SIL level is computed as SIL 2 based on PFDavg, and the RRF is computed as 113 – slightly greater than the requirement of the opening hypothetical example. In Figure 2, the analysis is repeated but this time the SSI level is assumed to be 3. Now, based on PFDavg, the SIL performance is a SIL level 1 and the RRF is computed as 98 – insufficient for the requirement of the hypothetical example. This shows how the impacts of site activities and practices can be quantified and how effects of changes to these activities and practices can be measured.

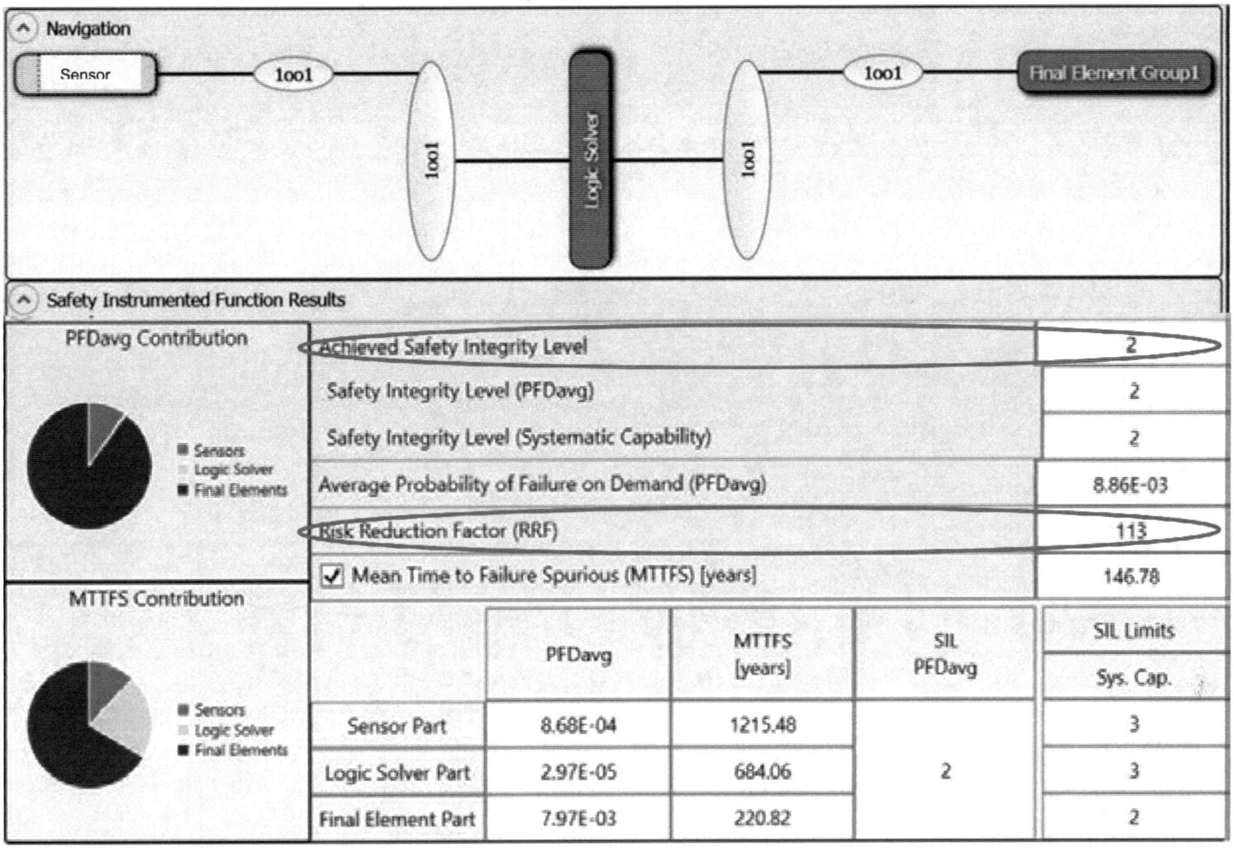

Achieved Safety Integrity Level	2
Safety Integrity Level (PFDavg)	2
Safety Integrity Level (Systematic Capability)	2
Average Probability of Failure on Demand (PFDavg)	8.86E-03
Risk Reduction Factor (RRF)	113
Mean Time to Failure Spurious (MTTFS) [years]	146.78

	PFDavg	MTTFS [years]	SIL PFDavg	SIL Limits Sys. Cap.
Sensor Part	8.68E-04	1215.48	2	3
Logic Solver Part	2.97E-05	684.06		3
Final Element Part	7.97E-03	220.82		2

Figure 1. SIL verification/RRF computations performed under assumption SSI = 4.

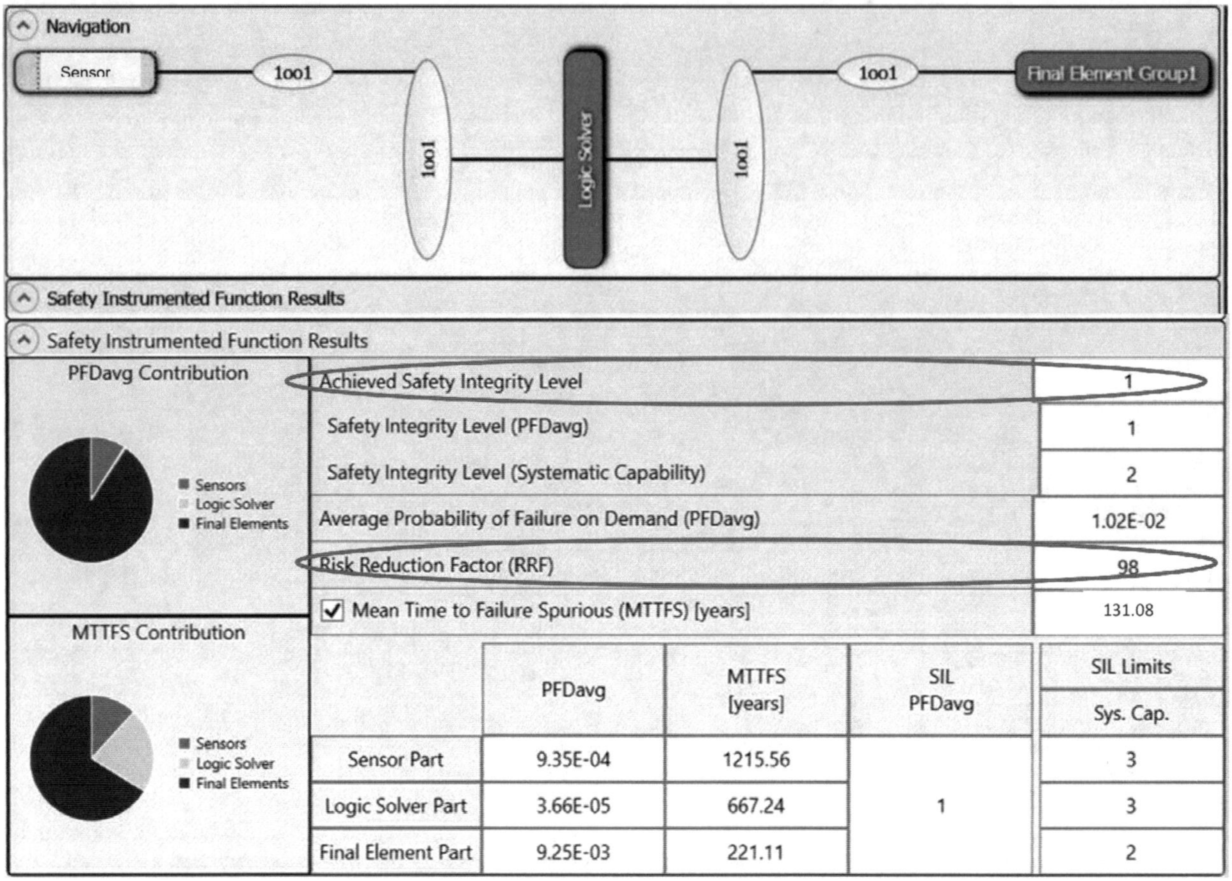

Figure 2. SIL verification/RRF computations performed under assumption SSI = 3.

5. Conclusions

The SSI provides a simple means to measure how well site-specific end-user activities and practices support the achievement and retention of SIF maximal safety performance. Further they can be used to adjust metrics of SIF safety performance from the ideal achievable to a more realistic measure of what is actually attained at a specific site given the site's activities and practices. This last concept is explored more fully in a separate paper [2].

6. References

[1] W. M. Goble and A. C. Brombacher, "Using a Failure Modes, Effects and Diagnostic Analysis (FMEDA) to Measure Diagnostic Coverage in Programmable Electronic Systems," *Reliability Engineering and System Safety*, Vol. 66, No. 2, Nov 1999, pp. 145-148.

[2] J. V. Bukowski and L. Stewart, "Quantifying the Impacts of Human Factors on Functional Safety Failure Rates," *Proceeding AIChE 12th Global Congress on Process Safety*, Houston, TX, April 2016.

[3] J. V. Bukowski, W. M. Goble, and I. van Beurden, "Product Failure Rates vs Total Failure Rates at Specific Sites: Implications for Safety," *Proceeding AIChE 11th Global Congress on Process Safety*, Austin, TX, April 2015.

[4] J. V. Bukowski, W. M. Goble, R. E. Gross and S. P. Harris, "The Effects of Maintenance Actions on the PFDavg of Spring Operated Pressure Relief Valves," *Proceedings of the ASME 2014 Pressure Vessels And Piping Division Conference*, Anaheim, CA, July 2014.

[5] W. M. Goble, "Maintenance capability: Does it matter in the long term?" Hydrocarbon Processing, June 1, 2013.

[6] IEC 61511, *Application of Safety Instrumented Systems for the Process Industries*, Geneva, Switzerland, 2nd edition, 2004.

[7] www.exida.com/SSI

[8] IEC 61508, *Functional Safety of electrical / electronic / programmable electronic safety-related systems*, Geneva, Switzerland, 2010.

441381 Fire Hazard Associated with Polymers and Application of Polymer Nanocomposites for Flame Retardancy

Abstract:

Due to advances in polymer science, their widespread application is found in daily products including vehicles, electronics, clothing, furniture *etc.* However, most of the items that first ignite in any fire scenario includes mostly polymeric materials according to NFPA reports which clearly elucidates the importance into looking for flame retardancy approach for polymers. Polymer nanocomposites is a relatively new approach undertaken for flame retardancy and has the potential to bring multi-functionality including improved thermal behavior, physical barrier effect, catalytic charring effect and better material performance. The objective of this paper is to initially provide the current state-of-the-art review on polymer nanocomposites in terms of their fire retardancy. The paper describes the hazards associated with burning of polymers, flame retardant mechanism, synthesis and characterization of polymer nanocomposites with focus on thermal analysis. Different kinetic model has been studied and applied for analysis of thermogravimetric analysis data. The kinetic parameters of the polymer degradation have been identified. Primarily polystyrene-silica nanocomposite has been prepared by in-situ method and were characterized for their morphology and thermal behavior using characterization methods, such as Scanning Electron Microscopy (SEM), X-Ray Diffraction (XRD), Thermogravimetric Analysis (TGA), and Differential Scanning Calorimetry (DSC). Initial results show nanocomposites with promising fire-retardancy properties.

441530 Applying QRA Study to Select and Design Effective Release Prevention Safeguards

Abstract:

In the past two decades, Quantitative Risk Analysis (QRA) has become a staple safety analysis in offshore oil and gas industry. A comprehensive QRA study could be a powerful tool if the results are understood and changes are implemented to improve safety in the engineering design. Conventional QRA study relies on generic leak frequency and parts count method. One major limitation of conventional QRA study is that it is focused primarily on the loss of containment of hazardous fluids and what happens when they are released. To reduce the risk to acceptable level, usually post-release mitigation measures are addressed. It is a challenge to tackle the effectiveness of release prevention safeguards. This paper presents an approach to address the dominant risk contributors identified in conventional QRA study. The aim is to increase confidence and facilitate decision making in the design selection of loss of containment prevention safeguards.

Sensitivity analysis is usually conducted as part of a QRA study to evaluate the effect of input parameters. However, in most cases, sensitivity analysis focuses on the mitigation measures, such as safety system locations or ratings, personnel distribution or modification to the QRA model itself. Means of release scenario prevention is rarely investigated to the point that it is beneficial to the operation or to the design project. This paper proposes that in order to maximize the use of QRA results, decision makers should investigate means of scenario prevention by performing reliability analysis (e.g., fault tree analysis, human reliability analysis) on the top contributing scenarios. This paper presents some examples of reliability analysis performed on top risk contributors. This paper also presents suggestions on how the results can potentially be applied for the benefit of the operation or design project. The outcome from the QRA study in combination with reliability analysis will increase decision makers' understanding of the top risk contributing scenarios. This approach can be utilized to identify opportunities for improvement in the prevention safeguards.

441536 Managing the Uncertainty in Blast Load Calculations with Enhanced Technical Safety Analysis

Abstract:

The safety of personnel working within onshore and offshore oil and gas processing facilities with respect to vapor cloud explosion (VCE) scenarios revolves around the risk associated with the event. Conducting a risk assessment is quite typical within the offshore and onshore loss prevention communities. The consequence of a VCE event is a very complex phenomenon however. It varies based on several blast load parameters. With improved understanding of blast loading parameters such as blast overpressure, impulse, time lag, net blast loading, etc. structural designers have gained increased confidence in their design and have reduced conservatism in their blast designs. However, there is a significant gap in the risk assessment of onshore and offshore facilities between the process safety and the structural design groups. Using structural insight in combination with rigorous process safety risk evaluation techniques we can reduce the gap between the risk and consequence of a blast design, providing the stakeholders with a more accurate understanding of the risk and consequence of a blast event.

This paper will attempt to address the impact of uncertainty in pressure exceedance curves and;

- how the confidence level impacts the design blast loads;
- how the slope of the exceedance curves impacts the blast design loads;
- two different blast events can result in similar design blast loads for a given risk;
- how a small change is frequency can lead to a significant change in design blast loads;
- how the negative phase of a blast scenario could lead to failure of some specific type of structures;
- how the response of long wall or piping structure is impacted by time-lag of the design blast load;
- how the design blast pressure and design blast impulse impacts a structure;
- how non-structural members, typically not considered in blast design, impact the blast design;
- how simultaneous application of blast load on different parts of structure, taking into account time-lag, affects the blast design.

Currently for a blast event, both blast pressure and blast impulse are defined as independent variables whereas in reality they are coupled. The correlation between the impulse and pressure will also be discussed in the paper. Defining appropriate pressure time-history is critical for structural design. Rise time, positive-phase and negative-phase duration are unknowns in a blast design. How an appropriate time-history is established will also be addressed.

Prior knowledge of implications of a blast event for an associated risk is helpful during structural blast design. Early involvement of structural engineers within the process safety procedure is a recommended solution to such questions. Enhanced loss prevention of offshore and onshore structures can only be established by always keeping the consequence for a blast event for a given risk in mind.

441537 Occupant Vulnerability for Process Plant Building Design Due to Secondary or Domino Events

Abstract:

Statistics indicate that it is in fact not uncommon to have an accidental explosion on an offshore platform, refinery or petrochemical installation. The focus for several years has been on occupied buildings and their resistance to blast loads from accidental releases of hydrocarbons. Common practice dictates that domino effects may not need to be considered given that a large catastrophic release is equivalent in consequence. However, the catastrophic events are filtered out in a risk assessment due to their unlikelihood and as a result, domino effects are not captured. For example, a small release of highly reactive fuel, can lead to a small explosion which can easily damage nearby piping and vessels which can in turn become catastrophic events. The design and assessment of piping systems and vessels against blast loads are of paramount importance and require advanced analysis capabilities. In order to have a safe, i.e. blast-resistant process piping system design, an advanced nonlinear analysis in time domain is required so that we may properly capture the dynamic response of the piping and vessels subjected to blast loading. This type of analysis, in general, should take into account (1) Blast drag load and blast overpressure; (2) Nonlinear material properties, i.e. thermal and strain-rate dependence; (3) Effects of non-structural masses and adjacent piping systems; (4) Pipe supports; (5) Failure of flanges and piping detail; (6) Effects of Operating Temperature on Material Properties; (7) Blast direction/Ignition location/Attenuation sensitivity; (8) Shielding effect, e.g. pipes behind large objects such as vessels; and (9) Effects of pipe insulation, i.e. increased pipe diameters. This paper first discusses the methodology/techniques to account for such effects in finite element analysis for a safe design. The paper then uses case studies to demonstrate the methodology we develop to show how advanced simulation techniques are applied as tools for piping blast response to minimize the risk of secondary failure due to initiating events. These case studies involve a piping system that is simulated from global to detail modeling. The goal of the paper is to discuss how to achieve a safer design for process piping systems by using advanced analysis.

441629 Assessment of the Mechanical Integrity of a Pipeline with Metal Loss, Using Finite Element Method

Abstract:

Following the guidelines proposed by the ASME B31G and API579 standards, for the assessment of systems that compromise its mechanical integrity, given the existence of defects associated with the presence of corrosion damage, the probability of failure of a pipeline, which transports hazardous materials associated with the Oil & Gas industry, and is exposed to pitting corrosión damage was calculated when it overcomes its yield stress. For this, a methodology that combines the corresponding structural analysis of the system and subsequent reliability analysis was proposed. Through the first, the calculation of the von Mises stress present in areas where the pitting defects was made; for this, the corrosion growth model to be used in the present work was defined by implementing the correlation proposed by the Southwest Research Institute (SwRI) to find out the corrosion rate growth in annual terms. Subsequently the pipeline section where the study would be carried out was defined, based on the results obtained in a previous study, which analyzed information related to corrosion damage present on the pipeline, collected through the In Line Inspection (ILI) technique. This allowed to restrict the test area to a total length of 80 cm and 1/8 section of pipe, lastly simulating the system and obtaining the loads associated with the von Mises stress affecting corrosion defined areas. For reliability analysis it was necessary to find the probability distribution function associated with the von Mises stress on the system studied; defining an experimental design which sought to evaluate different values of the random variables defined on the model proposed by SwRI, thus simulating the system a sufficient number of times in order to find the distribution of von Mises stress for this case study. With the distribution function defined, system reliability analysis was carried out through Monte Carlo simulations; thus, it was established that the limit state function would have a conventional structure comparing resistance with loads. Thus, the resistance was set as the yield stress of the material, while the load would be the von Mises stress, and thus the random variable from which random numbers governed by the probability distribution found previously were generated. As mentioned previously, it was defined that the present system fails when the von Mises stress exceeds the yield stress of the material; finding that for the time range evaluated (year one, five, ten and fifteen), the probability of failure is behaved as expected, increasing year by year if no maintenance is not performed to the pipeline section evaluated.

441539 An Improved Risk-Based Approach for Accidental Loadings

Abstract:

Accidental loading such as fire, blast, ship impact, and dropped objects constitute the great majority of potential and actual fatalities in both onshore and offshore operations. In order to help prevent human loss and for increased safe design or assessment of an onshore/offshore asset, the risks of these accidental loadings should be quantified, in terms of both probability/frequency and consequence aspects. In this paper we propose a unified risk-based approach for the assessment of all accidental loadings mentioned above.

This approach will address both human and asset risk assessment against accidental loadings. Technical safety and process safety inputs such as hazard curves, frequency exceedance curves, and return periods are where the process all begins. Based on these inputs, an event frequency analysis is carried out to determine how often one should expect the event to happen, i.e., the frequency aspect of the assessment. Depending on the type of the accidental loadings and the subject of the risk assessment (human or asset), the output can include a fragility curve, damage frequency matrix, and individual risk matrix. Additionally, a consequence analysis is carried out to determine the severity levels of the event. In a risk-based consequence analysis, the severity levels are fully determined based on the risk associated with the event. Based on both event frequency and consequence, the risk ranking is identified using a risk matrix. If the risk is not acceptable according to the risk matrix, mitigation/repair/strengthening options are then developed.

As an example, a risk-based structural fire analysis is compared with the conventional approach which is widely used in the industry. In a conventional Passive Fire Protection (PFP) analysis using Ductility Limit Analysis (DLA) according to API RP 2FB, fire loads are deterministically applied to a structure whose response is then analyzed. The initial PFP scheme is developed based on the analysis and then optimized based on input from all disciplines (e.g. technical safety, operations, structures, etc.). This approach is sometimes misinterpreted as a "risk-based" approach; however, it does not take into account the frequency aspect of the risk assessment. In a risk-based PFP analysis using DLA, fire scenarios are developed in a particular target zone based on exceedance curves (i.e., input from technical safety discipline). The ductility level analysis is performed to determine the structural consequence. If personnel safety is of interest, the consequence of the structure is then linked to individual risk to determine fatalities. The amount of PFP to be applied on the structure is fully based on the risk that is produced by the fire scenarios in the target zones or areas.

Followed by the approach, a new perspective on safe design of onshore/offshore structures for accidental loadings is outlined to estimate the associated risk to potential targets such as human personnel as well as asset components. The proposed assessment methodology will contribute towards identifying the mitigation measures and safety-critical procedures and equipment, and toward a safer design.

441540 Occupant Vulnerability for Temporary Blast Resistant Modular Buildings

Abstract:

Human fatality and injury in the process facility due to overpressure effects can be categorized as (a) direct injury to body due to pressure change, (b) secondary injury resulting from fragment or debris impact, and (c) tertiary injury as a result of body being thrown by explosion drag on stationary objects. Secondary and tertiary effects are typically the dominant causes of fatalities in the process facilities. Direct injuries may result from change in internal pressure as a result of wall/roof response or penetration of blast overpressure into the building. Existing studies allow for an estimation of occupant vulnerability to blast on typical default building construction types. There exists no such building occupant vulnerability calculation method for modular buildings. This paper seeks to develop an improved understanding of the impact of blast loads on modular building and a more comprehensive understanding of the vulnerability to personnel who are within this type of building if they are subjected to a blast load. The study will be carried out in analytical basis which will include basic and advanced numerical methods along with engineering judgment and accident investigation for estimation of vulnerability. A numerical model can include deformable crash test dummies as well as many types of non-structural components in order to estimate occupant vulnerability and maximum individual risk. The outcome of this study will include new response criterion limits related occupant vulnerability in modular buildings as well as design and retrofit guidelines for critical non-structural components.

441541 Occupied Building Blast Retrofit and Upgrade for Personnel Safety

Abstract:

Accidental blast events at petrochemical facilities such as BP Texas City have shown that the occupied buildings near the process areas can be vulnerable to blast and collapse even under low blast overpressures. Additionally, existing permanent and portable buildings may exhibit insufficient capacity to resist blast loading during facility siting studies per API RP 752 and 753. Although it is not possible to completely eliminate risk from blast loading, it is possible to minimize the individual risk, which is a probability an occupant will experience fatality due to an explosion event, by applying retrofit and upgrade strategies to increase the blast-resisting capacity of an occupied building and meet the safety performance criteria.

There are several ways to decrease individual risk due to explosion with respect to occupants in a building:

1. Moving the building from a high blast region near the process area to a lower blast region, and design a new building;
2. Retrofitting existing occupied building;
3. Replacing existing occupied building with a blast resistant modular building; and
4. Replace the existing occupied building with a new blast resistant building.

The first and forth options can be inefficient from construction cost and time perspective. Additionally, the first option may require purchasing an additional land and obtaining permit. The second and third options are significantly cost and time efficient, and causes less interruptions for the operations. However, these options require analysis and engineering judgment from a highly skilled and experienced engineer. In this paper, case study will be presented where a variety of retrofit and upgrade options will be discussed from perspectives such as design, constructability, cost and individual risk.

12TH GLOBAL CONGRESS ON PROCESS SAFETY

Implementing Engineered Controls to Reduce the Likelihood of High Severity Scenarios

Alek Hamparian
Kazarians & Associates, Inc.
100 West Broadway, Suite 970, Glendale, California
Alek.Hamparian@Kazarians.com

Mardy Kazarians
Kazarians & Associates, Inc.
100 West Broadway, Suite 970, Glendale, California
MKazarians@Kazarians.com

Prepared for Presentation at
American Institute of Chemical Engineers
2016 Spring Meeting
12th Global Congress on Process Safety
Houston, Texas
April 11-13, 2016

Implementing Engineered Controls to Reduce the Likelihood of High Severity Scenarios

Alek Hamparian
Kazarians & Associates, Inc.
100 West Broadway, Suite 970, Glendale, California
Alek.Hamparian@Kazarians.com

Mardy Kazarians
Kazarians & Associates, Inc.

Keywords: Hazard Identification and Risk Analysis, Process Safety

Abstract

The Center for Chemical Process Safety (CCPS) *Guidelines on Inherently Safer Chemical Process A Life Cycle Approach* (2009) defines a hierarchy of controls that can be used to decrease the risk of a system. These guidelines include a hierarchy of layers of protection that can be used in Process Hazard Analyses (PHAs) to determine adequacy of the existing safeguards. This approach can be particularly useful for companies that have potentially high severity scenarios for which only a procedural layer of protection is available.

Releases of hazardous materials have occurred that could have been prevented by implementation of engineered safeguards. From a review of these incidents, it had become evident that the owners and operators of the affected facilities relied on procedural safeguards to prevent occurrence of those events. Examples of such cases have involved gasoline, chlorine, and anhydrous ammonia, which are three commonly used hazardous materials at facilities in the United States.

Process Hazard Analysis (PHA) provides a thorough and efficient method for systemically reviewing complex systems for safety concerns. A commonly applied PHA method is the Guide-Word Style Hazard and Operability (HAZOP) approach. HAZOPs are a team effort and risk determination varies based on the experiences and safety culture of team members. Some teams may determine that procedural safeguards are adequate to prevent against a scenario that could lead to a severe event (e.g., liquid release of a volatile toxic material such as chlorine). However, accidental releases have been experienced demonstrating inadequacy of procedural safeguards to minimize the risk of a severe event.

The above noted CCPS guidelines consider procedures as the least effective protection layer. An active engineered safeguard is considered a more reliable protection layer. The purpose of this paper is to point out the potential for a pitfall in risk acceptance where a high severity consequence scenario is concluded to be of acceptable risk based solely on operator action. It is recommended for owner and operators of such facilities to require mitigation feature other than procedural controls for high severity scenario regardless of the risk ranking.

1 Introduction

Releases of hazardous materials have occurred in a wide range of systems with safeguards in place. A number of these events have occurred in systems that relied on procedures and operator intervention as the primary safeguard. The Center for Chemical Process Safety (CCPS) [1] and the American Industrial Hygiene Association (AIHA) [2] have each published guidelines on the topic of hierarchy of controls which can be used as a qualitative measure to establish the effectiveness of risk reduction measures. Both CCPS and AIHA have ranked engineered controls as a higher level safeguard than human intervention. The October 23, 2009 explosion at the Caribbean Petroleum Corporation's facility in Puerto Rico shows the importance of implementing engineered controls in addition to human intervention. The Chemical Safety Board (CSB) [3] investigated that event and recommended for similar facilities to add engineered controls such as "automatic overfill protection systems" to prevent future accidents.

Facilities that contain certain hazardous materials above a certain quantity must comply with the Process Safety Management (PSM) [4] and Risk Management Plan (RMP) [5] process safety regulations that require a Process Hazard Analysis (PHA). The purpose of a PHA is to identify potential hazards posed by a system, the safeguards that are in place to minimize the likelihood or severity of the event and risk rank the scenarios that can lead to adverse consequences. In some instances, the probability qualitatively assigned to an adverse scenario with only procedural safeguards may be underestimated. This underestimation of probability can lead to a lack of urgency to implement engineered safeguards to prevent an adverse consequence.

In addition to PHAs, some companies voluntarily perform Layer of Protection Analysis (LOPA), which is a more detailed analysis than PHA. LOPA is used to semi-quantitatively assess whether sufficient safeguards exist to meet a target risk level. In some instances, teams may use LOPA in such a way as to show that adverse consequences are unlikely by crediting conditional modifiers that may not be applicable to the scenario. This can lead to a determination that additional safeguards are not necessary as risk tolerance has been met.

It may be prudent to establish rules to implement engineered safeguards in potentially severe scenarios that rely only on procedural safeguards. The purpose of this paper is to describe the establishment of a criterion in which potentially high severity scenarios are identified that rely primarily on procedural safeguards and how to add an engineered safeguards to reduce the likelihood of the scenarios occurrence.

2 Industry Incidents

Many incidents have occurred in the chemical process industry where a procedural safeguard was the main layer of protection.

The CSB has performed many incident investigations and provides recommendations to different organizations with the main purpose of learning from past events to enhance safety. A recent investigation focused on the CAPECO incident of October 23, 2009 when a gasoline storage tank was overfilled leading to the formation of a vapor cloud and explosion [3]. In Appendix B of that investigation report [3], CSB lists twenty-two major accidents that have occurred in the process industry since 1962 that involved hydrocarbon storage tank overfill. A majority of these accidents involved processes that did not have an automatic overfill protection system and only relied on human intervention to prevent an overfill.

Similar events have occurred at facilities that handle anhydrous ammonia, a highly hazardous material. Ammonia is a toxic material with a low boiling point that vaporizes upon release into the atmosphere. Inhalation of ammonia vapors can result in irritation of the mucous membranes of the nose, throat, and lungs. If a plume of ammonia vapors with a concentration above 5,000 ppm is inhaled, sudden death may occur [6]. Many refrigeration systems use anhydrous ammonia as the refrigerant in the United States. In ammonia refrigeration systems, compressors provide the motive force to circulate the ammonia. Compressors require oil for lubrication, which by design leaks into ammonia side of the system. Entrained oil is removed from the system from the oil pots where oil separates from ammonia and settles in the pot. In order to ensure proper ammonia refrigeration system cooling efficiency, these oil pots must be drained periodically to remove the oil on the ammonia side of the system. This is achieved by opening a drain valve into an open pale. System pressure moves the oil. The operators are required to be vigilant and close the valve as soon as oil composition starts to change indicating that the pot is drained of oil and ammonia may follow.

Releases of ammonia have occurred from ammonia refrigeration systems due to oil drain operations. An engineered safeguard that could prevent this type of release is a self-closing emergency stop valve. It is designed such that an operator must hold the valve open by holding on to it and when operator's hand is removed the spring loaded self-closing emergency stop valve immediately closes. If an emergency stop valve is installed on an oil drain valve, it can limit the amount of ammonia released to atmosphere in the event that the drain operation is performed too long and ammonia starts to release once the oil has been drained from the pot.

The accident at CAI, Inc. and Arnel Inc., [7] is another example of an accident involving a system without an engineered safeguard. The CSB performed an incident investigation on this accident and issued recommendations [7]. According to the CSB, an initiating cause of the explosion was determined to be that a steam valve was left open to a tank containing flammable liquid. This steam valve was periodically used to heat the contents of a tank that stored ink and then manually shutoff by an operator once the desired temperature of the material was reached. On November 22, 2006 the steam valve was forgotten open and the ink in the storage tank overheated and vaporized. The vapor cloud found an ignition source and a large explosion occurred. The CSB recommended that CAI, Inc. install safety controls that would prevent overheating of flammable or combustible liquids.

3 Hierarchy of Controls and its Application

3.1 Hierarchy of Controls

CCPS has published "Inherently Safer Chemical Processes, A Life Cycle Approach", which describes the importance of making chemical processes inherently safer [1]. The CCPS states the following, "A chemical manufacturing process is inherently safer if it reduces or eliminates the hazards associated with materials and operations used in the process and this reduction or elimination is permanent and inseparable" [1]. The CCPS has described methods of making chemical processes safer and an important concept of the guidelines is levels of inherent safety. The CCPS [1] describes the levels of inherent safety as follows from most effective to least effective:

First Order Inherent Safety Measures –
  Avoid Hazards
Second Order Inherent Safety Measures –
  Reduce Severity;
  Reduce Likelihood
Layers of Protection –
  Apply Passive Safeguards;
  Apply Active (Engineered) Safeguards;
  Apply Procedural Safeguards

It is important to note that the least robust layer of protection listed by the CCPS is the application of procedural safeguards. This means that the application of active safeguards is a more robust layer of protection than procedural safeguards. The highest or first order of inherent safety listed is avoiding hazards. This is defined as elimination of a hazard. The two second orders of inherent safety listed are reducing severity and reducing likelihood. The severity of a hazard can be reduced by a number of methods including reducing the quantity of the material in a system, reducing the pressure or temperature within a process, substituting the material for a less hazardous material, etc. The likelihood of a hazard's occurrence can be reduced by a number of methods including using more corrosion resistant material, redesigning a process from four vessels to requiring two vessels, etc. The next order of inherent safety after the first and second orders are layers of protection. Layers of protection are ranked in order as passive safeguards, active safeguards, and procedural safeguards with procedural safeguards being the least robust and passive safeguards being the most robust. An example of a passive safeguard is designing the vessel within a system with a maximum allowable working pressure that is higher than the highest pressure that can potentially be reached within the system. An example of an active safeguard is an automatic overfill protection system that shuts off flow to a tank when high level is reached in the tank. An example of a procedural safeguard is relying on operators to shut off the flow to a tank when a certain level is reached.

The AIHA also addresses hierarchy of control from the standpoint of occupational safety. The "Z10-2012: Occupational Health and Safety Management Systems" Standard, defines hierarchy of controls which can be summarized as follows:

A. Elimination;
B. Substitution of less hazardous materials, processes, operations, or equipment;
C. Engineering controls;
D. Warnings;
E. Administrative controls; and
F. Personal protective equipment

There are many similarities between the CCPS's inherent levels of safety and the AIHA hierarchy of controls. Similar to the CCPS, the AIHA hierarchy of controls considers engineering controls as more robust than human intervention, warnings and administrative controls. The AIHA hierarchy of controls is written for occupational safety whereas CCPS's hierarchy of controls is written specifically for process safety applications. In both hierarchy of controls, the first item listed is eliminating a hazard. The second order of inherent safety listed by the CCPS is "reduce severity, or reduce likelihood", which is very similar to the second hierarchy of control listed by AIHA. Substituting a less hazardous material translates to reduced severity. Substituting less hazardous operations can be considered as a measure that reduces likelihood. The third hierarchy of control listed by AIHA is "engineering controls", which can be considered the same as the first two layers of protection listed by the CCPS, "Apply passive safeguards, Apply active safeguards". The third layer of protection listed by the CCPS, "Apply procedural safeguards" can be considered a combination of the fourth and fifth hierarchy of controls listed by AIHA, "Warnings; Administrative controls". The one major difference is that the CCPS does not call out alarms specifically as the AIHA does with the *warnings* hierarchy of control. The CCPS's inherent safety measures refer to alarms in the procedural controls as all alarms require personnel intervention.

An example of an engineered control referred to by AIHA in regards to occupational safety is a shutoff system for an automated laser welding machine that shuts off immediately upon detection of limbs within the welding area. This is an example of an engineered control designed to prevent personnel exposure to laser burns. An automatic overfill protection system that is considered an engineered safeguard by the CCPS for process safety applications is also considered an engineering controls according to the AIHA hierarchy of controls.

Regulatory bodies involved with PSM and RMP have taken notice of the importance of hierarchy of controls. Current drafts of California's Process Safety Management [8] and Risk Management Plan [9] regulations for refineries will specifically require hierarchy of hazard controls to be formally addressed. The draft regulations state that an analysis must be performed on all existing processes at a refinery that pose a process safety hazard. The analysis team must document the possible inherent safety measures for each hazard that can be implemented to reduce risk. The analysis team must review the feasibility of each level of inherent safety and must document why an inherent safety measure cannot be implemented for the process if that decision is made.

3.2 Application of Hierarchy of Controls

The most preferable time period to apply the hierarchy of controls to a system is during its design phase when changes can be made without severe economic impact. In addition to the design phase of a system, the hierarchy of controls can be applied throughout a system's life cycle when changes are considered. However, the upcoming regulations in California will be requiring application of the concept to built and operated systems.

To be able to apply hierarchy of controls to a system, it is only rational to establish what hazardous scenarios are possible and what safeguards are in place as protection. Process Hazard Analysis (PHA) provides a systematic method to identify and rank potential hazard scenarios and to identify the layers of protection in place to minimize the risk. By conducting the PHA, one can claim that all potential hazard scenarios that a team of experts could identify have been addressed systematically.

The hierarchy of controls can be used during a Process Hazard Analysis (PHA) to determine the adequacy of available safeguards to prevent each release scenario identified. This can be completed by identifying the available safeguards for scenarios that are postulated to have severe consequences and questioning whether or not inherently safer measures can be implemented. In addition, the hierarchy of controls can be applied when a recommendation is made during a PHA. The hierarchy of controls can act as guidance to choosing a higher level control measure to reduce the risk of a scenario. The hierarchy of controls can be reviewed while making any major changes to a system to assess if there are inherently safer options than the change being proposed. The hierarchy of controls can also be reviewed during an incident investigation to determine if a higher order of control should be implemented to prevent the recurrence of an incident.

The thought process of applying hierarchy of controls for a gasoline producer may be the following:

1. Currently there is a need for gasoline as vehicle fuel since a majority of automobiles sold in the United States require gasoline as the energy source. Therefore, this is a viable business venture that society at large needs. Eliminating gasoline from the line of products produced and sold by the company is not a prudent business decision because its safety risk is relatively well understood and has historically been accepted by society. Therefore, eliminating the hazard (i.e., moving out of gasoline business) as the very first item of the hierarchy of control is not a prudent option.
2. The next hierarchy of controls addresses reducing severity and likelihood. In order to reduce the severity of a gasoline storage tank related mishap (e.g., overfill and vapor cloud explosion), the hazard radius of a vapor cloud would need to be reduced. This can be done by reducing the pool surface of released material, as an example. Pool area is a function of the berm walls around the storage tanks and height of the walls. As another example, the owner may decrease the flow rate into a tank allowing the operators more time to discover and shutoff the inflow. These changes may not be feasible or the risk reduction gained from them may be smaller than other options (e.g., automatic shut-off on high level).

3. In order to reduce the likelihood of a gasoline storage tank overfill, the number of times the storage tank filling operation is performed can be reduced. This will require larger storage tanks which may increase the quantity of gasoline release in case of tank failure. Thus, the risk reduction gained from decreasing frequency of fills may be offset by the increase in risk due to larger tank size.
4. Engineered controls are the next available hierarchy of control. The first type of layer of protection based on CCPS's hierarchy of controls is a passive safeguard. One commonly used passive safeguard for large hydrocarbon storage tanks is secondary containment (a berm). The purpose of a berm is to contain a spill. However, in the case of gasoline, the light ends within the mixture would vaporize resulting in a vapor cloud formation with the potential for an explosion. As discussed above, reducing the pool size may reduce the evaporation rate but better risk reduction can be obtained by other means.
5. The next hierarchy of control is an active safeguard such as an automatic overfill protection system that will shut off the pump filling the storage tank when high level is detected. For a built and operated system, this option may provide the best risk reduction.
6. Procedural safeguards are the next hierarchy of control after active safeguards. The procedural safeguard for a gasoline storage tank filling operation would be operator monitoring of the level on the storage tank, installation of a high level alarm, and shutting off the feed pump manually at a certain level.

4 Process Hazard Analysis

4.1 Use of Process Hazard Analysis

A Process Hazard Analysis (PHA) is a hazard analysis tool that must be performed for systems covered under the PSM and RMP regulations [4 and 5]. PHA is a systemic tool for identifying the hazards of a system. PHA also provides a means to establish the potential severity of each hazard scenario, the available safeguards to prevent the hazard and likelihood of the scenarios occurrence. By applying risk ranking protocols, the risk gap of each scenario is identified in the PHA process.

The most common PHA method is HAZOP, which is typically performed in a brainstorming session with the participation of team members with experience in different fields necessary to properly analyze a system.

An important part of a PHA is establishing the risk of a hazard scenario. Risk is defined as "the combination of the expected frequency (events/year) and severity (effects/event) of a single incident or a group of incidents" [10]. Before a PHA study commences, the management responsible for the system must decide what risks they are willing to accept and what risks they will consider unacceptable. The acceptable and unacceptable risks are then translated into a risk tolerance matrix. The risk tolerance matrix contains severity and likelihood pairs that are acceptable and severity and likelihood pairs that are considered unacceptable and require risk reduction. This risk matrix is then used by PHA teams in all the PHA studies owned and operated by the management. The risk matrix provides a basis to identify risk gaps and whether or not system upgrades should be considered to improve safety.

4.2 Augmenting PHAs with Hierarchy of Controls

PHA results can vary from team to team. People working in the chemical process industry have different career experiences, which can result in different risk estimations during a PHA. It is possible and the authors' experience in leading PHA studies corroborates this observation that the likelihood of a severe consequence may be underestimated by PHA Teams. For example, operators who work with a system on a daily basis and have not experienced any incidents may be under the impression that the probability of a specific high severity scenario occurring is very low. Operators may feel that a system has been performing well for many years and if an issue did arise, the operators are knowledgeable enough to stop the progression to the bad event. Also, some PHA team members may believe that certain initiating causes of hazardous consequences are very unlikely based on their experience. Many of these issues can lead to underestimation of the risk of a severe consequence. Therefore, it is possible that the risk of a high severity scenario is estimated within the acceptable range allowing reliance on procedures based safeguards with no other safeguard. Due to the fact that risk ranking issues such as the one described above exist, it may be prudent for companies to use hierarchy of controls as the bases to require that all high severity scenarios to include at least one independent layer of engineered safeguard protection.

5 Layer of Protection Analysis

Layer of Protection Analysis (LOPA) is a tool that can be used to semi-quantitatively estimate the likelihood of a high severity scenario. Typically, LOPA is applied to the high severity scenarios identified during a PHA in order to enhance the accuracy of scenario likelihood. As part of the LOPA process, typically the LOPA Team estimates probabilities on an order of magnitude basis. The values that are estimated are the initiating cause likelihood, failure probabilities of the safeguards, occupancy factor, ignition probability, time at risk, and other conditional modifiers [11]. The initiating event likelihood is an estimate of how likely it is that the initiating cause occurs that can initiate the postulated chain of events. For example, according to the CCPS [12] the probability of a basic process control system (BPCS) failure is 0.1 per year. Also, if the safeguard has special features (e.g., it is designed to be of high reliability) the team may assign 0.01 probability of failure.

Occupancy factor refers to the amount of time personnel are within the hazard zone. For example, the CCPS [13] states that the probability of one person being present in a hazard zone is 0.1 for an area where operators spend 17 hours on average per week. Ignition probability refers to the probability that a released flammable material finds an ignition source resulting in a fire or explosion. Ignition probabilities will vary based on many factors such as the properties of the released material in terms of the released temperature, hot equipment near the release point, etc. Time at risk factor applies to non-normally operating modes of a system. Other conditional modifiers may also be used depending on the scenario. For example, some companies may take a conditional modifier stating that a catastrophic compressor seal failure due to liquid being sent to the suction side may occur once every ten times.

LOPA teams may credit additional conditional modifiers and enabling conditions resulting in a low risk level within the acceptable range. One example is crediting a time at risk factor for scenarios that can only occur during startup. Another likelihood reduction can be achieved through the application of occupancy factor for scenarios where generally personnel are not present within the hazard zone. Application of these factors raises an important question about the possibility of reliance on procedural safeguard as the only safeguard against a high severity consequence. Similar to the PHA, management may mandate that for all high severity scenarios, the LOPA team should consider an engineered safeguard regardless of the likelihood and risk ranking.

Startup is an interesting case as a safety risk contributor. The CCPS states that time at risk is not advised to be used during startup, "Use of a time-at-risk enabling condition in such a situation may lead to a conclusion that the average risk meets a facility's risk criteria; however, the peak risk imposed on persons exposed to the hazard may be inappropriately high" [13].

6 Limitations and Benefits

6.1 Increased Complacency Potentially Leading to Adverse Consequences

A potential adverse outcome of adding engineered safeguards to a system is an increase in operator complacency, which may lead to an adverse consequence. Operators involved in tasks such as storage tank filling may become complacent if an engineered safeguard is available as a backup in case they do not shut off the filling operation in a timely manner. Many engineered safeguards are meant to serve as a final barrier to prevent the occurrence of a release. The knowledge that a final barrier such as an engineered safeguard is available may lead to personnel relying on that safeguard and not being as attentive to system operation as before the addition of the engineered safeguard. Increased complacency may unintentionally lead to an adverse consequence as typical engineered safeguards have a reliability of approximately 90%. This means that approximately 10% of the time that they are required to prevent an adverse consequence, they may not work. If operators become complacent and rely on engineered safeguards to prevent incidents, an incident may occur if the engineered safeguard does not function as required. For this reason, it must be stressed to personnel prior to installation of engineered safeguards that they are not meant to be a replacement of a safeguard (i.e., engineered safeguard replacing a procedural safeguard). Adding engineered safeguards to a system must be understood to be a measure to reduce risk significantly, which it cannot accomplish if it only replaces a procedural safeguard.

6.2 Engineered Safeguards Are Not Perfect

Engineered safeguards should not be considered as means of eliminating the potential for an adverse consequence. Engineered systems are not guaranteed to work every time they are needed. Incidents have occurred in the chemical industry in which an engineered safeguard did not work. One such incident occurred on August 14, 2002 in which a chlorine hose ruptured during a chlorine transfer and resulted in a release of over 48,000 pounds of chlorine [14]. The chlorine transfer system did have automatic shutoff valves that were supposed to activate if chlorine was detected at nearby chlorine detectors. The chlorine detectors did sense chlorine and sent a signal to the automatic shutoff valves to close. However, all automatic shutoff valves did not close as they had experienced some internal corrosion. The automatic shutoff system was not tested properly as part of the Mechanical Integrity Program. Management at the site was not aware of the corrosion issue within the system and did not take proactive measures to ensure the automatic shutoff system activates properly in the event that it was needed.

Another example of an engineered safeguard failing to work is the explosion at the Buncefield Oil Storage Depot on December 10, 2005 [15]. A gasoline storage tank was overfilled and the automatic overfill protection system did not activate as it was not placed into service properly after its previous test. Maintenance errors such as this can happen and it is important for personnel working with engineered safeguards to understand its functionality, and test and maintenance procedures in order for it to activate when needed.

The above two examples of an engineered safeguard failing to activate should not lead to a conclusion that adding engineered safeguards to a system does not reduce risk. Engineered safeguards do indeed reduce risk if the Maintenance Program of the system to which it is added is updated to include the proper procedures and testing frequency recommended by the manufacturer of each piece of equipment associated with the engineered safeguard. It should also be noted that personnel must be aware of the functional details of the engineered safeguard system to ensure that it is operable so that it can prevent an accident when needed as it is designed to.

6.3 Resources

Adding engineered safeguards to a system requires resources from many departments. Engineered safeguards added to an existing system may need to be designed according to the Instrumentation, Systems, and Automation Society (ISA) Standard 84.00.01-2004 [16]. The ISA 84 Standard has many detailed requirements for installing Safety Instrumented Systems (SIS). In order for these requirements to be properly met, input is needed from many different groups within a company.

6.4 Financial Aspect

Adding engineered safeguards may prevent financial losses incurred due to major accidents. Some may argue against adding engineered safeguards to a system that has been found to be of low risk but high consequence severity due to the immediate financial strain it may have on a company. It is true that adding an engineered safeguard requires some financial investment and the system must be included in the periodic test and maintenance program. However, a major accident can cost millions of dollars in monetary losses and other non-tangible impacts (e.g., personnel injuries and reputation impact). For example, in 1999 an explosion occurred due to a gasoline storage tank overfill at the Thai Oil Company that cost $22.3 million and resulted in seven fatalities [3]. The storage tank only had a high level alarm, which was not heard by onsite personnel. The automatic overfill prevention system would have cost a small fraction of that monetary loss, not to mention the loss of life that was experienced.

An opinion that is increasingly being accepted by industry management and safety professionals is that safety is an investment, not a cost. According to OSHA's Business Case for Safety Webpage [17], "Employers that implement effective safety and health management systems may expect to significantly reduce injuries and illnesses and reduce the costs associated with these injuries and illnesses, including workers' compensation payments, medical expenses, and lost productivity." We can extrapolate this statement to process safety. "Z10-2012: Occupational Health and Safety Management Systems" Standard [2] is generally accepted as being a very good guidance document to implementing an effective safety and health management system. A very important section within that standard is the one on hierarchy of controls. Therefore, implementing a system in which engineered safeguards are added for high severity scenarios where only procedural safeguards exist would be a step towards improving process safety. This can result in a decrease in incidents at a chemical processing facility, which can have a direct positive impact on facility profitability.

Adding engineered safeguards to a system does not necessarily mean that future accidents will be prevented. However, it will reduce the likelihood of a major accident occurring at a facility. The investment required to add and maintain an engineered safeguard can be significantly less than the costs associated with an accident.

7 Conclusion

In the chemical process industry, many potentially high severity tasks exist for which procedural safeguards are the main layer of protection. The hierarchy of controls guidance developed by the process industry ranks engineered safeguards as a more robust layer of protection than procedural safeguards. Multiple chemical process industry accidents have occurred in which the procedural safeguard layer of protection failed and an engineered safeguard did not exist. These accidents could have been prevented if the associated system had an engineered safeguard as a layer of protection in addition to the procedural safeguard.

The PHA and LOPA methodologies are tools that can be used to identify high severity scenarios and the existing safeguards. It is possible, to apply these methods properly and arrive at a low risk conclusion for high severity scenarios that solely rely on a procedural safeguard. It is recommended in this article that these methods be augmented by owners and operators responsible for highly hazardous operations with a mandate that all high severity scenarios to include at least one engineered safeguard. This should ensure that regardless of the estimated likelihood of the scenarios, the operators are not the only safeguard preventing the high severity release.

References

1. *Inherently Safer Chemical Process: A Life Cycle Approach.* Center for Chemical Process Safety, American Institute of Chemical Engineers, New York, NY, 2nd Edition, December 2008.

2. *Z10-2012: Occupational Health and Safety Management Systems*. American Industrial Hygiene Association, Falls Church, Virginia, June 2012.

3. *Final Investigation Report – Caribbean Petroleum Tank Terminal Explosion and Multiple Tank Fires, Bayamon, Puerto Rico, October 23, 2009*. U.S. Chemical Safety and Hazard Investigation Board, Report 2010.02.I.PR, October 2015.

4. *Process Safety Management of Highly Hazardous Materials*. Occupational Safety and Health Administration, Federal Code of Regulations, 29 CFR, §119.119.

5. *Risk Management Programs for Chemical Accidental Release Prevention*. U.S. Environmental Protection Administration, 40 CFR Part 68.

6. *Safety Data Sheet (SDS) for Anhydrous Ammonia*. Hill Brothers Chemical Company, HBCC SDS No. CA10000, Revision Issued: 5/14/2015.

7. *Investigation Report – Confined Vapor Cloud Explosion – CAI, Inc. and Arnel Company, Inc. – Danvers, Massachusetts*. U.S. Chemical Safety and Hazard Investigation Board, Report 2007-03-I-MA, May 2008.

8. *Process Safety Management for Petroleum Refineries*. California Occupational Safety and Health Administration, California Code of Regulations, Title 8, §5189.1, Draft Version 5.0, September 14, 2015.

9. *California Accidental Release Prevention (CalARP) Program*. California Governor's Office of Emergency Services. Title 19 of CCR Division 2, Chapter 4.5, Draft, September 24, 2015.

10. *Guidelines for Hazard Evaluation Procedures*. Center for Chemical Process Safety, American Institute of Chemical Engineers, New York, NY, 3rd Edition, April 2008.

11. *Layer of Protection Analysis – Simplified Process Risk Assessment* Center for Chemical Process Safety, American Institute of Chemical Engineers, New York, NY, October 2011.

12. *Guidelines for Initiating Events and Independent Protection Layers in Layer of Protection Analysis*. Center for Chemical Process Safety, American Institute of Chemical Engineers, New York, NY, February 2015.

13. *Guidelines for Enabling Conditions and Conditional Modifiers in Layer of Protection Analysis*. Center for Chemical Process Safety, American Institute of Chemical Engineers, New York, NY, November 2013.

14. *Investigation Report – Chlorine Release – DPC Enterprises, L.P.* U.S. Chemical Safety and Hazard Investigation Board, Report No. 2002-04-I-MO, May 2003.

15. *Buncefield: Why Did It Happen?*. United Kingdom Health & Safety Executive, February 2011.

16. *ANSI/ISA-84.00.01-2004: Functional Safety: Safety Instrumented Systems for the Process Industry Sector*. Instrumentation, Systems, and Automation Society (ISA), September 2004.

17. https://www.osha.gov/dcsp/products/topics/businesscase/. Occupational Safety and Health Administration, Business Case for Safety, Accessed on February 8, 2016.

12TH GLOBAL CONGRESS ON PROCESS SAFETY

Maintaining Consistency in the Determination of Consequence Severity Rankings for Process Hazard Analyses

Kirk Busby
Kazarians & Associates, Inc.
100 West Broadway
Glendale, CA 91210
Kirk.Busby@kazarians.com

Prepared for Presentation at
American Institute of Chemical Engineers
2016 Spring Meeting
12th Global Congress on Process Safety
Houston, Texas
April 11-13, 2016

Maintaining Consistency in the Determination of Consequence Severity Rankings for Process Hazard Analyses

Kirk Busby
Kazarians & Associates, Inc.
100 West Broadway
Glendale, CA 91210
Kirk.Busby@kazarians.com

Keywords: Hazard Identification and Risk Analysis, Process Safety, Consequence Modeling

Abstract

A Process Hazard Analysis (PHA) is an important tool for identifying and managing the hazards associated with a chemical process. One of the most critical aspects of the analysis is the risk ranking of scenarios, which is used to determine which events pose a high risk and may require further mitigation. The risk ranking of a scenario is built on two factors – severity and likelihood. The severity of a scenario is a measure of the potential worst-case impact on facility operation, facility personnel, the public, or the environment, while the likelihood is an estimate of the probability with which an event may occur, given the presence of the safeguards acting to prevent or limit the severity of the event.

Severity of a scenario is generally expressed in qualitative terms. While a typical approach to assigning consequence severity rankings involves categorizing events by impact (e.g., recordable injury, permanent disability, or fatality), this approach relies on the subjective judgment of the analysis team in determining the exact impact of a specific event. Experience has shown that, because of the subjective nature of this step, different PHA study teams have had varying opinions on the potential severity level of a given scenario. It is the aim of this study to provide guidelines to improve consistency of severity rankings both within a single analysis, and across multiple analyses with separate PHA study teams. It is possible to increase the consistency in consequence severity rankings through application of standardized guidelines. A method has been developed to aid PHA facilitators based on dispersion modelling of various conditions typically encountered in a PHA. The underlying data is based on groups of chemicals with similar properties and other conditions (e.g., break size, break location, etc.).

1. Introduction and Background

Process Hazard Analysis (PHA) is an important tool for identifying and managing the hazards associated with a chemical process. The goal of a PHA is to use a systematic methodology to identify each potential hazard source within the process. A typical PHA methodology is the guideword-style Hazard and Operability (HAZOP) study. To ensure completeness, the HAZOP methodology breaks down a complex system into individual nodes, making it easier to thoroughly analyze each section of the process. Further, within each node, hazards are postulated according to a standard combination of guidewords and process deviations. For example, when the guideword 'No' is paired with the deviation "Flow," all scenarios within the node which lead to a loss of flow are considered.

Once a scenario is identified, the risk posed to the facility must then be determined. This is normally achieved through the risk ranking process. To determine the risk ranking, each scenario is assigned a severity and a likelihood level. This pair is then used in a risk ranking matrix, which defines the risk targets developed by company management prior to the start of the PHA. Assigning a risk ranking is one of the most important facets of the PHA process, as it is used to determine which risks are acceptable given the safeguards currently in place, and which require further actions to ensure that the risk is maintained at a tolerable level.

Currently, there is a large potential for variance in risk ranking determination, specifically with regard to consequence severity. While severity determination is meant to be a qualitative ranking, there are a number of benefits to improving the consistency of severity rankings among similar release scenarios. Kazarians & Associates, Inc. has developed a method that has proven effective in reducing variance among consequence severity rankings through the use of pre-determined dispersion modeling results.

2. Current Methods for Consequence Severity Determination

It is common practice to estimate the worst case consequence severity for the identified potential release scenarios. For this, it is assumed that all existing active safeguards are unavailable. Typically, the severity ranking is a qualitative determination made by the PHA study team based on their understanding of the potential impact a given scenario may have on the safety of facility personnel and the public, as well as the environment or business operations. The text provided in defining each severity level is generally used to determine the consequence severity. Table 1 provides an example of severity level definitions available to the PHA study team.

Definitions such as those in Table 1 can provide a good starting point for consequence severity selection. However, use of these limited definitions can result in inconsistent severity rankings, both within an individual PHA, as well as among PHAs. This variation stems from the fact that the PHA study team generally carries the chain of events to the release condition (i.e., flange leak, pipe rupture, etc.) whereas the definitions of consequence severity level is often in terms of level of harm (i.e., first aid injury, fatality, etc.). This then requires the study team to envision how the release will ultimately result in harm, which means that they have to take into consideration factors such as source conditions, dispersion, and ignition, which are often complex phenomena by themselves. Therefore, this leads different PHA study teams to hold varying opinions of what the impact of a certain event may be. Typically, these opinions are based on the experience of the personnel present during the PHA.

Table 1. Typical Consequence Severity Guidance

Consequence Severity	Impact on Personnel	Impact on the Environment	Impact on business
1	Minor first aid injury	Slight spill onto soil, release to flare	< $10,000
2	Recordable injury	Small spill onto water, moderate spill onto soil	< $100,000
3	Permanent disability	Moderate spill onto water, large spill onto soil	< $500,000
4	Fatality	Moderate spill into an environmentally sensitive area	< $1,000,000
5	Multiple fatalities	Large spill into an environmentally sensitive area	< $5,000,000

The following example is taken from PHAs conducted for chlorination systems at two facilities operated by the same organization. The example demonstrates the difference operational experience can make in the severity ranking of a given scenario. During PHA sessions for the two systems, the PHA team encountered the same scenario in which a storage tank is exposed to an external fire, resulting in a pressure increase within the tank, ultimately resulting in tank rupture. In this scenario, the two teams have arrived at an identical endpoint (i.e., release scenario). While the release conditions (i.e., temperature and pressure) will be the same for both cases, the first team concludes that a single fatality may result, the second believes that multiple fatalities may result. The two teams may record the scenarios according to Figure 1.

In each case, the PHA study team has arrived at the same chain of events leading up to an integrity failure of the storage tank, and release of chlorine into the atmosphere. However, the final consequence statement and therefore risk rankings vary based on the estimated severity of the release.

Inconsistency in severity rankings can cause a number of problems following a PHA. Most importantly, they can lead to an over- or under-estimation of the risk a scenario poses. A consequence with the potential to cause severe injury that has only been identified as a first aid issue could result in insufficient safeguarding to protect against the event. This could ultimately lead to major safety incidents during the operating lifetime of the process. Safeguards are typically added when the risk ranking of a scenario does not meet the pre-determined corporate risk target. A lower severity ranking may indicate that the appropriate safeguards are in place, when in reality an additional protection may be necessary.

Team 1:

Causes	Consequences	Safeguards	S	L	RR	HAZOP Recommendations
Greater than 140°F in the Chlorination Room caused by a fire	▪ Pressure buildup in the Chlorine Storage Tanks. ▪ Potential for loss of containment of a Chlorine Storage Tank if the relief valve fails to lift. Release of chlorine into the Chlorination Room. Potential for personnel exposure.	▪ Relief Valves will lift to prevent overpressure. ▪ Detectors alarm locally and on SCADA. ▪ Scrubber activated by detectors at 3 ppm. ▪ Plant housekeeping practices prohibits storage of any combustible materials in the Chlorination Room.	4	D	M	

Team 2:

Causes	Consequences	Safeguards	S	L	RR	HAZOP Recommendations
Fire inside the Chlorine Building.	Potential for increased pressure in the storage tanks. Potential to rupture the storage tank. Potential for release of chlorine inside the building. Potential for personnel and public exposure.	1. PSVs on the storage tanks set at 225 psig. 2. Chlorine detectors alarm at 1 ppm. Scrubber system will start at 3ppm on any one detector. 3. Good housekeeping practices.	5	E	M	No recommendations necessary. Safeguards are considered adequate.

Figure 1. Identical PHA Scenarios with Differing Consequene Statements and Risk Ranking

A second and more subtle consequence of inconsistent severity ranking is the potential discrepancy between processes or facilities. If two PHA study teams assign different severity rankings to similar events on two similar systems, there is a potential for one system to require an extra safeguard to be installed which is not in place on the second system. This discrepancy between units could potentially lead to confusion later in the life of the plant. If a group of operators is responsible for both systems, there could be confusion as to which setup is correct, as well as the incorrect assumption that both systems are configured in the same manner. Additionally if an incident occurs in the system with fewer safeguards, there is a potential for increased regulatory scrutiny and fines due to the fact that the second system is not as well protected as the first.

3. Improving Consistency through Dispersion Modeling

To reduce variation among consequence severity rankings, it is possible to provide PHA teams with improved guidance on the potential impacts of a release. Dispersion modelling is a tool which is commonly used to determine the hazardous area resulting from a specific chemical release scenario. There are a number of sophisticated dispersion models and software packages currently on the market which can provide a picture of the hazard distance presented by various chemical release events. Many of the commercially available packages allow the user to input specific data such as the temperature, pressure, quantity, and release orifice size, and calculate the released quantity, pool size, vaporization rate, and resulting hazard distance. Depending on the model chosen, the user may be able to customize the scenarios to a greater degree, including such factors as terrain type, degree of confinement, explosive characteristics, etc.

Currently, some PHA teams already use dispersion modeling to assist in performing a more in-depth analysis for select scenarios with complex release characteristics. These studies are useful, but require a large amount of input data to be determined by the PHA team in order for them to generate accurate results. Therefore, it is much too time consuming to use such an analysis for every release scenario postulated during a PHA.

To combine the time and cost efficiency of PHA with the increased accuracy afforded by the use of dispersion modeling, the dispersion calculations can be performed generically prior to beginning the PHA discussions. To utilize this method, a database is developed for each hazardous chemical which may be released, and dispersion calculations are performed for the range of temperatures, pressures, phases (liquid or gas), and break sizes which may be encountered during the PHA. Every possible combination of parameters does not need to be considered, as the PHA team may quickly interpolate between results when needed. With a pre-developed matrix in front of them during the PHA sessions, participants can quickly relate a set of known release conditions to an approximate hazard zone, which will allow them to choose a more refined consequence severity. Some large chemical processing and manufacturing corporations have already begun to adopt such methods to improve the consistency of their PHAs.

The method developed by Kazarians utilizes PHAST modeling software by Det Norske Veritas [Reference 1] to create databases for both flammable and toxic chemicals. Specifically, the common alkanes (i.e., methane through hexane), hydrogen, hydrogen sulfide, anhydrous ammonia, and chlorine have been analyzed, in both liquid and gaseous form. In each case, a generic outdoor release event is postulated for varying temperatures, pressures, spill quantities, and release sizes. From this basic input data, PHAST is capable of generating information on the maximum dispersion distance, the flammable range, the range of explosive overpressure, and toxic effects. For each release event, the maximum distance of each hazard type is compared, and the greatest distance is selected. For example, for a specific set of release conditions, the maximum distance at which an injury may occur due to thermal radiation is 200 feet, while the maximum injury distance due to explosive overpressure is 250 feet. For these release conditions, 250 feet will be selected and recorded as the worst-case injury hazard distance. The final result is a set of eight tables for each chemical, with four liquid phase tables and four vapor phase tables. Tables 2 and 3 below show the four hazard distance tables for methane in the liquid phase.

Table 2A. Methane Liquid Injury Hazard Distance (ft)

	Source Temperature (F)			
Release (lbs)	**-250**	**-200**	**-150**	**-117**
10	110	100	140	160
100	350	230	320	350
1,000	610	510	710	780
10,000	1,520	1,130	1,570	750
100,000	4,060	2,480	3,350	3,350
500,000	7,690	4,260	5,360	5,360

Table 2B. Methane Liquid Injury Hazard Distance (ft)

	Source Pressure (psi)				
Break Diameter (in)	**50**	**150**	**300**	**400**	**450**
0.25	20	30	30	30	30
0.5	110	110	110	110	110
0.75	180	210	210	210	210
1.0	210	280	310	310	310
1.5	280	410	480	510	510
2.0	380	540	640	670	700
3.0	510	770	970	1,070	1,100
4.0	610	1,000	1,330	1,430	1,460
6.0	840	1,430	2,020	2,180	2,250

Table 3A. Methane Liquid Fatality Hazard Distance (ft)

	Temperature (F)			
Release (lbs)	**-250**	**-200**	**-150**	**-117**
10	40	40	60	70
100	240	90	130	150
1,000	530	210	310	340
10,000	1,390	470	690	1,690
100,000	2,990	1,070	1,500	1,500
500,000	5,290	1,860	2,400	2,400

Table 3B. Methane Liquid Fatality Hazard Distance (ft)

	Pressure (psi)				
Break Diameter (in)	**50**	**150**	**300**	**400**	**450**
0.25	20	20	20	20	20
0.5	40	50	50	50	50
0.75	110	140	140	140	140
1.0	140	210	240	240	240
1.5	210	340	400	440	440
2.0	300	470	570	600	630
3.0	440	700	890	990	1,030
4.0	530	930	1,260	1,350	1,390
6.0	760	1,350	1,940	2,110	2,170

Tables 2 and 3 present two sets of hazard distances. One represents the distance inside which an injury is expected to occur, and one inside which it is possible for fatalities to occur. Additionally, the sets have been subdivided into Tables A and B, which provide dispersion information for Release Quantity vs Temperature, or Break Diameter vs Pressure. The PHA team may elect to use either of the tables based on what information is available for a given scenario. For example, Break Diameter vs Pressure tables may be more useful for a pipe break scenario, while Release Quantity vs Temperature may be easier to use in the event of a seal failure or liquid overfill scenario.

There are a few notable trends which can be identified from Tables 2 and 3. Specifically, it can be seen that in the tables relating Break Diameter and Pressure, the maximum hazard distances reach a constant value as pressure increases. This is due to the release reaching what is known as "sonic velocity." [Reference 2] When a material under pressure is released through an orifice, the velocity of the release is initially dependent on the pressure within the vessel. Higher release pressure will typically result in a higher release velocity. However, once the release velocity has reached sonic velocity, further increases in pressure will no longer cause the velocity to increase. Put simply, no more material can fit through the release orifice, no matter how much force is behind it.

4. Defining Exposure Thresholds

The final dispersion distances reported in Tables 2 and 3 can vary greatly depending on what thresholds values are chosen for injuries and fatalities. This analysis used three potential sources of impact to the population: overpressure due to explosion, thermal radiation due to fire, and exposure to toxic vapors. Each threat has a unique hazard zone that is determined by its effects on both the human body and surrounding structures.

There is a great deal of research performed on the various effects on the human body caused by exposure to overpressure, thermal radiation, or toxic chemicals [References 3 through 10]. From this research, threshold values can be estimated to determine the exposure limits for injuries and fatalities. Table 4 presents the threshold values used in this article to differentiate between effects.

Table 4. Exposure Threshold Definitions

Explosion Overpressure	
Effect	**Threshold Value**
Injury	0.5 psi
Fatality	2 psi
Thermal Radiation	
Effect	**Threshold Value**
Injury	5 (kW/m^2)
Fatality	10 (kW/m^2)
Toxic Exposure	
Effect	**Threshold Value**
Injury	ERPG-2
Fatality	ERPG-3

It is important to note that these values represent conservative estimates of the potential impact an event may have on a person. For example, the threshold values for thermal radiation exposure are time dependent. Exposure to 10 kW/m^2 will not result in an immediate fatality, but prolonged exposure of approximately 60 seconds could result in death. This value is chosen as it represents the possibility of a fatality since it requires rapid evacuation from the hazard zone.

For each scenario modeled in PHAST, the distances to each threshold value described in Table 4 are compared, and that which results in the largest hazard distance is selected as the worst-case hazard zone. Toxic effects are only calculated for those chemicals which pose a significant risk when inhaled or contacted.

5. Using Hazard Distance Tables During PHA Sessions

Once developed, hazard distance tables can be used during PHA discussions to quickly determine an appropriate severity ranking based on the potential impact to facility personnel and the public. To convert the hazard distance tables into a form which may be more easily used during the PHA, the PHA facilitator may review the company risk matrix, as well as the facility layout prior to starting the PHA sessions. During this time, the facilitator should convert the dispersion distances in Tables 2 and 3 into charts providing guidance on the potential worst-case severity of different release events.

For example, a hypothetical company has developed a risk matrix with the consequence severity rankings described in Table 1. The system under analysis is an LNG storage tank and associated piping, which are located in an isolated part of the facility, but approximately 700 feet from the closest residences. Using this information, the PHA facilitator may translate the worst-case hazard zones in Tables 2 and 3 into a set of tables with suggested consequence severity rankings which are tailored to the system under review. Table 5 provides an example of these consequence severity tables for the specific LNG system described above.

Table 5A. Methane Liquid Consequence Severity Rankings

	Source Temperature (F)			
Release (lbs)	**-250**	**-200**	**-150**	**-117**
10	1	1	2	2
100	3	2	2	3
1,000	4	3	3	4
10,000	4	4	4	5
100,000	5	5	5	5
500,000	5	5	5	5

Table 5B. Methane Liquid Consequence Severity Rankings

	Source Pressure (psi)				
Break Diameter (in)	**50**	**150**	**300**	**400**	**450**
0.25	1	1	1	1	1
0.5	1	1	1	1	1
0.75	2	2	2	2	2
1.0	2	2	3	3	3
1.5	2	3	4	4	4
2.0	3	4	4	4	4
3.0	4	4	4	5	5
4.0	4	5	5	5	5
6.0	5	5	5	5	5

When translating the calculated hazard zone to consequence severity, it is important to note that the dispersion modeling calculations are performed generically, and their results may need to be adjusted based on the specifics of the system under review. The relationship between hazard distance and similar consequence categories (i.e., first aid injury, reportable injury, and permanent disability) should be developed with the primary decision maker for the facility (i.e., company management). To accomplish this, the facilitator should record the decision making process used to relate hazard distances to consequence severity levels, and present these decisions to company management prior to beginning the PHA sessions, and adjust the severity tables based on comments received. .

Additionally, the model calculations make a number of assumptions which may or may not apply to the system being analyzed. Most importantly, the calculations assume that releases occur outdoors, and outside of any secondary containment areas. Indoor releases can have a significantly reduced hazard zone outside of the building. This may depend on the ventilation capabilities of the building, or doors normally kept open. Toxic material may build to higher concentrations within the building than those reported in the hazard distance tables. Similarly, the presence of secondary containment limits the surface area of a liquid pool, which may result in a smaller vapor cloud than an uncontained liquid spill. To adequately assess the severity of a chain of events, these deviations from the assumed model must be accounted for. In cases such as these, the hazard distance tables should be used as a starting point, but adjust their consequence severity ranking according to the physical systems present in the field.

During the PHA discussions, once a scenario has been identified and the release conditions are established, the consequence severity tables can be consulted to assist in selecting the most appropriate consequence severity ranking. To use the tables, a process engineer or other person familiar with the design of the system under consideration should determine the physical parameters of the release. It must first be determined whether the released material is in the liquid or the vapor state. Subsequently, the release conditions must be specified in terms of temperature, pressure, quantity spilled, or breach diameter, depending on what information is most readily available to the team. This determination is not intended to be a rigorous calculation, but rather a quick estimation of the conditions of the release.

Once the release parameters have been agreed upon, the team may then use the consequence severity tables to determine the worst-case severity ranking. For example, in the LNG system described above, it is postulated that a full break occurs in a one inch diameter pipe containing LNG at 170 psig. Given the information at hand, the team consults Table 5B, and identifies 150 psig as the closest pressure rating found on the tables. From the Table, the team selects a severity ranking of 2, indicating that a recordable injury is the anticipated worst-case consequence.

6. Conclusion

Kazarians & Associates, Inc. has developed a method which uses pre-developed consequence severity tables to refine the judgement of PHA teams with regard to the impact that a scenario may have on personnel. This method has been applied, resulting in a more accurate determination of consequence severity. When applied across multiple teams and studies, this increase in accuracy will lead to consistency in severity ranking among similar PHA scenarios. In turn, this will lead to consistent application of safeguards across multiple systems, resulting improved plant operability and avoiding the potential for confusion among operating staff.

7. References

[1] Phast Modeling Software, version 7.11.33.0. DNV GL AS.

[2] *Guidelines for Hazard Evaluation Procedures*, published by the Center for Chemical Process Safety of the American Institute of Chemical Engineers, 3rd Edition, April 2008.

[3] *OGP Risk Assessment Data Directory: Vulnerability of Humans*. The International Association of Oil and Gas Producers. Report No. 434-14.1. March 2010.

[4] Toups, Harry J., *Fires and Explosions - Fundamentals and Design Considerations* [PowerPoint Slides]. Retrieved February 26, 2015 from www.sache.org / links / Pike21Jul2004 / Fires%20and%20Explosions.ppt.

[5] Clancey V J, 1972. Diagnostic features of explosion damage, 6th Intl. Meeting on Forensic Sciences, Edinburgh, Scotland.

[6] ALOHA 5.4.3 Modelling Software, Office of Emergency Management – U.S. EPA and Emergency Response Division – National Oceanic and Atmospheric Administration.

[7] *Handbook of Chemical Hazard Analysis Procedures.* Federal Emergency Management Agency, U.S. Department of Transportation, U.S. Environmental Protection Agency.

[8] Hockey, S M, and P J Rew. *Review of Human Response to Thermal Radiation.* HSE Contract Research Report No. 97/1996. Health and Safety Executive. July 2, 1996.

[9] Raj, Phani K. *A Review of the Criteria for People Exposure to Radiant Heat Flux from Fires*. Technology & Management Systems, Inc. October 25, 2006.

[10] National Fire Protection Association. (2013). *NFPA 59A: Standard for the Production, Storage, and Handling of Liquefied Natural Gas (LNG).*

441564 Application of Open Source Software Tools in the Process Safety Education

Abstract:

Process safety professionals have used, developed, and contributed to open source software tools. They have fascinated more and more notice from the society. The free software will enhance the collaboration and motivate the scientific and technical development. The software users can use and even modify the software modules. A lot of applications can be found in the real industrial practice so the engineering learning process should follow this trend. Before starting the career, the students should get to know how to take the full advantage of these tools in their future jobs because the future engineers should know how to find the free software packages in order to solve process safety problems.

This presentation will tell the audience how to integrate the open source software into process safety education or learning practice. The examples of modeling projects which can be practiced by the students in the classroom will also be introduced. The modeling will give the students more opportunities to recall their knowledge of mass and energy balance, thermodynamics, heat transfer, reaction kinetics, transport phenomena, process safety, statistics and numerical methods. These projects will also help them to learn how to conduct the risk assessment and process safety management through different kinds of model development. These chemical engineering education activities will allow them to explore computer aided tools such as computing and process simulators throughout the curriculum.

441688 Beyond Quantitative Risk Analysis Results - Part I: Explosions and Blast Phenomena Characterization

Abstract:

The present paper illustrates the results of a Quantitative Risk Analysis (QRA) case study of a refinery. The main purpose of a QRA is to evaluate the risk levels (both individual and societal risk) of a process due to Loss Of Containment scenarios (LOCs). A QRA can be used to optimize the risk reduction measures to be implemented to comply with a given tolerable risk level. Moreover, the development of a detailed QRA is the best opportunity to perform a specific analysis addressing structural response due to overpressure impacts caused by explosions and also estimating the Time To Failure (TTF) of structures and equipment due to thermal radiation of fires.

Part I of this paper is focused on illustrating how a detailed analysis of QRA results can provide valuable information for characterizing the blast phenomena parameters for a given location. The analysis gathers three key sources of information: (1) peak overpressure, maximum and minimum positive phase durations and impulses as a function of the cumulative frequency of occurrence; (2) total number of LOC scenarios which contribute to the impact of explosions (including vapor cloud explosions due to both immediate and delayed sources of ignition); and (3) the identification of each LOC scenario and associated individual frequency of occurrence. This information is valuable for ensuring the blast loading function for a subsequent structural response analysis (e.g., equivalent elasto-plastic Single Degree Of Freedom analysis), and also provides specific information regarding which LOCs contribute most on overpressure impact.

441690 Beyond Quantitative Risk Analysis Results - Part II: Fires and Domino Effect Characterization

Abstract:

The present paper illustrates the results of a Quantitative Risk Analysis (QRA) case study of a refinery. The main purpose of a QRA is to evaluate the risk levels (both individual and societal risk) of a process due to Loss Of Containment scenarios (LOCs). A QRA can be used to optimize the risk reduction measures to be implemented to comply with a given tolerable risk level. Moreover, the development of a detailed QRA is the best opportunity to perform a specific analysis addressing structural response due to overpressure impacts caused by explosions and also estimating the Time To Failure (TTF) of structures and equipment due to thermal radiation of fires.

Part II of this paper is focused on illustrating how a detailed analysis of QRA results can provide valuable information for characterizing potential domino-effects due to thermal radiation of fires at a given location. The analysis gathers three key sources of information: (1) thermal radiation as a function of the cumulative frequency of occurrence; (2) total number of LOCs which contribute to the impact of fires; and (3) the identification and characterization of each LOC scenario and associated individual frequency of occurrence. This information not only is valuable for ensuring the heat radiation for a subsequent analysis intended to estimate the TTF and associated conditions remaining in the system (e.g., Wall Dynamics Analysis), but also provides specific information of which LOCs contribute most to the heat radiation impacts. This information ensures which LOC scenarios need to be addressed for optimizing the risk reduction decision-making process. Additionally, the accurate estimation of potential equipment failure due to thermal radiation provides added value for predicting when and how impacts such as BLEVEs (Boiling Liquid Expanding Vapor Explosions) are expected to occur.

441700 A Leadership Vision for Accelerating Learning from Past Lessons to Create a Future Free of Incidents

Abstract:

The practice of process safety management (PSM), involves the analysis and understanding of what can go wrong, what the effects could be, and what should be done to prevent or mitigate an incident. That is: (1) what are the causes (human errors and equipment malfunctions)? (2) What are the potential consequences and impacts (to people, property, process, reputation)? and (3) what are the risk controls needed to prevent, detect, control, mitigate, or respond to such events (safeguards, layers of protection)?

Lessons learned from industrial disasters, as well as successes by industry leaders in preventing such incidents, form the basis for a practical framework for a PSM system as a holistic management tool for managing hazards and their risks. Quality design/development and effective implementation of a robust risk management system driven by a proactive culture of leadership to reduce operational risks and improve performance, enables organizations to avoid future incidents.

However, a conventional approach of longer-term continual improvement of operational risk performance could easily be outpaced by the rapid changes in technology advances, organizational and human factors, and the dynamics of the market. Operating companies, service providers and regulators alike are challenged by such changes and their potential impacts. Lessons of disaster investigations from the process industry have been full of learning opportunities. In fact, many of the lessons and their root causes are common across most process safety incidents and near misses ... yet, 15-20 years has proven to be an unacceptable period for gradual learning.

This presentation will present some common lessons as a priority for establishing improvement strategies as organizations strive for operational excellence. Available proven risk management methodologies and best practices for operational risk reduction will be summarized. Given that the global industry has made remarkable progress toward developing proven solutions aimed at prevention, organizations must now commit to accelerating the adoption of these solutions before at-risk operations experience near future incidents.

The focus of the paper will be the challenge to accept a leadership position to resolve existing but perhaps unidentified risks, then to commit to action to proactively identify hazards and their associated risks, and implement risk reduction measures before losses occur. A framework of a robust risk-based management system, incorporating all of the necessary elements based on lessons learned and including supporting leadership and management processes to achieve risk reduction performance objectives, will be presented.

Attendees will benefit from practical takeaways of how each management system element can be developed with "leadership designed in" to control the process safety risks at three opportune times: before the incident (prevention), during the incident (emergency management) and after the incident (incident investigation, root cause analysis and lessons learned).

The presentation will pose a challenging question: "Knowing what we already know, why would we want to take years to gradually reduce our risks, when we could accelerate the learning from past lessons and prevent incidents using available best industry practices"?

This paper will draw upon the authors' extensive experience in designing, developing and implementing PSM systems.

441698 Analytics to Reveal Meaurable Factors to Process Safety Performance

Abstract:

Analytics on a Large Industry Data Set Reveal the Leadership and Organizational Factors that Drive Down Risk and Reduce Incidents

With access to a multi-year data set of over a million incidents from 1000's of operating facilities worldwide in the Energy and Chemicals industry, the research team mined the data set to explore the differences in risk management practices and behaviors which most effectively explain the variation in process safety outcome performance. By modeling and examining predictive relationships, the study identifies four attributes - worker engagement, process discipline, risk sensitivity, and learning-mindedness – as critical for reducing incidents.

Findings from this study support a progressive and integrated approach and suggest that companies that are proficient in driving process safety performance are companies that, to a greater degree, have successfully engaged their workforce, strengthened their operating discipline, sharpened their risk sensitivity, and become learning-minded. A learning-minded organization manages each incident and responds to each failure deliberately to identify and fully leverage improvement and learning opportunities.

While this study provides statistical evidence to support the strong correlation between leadership, organizational behavior attributes and safety performance, it also emphasizes the organizational factors that are necessary to enable and sustain effective risk mitigation efforts. By deriving these insights from a data set of routine field-level activities, operational measurements may be created to measure progress and drive focus on engagement, discipline, risks, and learning.

This paper will conclude by sharing examples of how companies have applied these findings to establish quarterly and monthly measurements to drive leadership and organizational focus in these critical areas and ultimately to improve outcome performance.

441960 Coupling Dynamic Pressure Relief System Design with Experimental Vent Flow Regime of Dicumyl Peroxide / Toluene System

Abstract:

Abstract

A safe process design requires knowledge of the chemical reactivity of the desired reactions as well as the undesired reactions due to process upset conditions. Techniques used for sizing pressure relief systems for reactive systems under thermal runaway conditions include direct empirical scaling of experimental data obtained from a vessel with very low thermal inertia, analytical methods, and computer simulations using reaction kinetic models developed from adiabatic calorimetry data.

Pressure relief size estimation is based on the required flow capacity through the relief device at a specific set pressure, in order to keep the pressure in the vessel below the maximum allowable accumulated pressure. The flow capacity of the relieving device is highly dependent on the quality of the vent flow regime, whether it is liquid, all-vapor/gas, or two-phase. This requires knowledge of the physical properties of the contents inside the vessel at the time of venting. For a reactive system, these properties can be difficult to determine.

In this paper, the pressure relief size of a reactive system containing 50% dicumyl peroxide in toluene will be estimated using the dynamic simulation method. The vent flow regime of the reactive system will be determined from measured data, according to the methods described in the DIERS (Design Institute of Emergency Relief Systems) project manual. An automatic pressure tracking adiabatic calorimeter (APTAC^{TM}), which is a low thermal inertia calorimeter, will be used to conduct vented tests. The pressure relief size estimates of the above dynamic simulation method will be presented and compared with the DIERS simplified method.

441701 We Did a Quantitative Risk Assessment Study. Now What Do We Do with It?

Abstract:

Quantitative Risk Assessment studies have been extensively used in risk based regulatory regimes and they are becoming more common as a replacement for facility siting requirements in Process Safety Management (PSM) covered regulatory regimes and as a corporate risk management tool. Besides assessing risks to society and personnel, it can also assist in practical risk reduction implementation in design projects and operating assets. This paper intends to discuss these practical risk reduction implementation to show value to projects and operating assets, and how implementation can be achieved to using specific examples. The following examples will be discussed in the paper and will outline a broader set of QRA implementation improvements:

- Operating modes input to isolatable inventories and failure criteria
- Failure mechanisms from Hazards and Operability studies, and corrosion or mechanical integrity surveys
- Survivability criteria for manned and unmanned buildings for upgrades or siting challenges
- Specialized consequence analysis such as uncommon release material, indoor releases, and elevated receptors.
- Spill containment application
- Thermal protection and passive fire protection application
- Personnel and structural cryogenic protection application
- Risk based fire and gas detection mapping
- Escape, evacuation and rescue impacts
- Operational risk reduction using performance standards

The objective of this paper would be to better understand how to utilize QRA studies by improving their content and by practical implementation (as necessary) of mitigating risk reduction measures in design projects and to existing assets.

442063 Effective Implementation of PSM to Reduce Operational Risks

Abstract:

Ultimate success of process safety management (PSM) is measured through effective operational risk reduction. Usually, there is a lot of emphasis on putting the various elements of PSM in place through adoption of established international standards and best practices. Many of these standards and best practices are very well documented and element leaders/champions are appointed. Unfortunately, many of these elements and element champions are in silos and thereby not contributing to the overall reduction in operational risk. The risk is not reduced until the documentation is converted into reality on the ground.

Also, there is a lot of emphasis on doing high quality technical studies and quantitative process hazards analysis but very little is done to follow up on implementing the recommendations and results of these studies. The operational risks are not reduced until the recommendations are actually implemented. So, despite significant amount of effort, the improvement in the effectiveness of PSM implementation is limited.

Learning from this experience, SCG Chemicals has adopted a practical, bi-focal approach that is driven with a specific purpose in mind — operational risk reduction. The short term focus was to address deviations from design intent and established procedures in the field. If these items are not addressed today, they can contribute to a potential catastrophic incident tomorrow. This enhanced operational discipline and prevented 'normalization of deviations'. The longer term approach focuses on redoing structured PHAs and using the risk matrix to establish the levels of acceptable/unacceptable risks. Management decisions were made to address priority of the unacceptable risks with a rigorous follow up on implementing recommendations. This process has contributed to better outcomes in increased effectiveness of PSM while optimizing the money, time, and effort required. This paper will share the experience of SCG, along with DuPont as a partner, and the journey in implementing PSM effectively to reduce operational risks.

442151 Probit Models for the Determination of Damage Probability of Storage Tanks Due to Flooding Impact

Abstract:

Global climate changes may significantly influence temperatures and precipitation patterns, increasing the frequency of severe meteorological events which may lead to flooding. Flooding may trigger severe accidents on population among urban as it may do on chemical process industries too, causing the damage of equipment and the loss of containment of hazardous materials which might transform into toxic dispersions, fires, explosions or severe environmental contamination. This type of events are known as Natech events (Natural-Technological). This type of events may lead to extremely severe consequences bringing up the need to develop approaches for the risk assessment of Natech scenarios. Recently important efforts have been dedicated to develop tools that include Natech events in the quantitative risk assessment framework. In order to allow a quantitative risk assessment of a Natech event, a key point is the definition of tools for the estimation of equipment damage probability on the basis of intensity parameters of flooding. Past accident data analysis evidenced that horizontal and vertical storage tanks were often damaged in Natech events triggered by floods. Because of this the present study presents specific probit models for the damage probability of horizontal and vertical storage tanks due to flooding intensity parameters (flooding water height and flooding water).

Probit models in this study were derived on the basis of a mechanical model used for the assessment of the vulnerability of storage tanks due to the impact of flooding. For the determination of damage probability Montecarlo simulations were applied over the mechanical model in order to obtain a more accurate value and a probabilistic distribution of this parameter in dependence of the flood impact vector. Once the damage probability distribution is determined the probit method was applied using a logarithm approach with the impact flood vector (water height and water velocity) as the dose variables and the probability of damage as the response variable. A logarithm regression using the minimum square method was utilized to determine the probit constants for the model. Since the mechanical model used as basis has a huge dependence on the geometry of the storage tank being studied, it was find the damage probability will be affected and its value will change. To make sure this affections will be taken into account in the probit modeling a relationship between the geometry of the tanks and the probit constants was established using a polynomic regression. Once again the minimum square method was used and these two variables were related using a second order polynomic approach. Probit constants were determined for two substances (gasoline and LPG) and two types of storages tanks (vertical and horizontal).

The results of this study are a useful tool for the development of the quantitative risk assessment of Natech events due to flood impacts in storage parks. Also contributes to the reduction of the uncertainty in the framework for the QRA of Natech events triggered by floods.

442178 Analysis of Fire Extinguishing Methods in Engine Control Rooms

Abstract:

The study of electrical fires is a major concern for fire safety in the industry and more particularly for oil industry. In petrochemical facilities, the fire hazard may cause severe consequences on safety functions inside the fire compartment and furthermore may lead to the loss of one or several compartments, as just as structural damage on the facilities. From the OECD Fire Database a majority of the fire ignition is due to electrical sources (close to 50%, with a significant contribution attributed to electrical or electronic cabinets, with an occurrence of more than 10% of all fire events). Because of the above mentioned, it is necessary to implement tools to study these phenomena without using an experimental method. For example, the simulation of the dynamics of combustion and fire spread, allows an estimation of the effects on the structure of the case study preemptively. This project addresses the comparison of different fire protection philosophies in engine control rooms, using Fire Dynamics Simulator as support tool. The project studies three different cases, one of them without a protection system, another a clean agent as extinguishing tool and a final one, which implements an inherently safe design. For the first case study, the obtained data for the room temperature and the heat release ratio indicate a potential risk for the personnel and the equipment of the facility. It is expected that the study of the remaining cases show the significant advantages of each fire extinguishing philosophy.

12TH GLOBAL CONGRESS ON PROCESS SAFETY

The Study of HAZOP Revalidation Approach

Jing Xu
SINOPEC Research Institute of Safety Engineering
No.218, Yan'an san Road,Qingdao,China
xuj.qday@sinopec.com

Chuntao Ge
SINOPEC Research Institute of Safety Engineering
No.218, Yan'an san Road,Qingdao,China

Yi Zhang
SINOPEC Research Institute of Safety Engineering
No.218, Yan'an san Road,Qingdao,China
Zhangyi.qday@sinopec.com

The Study of HAZOP Revalidation Approach

Jing Xu
SINOPEC Research Institute of Safety Engineering
No.218, Yan'an san Road,Qingdao,China
xuj.qday@sinopec.com

Chuntao Ge
SINOPEC Research Institute of Safety Engineering

Yi Zhang
SINOPEC Research Institute of Safety Engineering

Keywords: HAZOP, HAZOP revalidation, update HAZOP,PHA, MOC

Abstract

Abstract: HAZOP evaluation method is used to analyze the risk of the process by the worldwide chemical enterprises and design units. However, Standard Specification for many countries stipulate the enterprise needs to carry on the process hazard analysis once again (revalidating) in several years later. The goal of periodic revalidation is to identify the cumulative changes of risk over time and to improve understanding of risk. The ultimate goal is to avoid accidents. But there is no systematic approach to guide the HAZOP revalidation at present. If we redo HAZOP blindly (performing a new HAZOP similar to the initial one) will only be time-consuming. This paper introduces the concept of HAZOP revalidation which is based on the analysis of the initial HAZOP and discusses many of the factors that need to be considered for the revalidation such as the change management, accident analysis and HAZOP report quality. At the same time, the specific procedure and approach of the HAZOP revalidation are proposed. The panel of experts scoring method is used to determine the revalidating approach and then decides whether to redo or update the initial HAZOP. If the scoring results suggest redo the HAZOP, and then conduct the Hazard and Operability Analysis in accordance with the complete HAZOP procedures; if the scoring results suggest update the initial HAZOP and then modify and supplement previous HAZOP analysis results to meet enterprise current status. This approach is not only time-saving and more targeted which can guide enterprises to conduct better and faster HAZOP revalidation.

1 Introduction

Hazard and Operability Study (HAZOP) is a kind of process safety analysis methods which is invented by Britain's imperial chemical industries (ICI) in the 1960s[1]. It is carried out by a suitably experienced multi-disciplinary team (HAZOP team) during a set of meetings to identify potential risk and operational problems in the system.

However, the safety status of enterprise is not invariable after the HAZOP.Many changes may occur in the later production process inevitably. The safety status of enterprise will change if the technological process, equipment, safety equipment, operating procedures, personnel, organization structure, and environment changes in the entire life cycle. So getting through the HAZOP analysis is not once for all. HAZOP Revalidation is to conduct a comprehensive evaluating for process hazards situation of the system during the next few years after the first time HAZOP analysis. To eva`luate the validity of the analysis results and determine whether meet the requirements of laws and regulations,and improve the process safety management level of enterprises at the same time. The goal of regular revalidation is to identify the change of the risk as time goes on and to improve the understanding of the risk. The ultimate goal is to avoid the accident. However, HAZOP analysis of a typical chemical process takes a few weeks to a few months. When in the HAZOP revalidation, whether can take advantage of the existing HAZOP analysis report to make the HAZOP revalidation more simplified and efficient and guarantee the results conform to the enterprise current situation. More and more companies try to simplify the method of revalidation, rather than redo HAZOP blindly[2].

2 Status analysis of HAZOP revalidation

HAZOP technology has developed more than 50 years getting a lot of project application, and become one of the general evaluation method on process safety domain at present. Petrochemical companies and design companies worldwide use HAZOP method to analyze the risk of petrochemical process. Such as Dow chemical, Rhodia company, BP, Bayer, Royal DSM, SINOPEC, CNPC, CNOOC and so on[3].

OSHA(the Occupational Safety & Health Administration) put forward this revalidation concept at Process Safety Management in 1992. Claim that "After the first process hazard analysis, revalidation should be done at least every five years, to ensure consistency of process hazard analysis and the current process" [4]. In the risk management procedures of EPA(Environmental Protection Agency) and aEuropean SEVESO II have corresponding requirements in the instruction. CCPS(Center of Chemical Process Safety) published a book Revalidating process hazard analyses Defined that the revalidation process attempts to protect this investment by identifying and building upon the still pertinent portions of the prior PHA .Corrections are made and new content is added to the PHA as required,and the result are documented to serve as the new PHA[5]. In the United States, the large-scale petrochemical enterprises have done 2 ~ 3 times HAZOP, identifying the potential risk in the system, found a lot of potential safety problems. The study of HAZOP technology started late in China. Now a lot of enterprises done HAZOP before will conduct HAZOP revalidation. Guidelines for process safety management of petrochemical corporations (AQ/T3034-2010) standard requires: Guidelines for process safety management of petrochemical corporations (AQ/T3034-2010) standard requires: Companies should conduct a process hazard analysis in the process during construction and should to finish the process hazard analysis to confirm again and updated every three year[6]. The difference between Revalidation and first HAZOP is the former is conduct on the basis of the practical operating experience. The key for the first time HAZOP is to perfect the design and the retrieval permissions[7]. HAZOP Revalidation is to compare the results of the first HAZOP whether still applicable to the current environment, more dependent on site operation actual situation and record filing, etc. There is no systematic method to guide how to HAZOP review again. However, blindly redo HAZOP is not only time-consuming, also need a lot of resources. More and more companies expect to revalidate on the basis of the first HAZOP analysis.It can be more simple and efficient, and at the same time guarantee the quality of process hazards analysis.

3 The Study of HAZOP Revalidation Approach

3.1 The key factor of HAZOP revalidation

HAZOP revalidation using redo or update way is decided by many factors, such as whether there is a big defect of the previous process hazard analysis, whether change management system is perfect, whether there is a major accident happened, whether the information is complete, new standards, etc.

The way for HAZOP revalidation depends on the following three aspects：

a. The design change number, the size of the problem, process complexity, change process hazard analysis quality and management of change since the last HAZOP analysis；
b. The quality of Previous HAZOP analysis, including process safety information comprehensiveness, the comprehensiveness of HAZOP analysis, HAZOP record content integrity, HAZOP results application effect, etc；
c. Previous accidents and device status: the type of process safety accidents, statistical accident frequency, Consequences severity and the reason for the accident , etc.

3.2 The process of HAZOP revalidation

a. Establish revalidation team, the composition of team members are as follows[8]:

- the group leader: trained, rich experience in organization HAZOP analysis; Responsible for communication between HAZOP team and project management personnel; Review plan
- Process equipment technician: familiar with the past project, past and present design/operational problems and change the content
- Site operator: familiar with the situation, the operation experience
- owner (user) : familiar with the operating environment, the consequences of deviation and the degree of danger of deviation
- Experts: providing related professional knowledge. Technology, automatic control experts shall be invited to assist analysis group for part analysis
- Recorder: review record, record the related changes, and update or supplement HAZOP records, to assist the analysis team leader for planning
- Maintenance personnel (if required)

b.Collect information, including the HAZOP report, change management (MOC) data, the previous accident report, HAZOP recommendations, process engineering and safety information, operation procedures, etc

c.Expert assessment team scoring method. Details see Table 1 - the marking table for determining HAZOP revalidation methods and Table 2 - HAZOP revalidation method
d.According to the examination way to implement. Redo HAZOP or update HAZOP.

The flow to determine the specific revalidation way is shown in figure 1

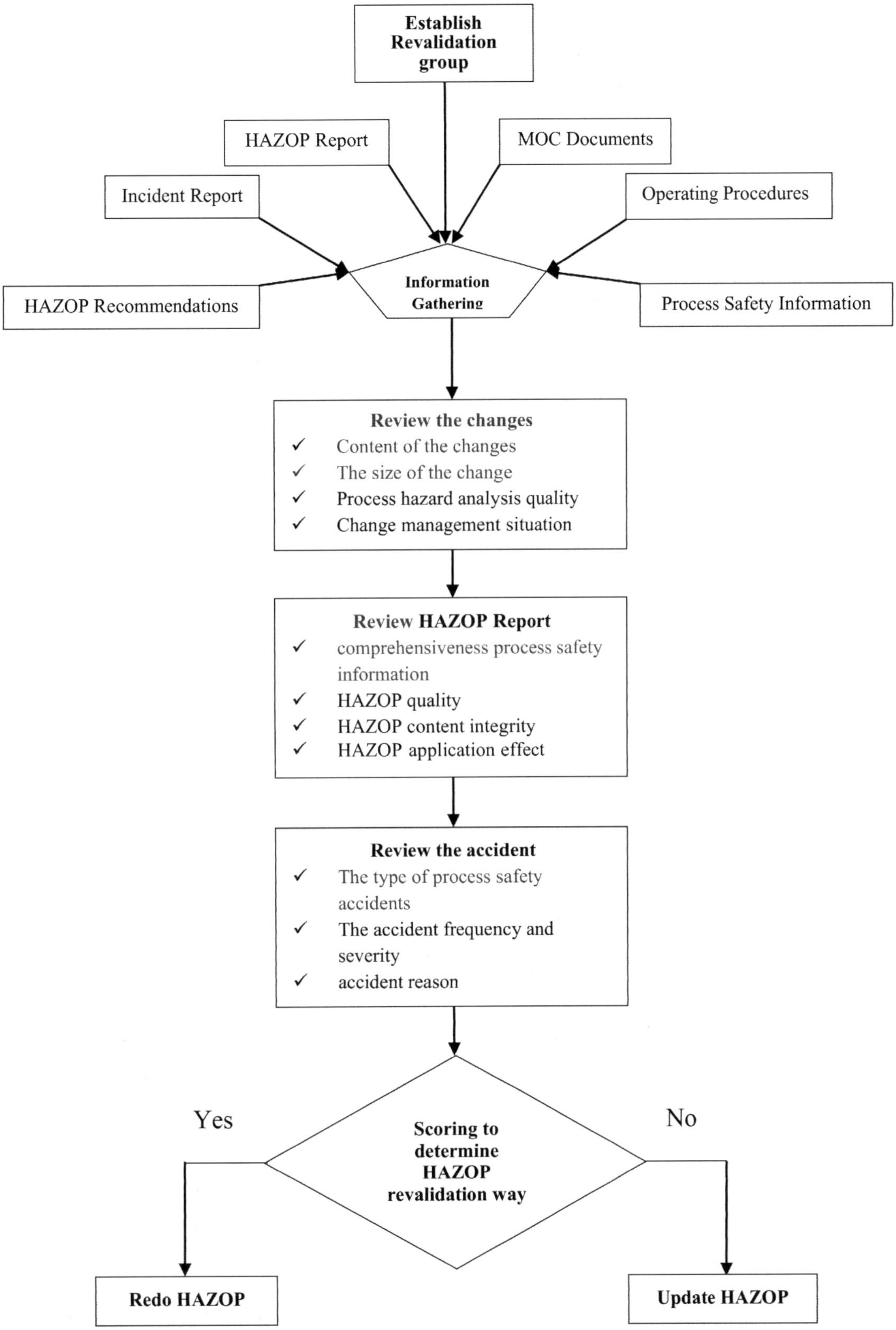

Fig.1 HAZOP revalidation approach decision diagram

Redo HAZOP is not introduced in the paper, the most effective way to update the HAZOP is to collect all data and accident information related to the change， to review the preliminary HAZOP Suggestions， to view the process pipeline flow chart meters (PID), understanding the changes and the defects of the system[9]. Such as process flow, equipment, safety equipment, operating procedures, operating conditions, location and environmental changes, and keep record by the recorder. According to the situation of change size, change management and change of process hazards analysis quality lead revalidation group to carry out different levels HAZOP for change unit by the team leader. Update the HAZOP results that conform to the enterprise the status quo. Update the HAZOP is time saving and maximum use of existing resources compared to redo HAZOP.

Table 1. HAZOP revalidation reference score table

Review items		**Index description**	**Reference scores**
Change Review	Change content （max）	Change of Process/tank type/operating procedure	15*
		Change of safety facilities	10
		Change of pipeline, pump, valve	5
		No change	0*
	Scale of change	Change of working hours more than one month/ 5 types of job post	10*
		Change of working hours about 2 weeks/3 types of job post	5
		Change of working hours about one week/2 types of job post	2
		No change	0*
	Quality of hazard analysis of change	No hazard analysis of change	10*
		Bad hazard analysis of change(lack of pertinence, record, etc)	5
		Good hazard analysis of change(More targeted, detailed record)	2
		Eexcellent hazard analysis of change(effective method, comprehensive analysis)	0*
	Management of change	Records and files are deficient and not combined with operating procedure or job training partly after change	5*
		Records and files are complete and combined with operating procedure or job training partly after change	2
		Complete and electronic records and files,which combined with operating procedure or job training after change	0*
HAZOP quality review	Comprehensiveness of process safety information	Deficient data or big change of site layout	5*
		Partly complete data or small change of site layout	2
		Complete data and less or no change of site layout	0*

	Integrity of HAZOP record	Deficient node description, design intent, drawing name, date, participant and the content of the standard form, bad reference	5*
		Partly complete node description, design intent, drawing name, date, participant and the content of the standard form, good reference	2
		Complete node description, design intent, drawing name, date, participant and the content of the standard form, wonderful reference	0*
	Rationality of HAZOP	Deficient deviation and guide word, unreasonable nodes, bad suggestions and measures	15*
		Partly complete deviations and guide words, reasonable nodes, common suggestions and measures	10
		Complete deviations and guide words, more reasonable nodes, good suggestions and measures	5
		Perfect deviations and guide words, most reasonable nodes, appropriate suggestions and measures	0*
	Application effect	Weak effect, less use of risk identification	5*
		Good effect, general use of risk identification	2
		Perfect effect, effectively use of risk identification	0*
Review of accidents		Vast or serious process safety accidents	30*
		Some specific process safety accidents	15
		Some accidents which has nothing to do with the process	5
		No accidents	0*
* reference point, according to the actual situation within the range of its rated arbitrary integer (interval)			

Table 2. HAZOP revalidation approach determination table

Scores	Review way
⩾60	Redo HAZOP
<60	Update HAZOP

4 Application examples

A spherical tank area of PetroChina has four 3000 m3 spherical tank, two 5000 m3 spherical tank, storage for butene 1, butadiene. The company has completed the HAZOP analysis in 2010. Device in service conduct HAZOP analysis every four years according to the company policy. HAZOP revalidation use the proposed method, completed HAZOP analysis efficiently.

Table 3. Application of HAZOP revalidation approach determination table

Review items		practical situation	Scores
Change Review	Change content (max)	No change	0
	Scale of change	No change	0
	Quality of hazard analysis of change	Good hazard analysis of change(More targeted, detailed record)	2
	Management of change	No change	0
HAZOP quality review	Comprehensiveness of process safety information	Partly complete data or small change of site layout	2
	Integrity of HAZOP record	Perfect deviations and guide words, most reasonable nodes, appropriate suggestions and measures	0
	Rationality of HAZOP	Complete deviations and guide words, more reasonable nodes, good suggestions and measures	5
	Application effect	Good effect, general use of risk identification	2
Review of accidents		No accidents	0
Scores			11

5 Conclusion

This paper introduces the concept of HAZOP revalidation which is based on the analysis of the initial HAZOP and discusses many of the factors that need to be considered for the revalidation such as the change management, accident analysis and HAZOP report quality. At the same time, the specific procedure and approach of the HAZOP revalidation are proposed. The panel of experts scoring method is used to determine the revalidating approach and then decides whether to redo or update the initial HAZOP. If the scoring results suggest redo the HAZOP, and then conduct the Hazard and Operability Analysis in accordance with the complete HAZOP procedures; if the scoring results suggest update the initial HAZOP and then modify and supplement previous HAZOP analysis results to meet enterprise current status. The method has been applied for a spherical tank area of PetroChina. Conclusion is spherical tank area further implement the last countermeasures of HAZOP in view of the actual operation situation, do not need the modified or added. To satisfy the needs of the enterprise for better and faster process hazards analysis.

6 References

[1]Dunjó J, Fthenakis V, Vílchez J A, et al. Hazard and operability (HAZOP) analysis. A literature review[J]. Journal of Hazardous Materials, 2010,173(1):19-32.

[2]Smith K E, Whittle D K. Six steps to effectively update and revalidate PHAs[J]. Chemical engineering progress, 2001,97(1):70-88.

[3]WAN Gujun, LI Qi, ZHANG Guangwen, et al. Implementing HAZOP to Control Major Accidents (Ⅰ)[J]. Safety Health & Environment, 2009,9(011):2-4.

[4]ZHAO Jianmin. Discussion About HAZOP Revalidation [J]. Modern Occupational Safety, 2012(134):83-85.

[5]L. F W, K. W D. Revalidating process hazard analyses[M]. Center for Chemical Process Safety, 2001.

[6]State Administration of Work Safety. AQ/T 3034-2010 Guidelines for process safety management of Petrochemical corporations[S]. 2010.

[7]FU Jianmin, ZHAO Dongfeng, CHEN Guoming, et al. Quantitative study over the probability of HAZOP analyzing technique in petro-chemical plants[J].Journal of Safety and Environment,2008,8(6):130-134.

[8]Herbert I. Learning the lessons-retrospective HAZOPs[J]. Offshore Europe, 2011.

[9]Wagner T, Champion J. A work process for revalidating LOPAs and other risk analyses[J]. Process Safety Progress, 2012,31(2):122-129.

The Importance of Process Safety:

A Young Engineer's Perspective

Waheed Wakil
Smith & Burgess
Waheed.Wakil@smithburgess.com

Travis Bruner
Smith & Burgess
Travis.Bruner@smithburgess.com

Prepared for Presentation at
American Institute of Chemical Engineers
2016 Spring Meeting
12th Global Congress on Process Safety
Houston, Texas
April 11-23, 2016

The Importance of Process Safety: A Young Engineer's Perspective

Waheed Wakil
Smith & Burgess
Waheed.Wakil@smithburgess.com

Travis Bruner
Smith & Burgess
Travis.Bruner@smithburgess.com

Keywords: Process Safety Management, Young Engineers, Overpressure, OSHA, LOPA

Abstract

Most young engineers experience a defining moment or moments in their career that shape their attitude towards importance of Process Safety Management (PSM). Experiencing a plant upset first-hand knowing that there were actual capital or human life impacts or both from an incident at the site, one soon realizes early in their engineering career why process safety is so important. It becomes evident that the most successful corporations are those who are willing to invest large capital costs into PSM documentation projects or projects that have come as a result of findings identified during a study, allowing them to prevent incidents from occurring. The findings generated during the course of a study may result in simple documentation update action items, or they may require a multi-million dollar investment. For young engineers, this exposure puts into perspective the value of their work, and emphasizes the importance of a thorough and adequate study.

1 Introduction

Many young engineers have a point early in his or her career when process safety hits home. Different companies develop their own unique strategies towards safety that they are comfortable with, which results in considerably different safety cultures within industry. Unfortunately, it is all too common for engineers to treat process safety as an afterthought, with priority given to efficiency and profitability. However, as we will see, treating safety with the utmost importance can prevent costly and life-threatening incidents.

2 First Hand Experiences

2.1 Organic Peroxides Plant Background

Two major upsets at a chemical plant within a year may very well serve as a defining "moment" in a young engineer's career that forces him to reflect on and realize the importance of dedication to process safety. A fellow Process Safety Consultant at Smith & Burgess who was a process engineer at a plant that produces organic peroxides was fortunate enough to live through two such accidents unscathed. His story is a lesson and reminder that even the most disastrous accidents can be prevented with due diligence, a safety-focused company culture, and proper attention to basic Process Safety Management.

Based on the nature of the chemicals produced, a Process Hazard Analysis for this particular plant would have revealed scenarios with severe consequences. However, the safety culture at this plant was not adequately focused in preventing and mitigating upsets. Not only were poor preventative maintenance practices used, but lack of written emergency procedures and poor training prevented adequate mitigating protection layers after the initiating events. A few relevant OSHA requirements are worth noting here as failure to meet these requirements was likely the reason for loss of containment:

- Per OSHA 29 CFR 1910.119(f)(1), "The employer shall develop and implement written operating procedures that provide clear instructions for safely conducting activities involved in each covered process consistent with the process safety information...".
- Furthermore, per OSHA 29 CFR 1910.119(j)(2), "The employer shall establish and implement written procedures to maintain the on-going integrity of process equipment."
- Lastly, per OSHA 29 CFR 1910.119(n), "The employer shall establish and implement an emergency action plan for the entire plant in accordance with the provisions of 29 CFR 1910.38. In addition, the emergency action plan shall include procedures for handling small releases."

If these basic elements of PSM would have been implemented at this plant, both accidents may have been prevented or mitigated.

Organic peroxides are highly unstable – their O-O bond breaks very easily, and that is why their decomposition properties are so useful in initiating polymerization reactions. This useful property can also be very dangerous. When an organic peroxide is in its pure form, the rate at which its exothermic decomposition generates heat is likely to be greater than the rate at which the heat can be dissipated to its surrounding environment if proper heat transfer equipment are disabled, resulting in a situation called self-accelerating decomposition in which the reaction continues to accelerate due to its own generated heat.

The organic peroxides produced in this particular plant are characterized by very low Self Accelerating Decomposition Temperatures (SADT), defined as the minimum temperature at which an organic peroxide would experience self-accelerating decomposition after one week of storage. As such, glycol cooling jackets on vessels and piping in this plant maintained operating temperatures in the ballpark of -30 °C (-22 °F) to prevent the possibility of self-accelerating decomposition. For this heat transfer to work properly, the organic peroxides must be continuously flowing in the process piping and equipment and cannot become stagnant. At the same time, the glycol that cools the organic peroxide must also continuously flow. Stoppage of flow for either component may be disastrous. As we will see in the following incident descriptions, loss of this cooling results in rapid increase of temperature and decomposition reaction rate, resulting in a deflagration that cannot be mitigated by a typical relief device such as a safety valve or a rupture disk. To prevent loss of containment and subsequent damage to surrounding structures, this scenario must instead be prevented or mitigated using written operating procedures, regular inspection and maintenance of equipment, and employee training.

2.2 Organic Peroxides Incident #1

What went wrong?

Organic peroxides are so thermally unstable that precautions must be taken when packaging and storing the material. Packaging must occur continuously to prevent the material from becoming stagnant in piping and the temperature must be monitored and controlled at or below the recommended storage temperature. If material remains stagnant without sufficient cooling, organic peroxides can reach their SADT temperature and undergo a violent reaction, which usually results in rupturing the packaging, dispersing peroxide, as well as discharging liquid and gaseous decomposition products considerable distances. The heat generated during this reaction has the potential to auto ignite flammable vapors.

An operator was transferring organic peroxides to 55 gallon drums, which are immediately transported by forklift to an icehouse for storage. An incident occurred which caused the operator to abandon the task in the packaging building to provide assistance elsewhere. While the operator was away, the product remained stagnant in the filter where glycol cooling was inefficient. This caused the material to exceed its SADT. The material began to decompose as temperature continued to increase. When the operator returned the material suddenly exploded propelling the head of the drum through the air. The remains of the drum were embedded in the ground and lights in the facility were shattered. The operator was behind a blast shield, but still suffered severely burnt hands and damage around the ear. Although no life threatening injuries were inflicted, the plant suffered considerable downtime and costly repairs to the filter, packaging equipment, and process piping.

How could it have been prevented?

Proper implementation of PSM standards likely could have prevented this incident from occurring - in particular, OSHA 29 CFR 1910.119(f)(1), mentioned previously. Documented operating procedures that clearly state not to leave the packaging unit unattended during operation to avoid stagnant organic peroxides, inefficient cooling, and the potential for explosion could have prevented (or, at the very least, limited) the disastrous effects of this incident.

2.3 Organic Peroxides Incident #2

What went wrong?

This plant utilizes large plug flow reactors that are jacket-cooled. On one particular day, the site was hit by a thunderstorm with wind speeds high enough to break a tree branch that would damage a power line, resulting in total loss of electricity to the plant. Electric power was vital to this plant as it kept peroxide flowing through the reactors and piping and it kept glycol flowing through the jacketing to keep the peroxide temperatures down to safe levels. The plant had a backup generator, but only a single operator was aware of the location of its switch. It just so happened that it was this operator's day off.

In the event of an emergency such as this, interlocks were in place to purge out residing organic peroxides in the equipment using control valves with actuators regulated by instrument air. Once glycol and peroxide flow was lost in the event of a power failure, the peroxides would rapidly exceed their SADT. The purging action would dump the peroxides into waste water drains that are gravity fed to open tanks, where it is much safer for the peroxides to reach their SADT. However, the instrument air that controlled the valves was supplied from air storage, and during the purging, the instrument air ran out. Nitrogen was used as backup for air during emergency situations, but due to poor inspection and maintenance for cost savings, there was damage to the backup nitrogen header that had gone unnoticed. This damage caused a leak in the header when it was exposed to the nitrogen pressure, rendering the header useless as there was not enough pressure to actuate the purge control valves properly. At this point, purging was halted. The catastrophic failure was now inevitable. With stagnant peroxides in piping and equipment, and no jacket cooling, fluid temperatures rapidly sored past the SADTs and eventually resulted in a deflagration. The reactor top heads failed first, and the building in which the reactors were located was designed to fail at the ceiling and the outer wall. The building and its contents were destroyed, and the plant suffered 6 months of downtime. Fortunately, operators were protected on the other side of a blast wall, and no one suffered major injuries.

How could it have been prevented?

Looking at the number of Independent Protection Layers (IPLs) that were in place at the time of failure, namely the operator's response to activate the backup generator and interlocks that functioned to purge out residing peroxides, there were not enough layers of protection in place to prevent a deflagration. Table 1 summarizes the highest possible credits that could be taken for each IPL based on CCPS's LOPA book [1].

Table 1: IPL Credit Summary

Initiating Event	Initiating Event Frequency	Number of IPL Credits Required for Category IV Consequence	Maximum Number of IPL Credits Available
Loss of Cooling	Frequency $\geq$ 1X10^{-2}	**2.0**	Operator Response for Backup Generator 0.5
			SIL 1 SIF 1.0
			Total **1.5**

Like many such accidents, it is not uncommon for multiple independent failures to cascade into a catastrophic event, especially when a key to a successful emergency response from operators is proper training and a key to proper function of mechanical equipment is proper maintenance and inspection. To pinpoint PSM measures that would have likely prevented this accident, we can start by calling on OSHA 29 CFR 1910.119(n) referenced previously in this paper. The first broken link in the series of events leading up to the accident was the lack of training amongst operators with respect to activating the backup generator in the event of loss of power. This should have been a basic emergency response action that every operator at the plant should have been trained on and been aware of, but unfortunately, no one at the plant at the time of the failure was trained on how to activate the generator. The failures cascaded into the second phase of emergency response when the instrument air was exhausted at exactly the wrong time. To address the issue of the backup nitrogen header leak, we call upon OSHA 29 CFR 1910.119(j)(2), "The employer shall establish and implement written procedures to maintain the on-going integrity of process equipment." To cut costs, plant management had reduced regular inspection and maintenance of the backup nitrogen header, and it proved disastrous on this day when they needed it the most.

3 Conclusion

As illustrated in the incident descriptions, a poor safety culture is likely to result in poor preventative inspection and maintenance cycles and lack of clear emergency operating procedures and training. In most process plants, inadequacies in PSM elements can result in disastrous consequences. For a young engineer, it sometimes takes a real-life experience to change their perspective on PSM from one of passive compliance to one of active responsibility. Each element of PSM is important and critical, from the most obvious and simple, like written procedures on activating a backup generator, to the most expensive and extensive, like redesigning a flare system. It makes every detail in every process safety analysis that much more important and that much more rewarding when the benefits are apparent.

References

[1] CCPS. *Layer of Protection Analysis: Simplified Process Risk Assessment.* Center for Chemical Process Safety, American Institute of Chemical Engineers, New York, NY, 2001.

Scaling Up Safely: Making Smarter Decisions about Rate Dependency

Zubin Kumana, P.E.
Smith & Burgess LLC
7600 W. Tidwell Rd, Ste 600
Houston TX 77040
Zubin.Kumana@smithburgess.com

Venkata Badam, P.E.
Smith & Burgess LLC
Venkata.Badam@smithburgess.com

Achilles Arnaez
Smith & Burgess LLC
Achilles.Arnaez@smithburgess.com

Prepared for Presentation at
American Institute of Chemical Engineers
2016 Spring Meeting
12th Global Congress on Process Safety
Houston, Texas
April 11-23, 2016

Scaling Up Safely: Making Smarter Decisions about Rate Dependency

Zubin Kumana, P.E.
Smith & Burgess LLC
7600 W. Tidwell Rd, Ste 600
Houston TX 77040
Zubin.Kumana@smithburgess.com

Venkata Badam, P.E.
Smith & Burgess LLC
Venkata.Badam@smithburgess.com

Achilles Arnaez
Smith & Burgess LLC
Achilles.Arnaez@smithburgess.com

Keywords: Process Safety, Safe Design, and Venting and Emergency Relief

Abstract

Over the lifetime of a facility, it is almost a certainty that the throughput of a process will be increased, often repeatedly. Rate-dependency is frequently used as a way to identify pressure relief system sizing that would be affected by the scaling up of process throughput. This paper will define rate-dependency as it relates to pressure relief systems and describe the various ways in which rate-dependent classifications may be affected by the mechanical and thermodynamic constraints of the process and equipment. In addition, examples of when the "rate-dependent" designation may no longer apply due to common assumptions and methodological choices will be provided.

1 Introduction

In the refining and chemical processing industries, increasing throughput through a processing unit is often considered as a means of maximizing value with existing capital or minimal additional investment. Often, the throughput will be increased many times over the lifetime of the facility. The pressure relief system for a process is designed in part based on the throughput of the unit; therefore, the change in the throughput may affect the adequacy of the pressure relief system design.

Verifying the design of the process unit's pressure relief system for these increased rates is typically done by identifying overpressure scenarios as either rate-dependent or rate-independent. A rate-dependent overpressure scenario is one whose required relief rate varies with changes in process throughput. A common example is a blocked outlet scenario on a separator. With higher the process throughput, the quantity of mass that must be relieved to limit the pressure to the allowable accumulation is expected to increase.

Conversely, rate-independence refers to overpressure scenarios whose required relief rate is independent of process throughput. A common example is exterior fire, where the required relief rate depends on the heat of vaporization of the liquid mixture inside the vessel, the density of the vapor and liquid phases, and the wetted area of the vessel the fluid is residing in. Changing the feed rate to the vessel is not expected to result in a change to the required relief area for the exterior fire scenario. In some circles, these scenarios may also be termed "equipment-dependent". While the equipment and piping remains constant, the required relief rate remains constant. However, as rate scale-up increases, the equipment and piping may be changed to accommodate the additional capacity.

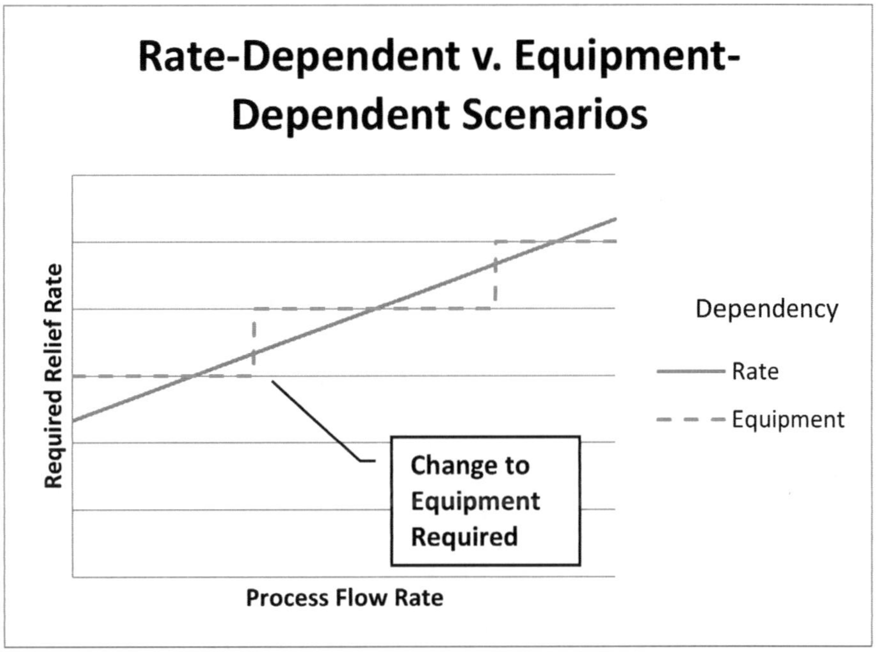

Figure 1: Rate-dependent scenarios increase in required relief rate (and therefore area) with process flow rate increase, while equipment-dependent scenarios are effectively constant until a change to equipment is required.

1.1 How Rate Dependency is used in Pressure Relief System Design

The common practice in pressure relief system design is to select a unit process basis that the user believes represents both the actual operation of the process as well as the worst case for the relief system design – specifically, the case that will result in the largest required areas and piping diameters. Since process units can have a range of operating modes and rates, selection of these cases tends to favor the following:

- Modes of operation with lighter feed, as this can result in more vaporization and therefore larger vapor and two-phase relief loads
- Higher process throughput rates; often the peak rate achieved, or some percentage above the peak rate

The conscientious user might separately specify a worst-case heavy liquid relief load basis in order to properly design the liquid handling capabilities of the disposal system.

As the pressure relief system design is completed, the engineer performing the design will typically identify overpressure scenarios as rate-dependent as the analysis proceeds. This designation is then used when the throughput rate is modified in the future. This allows engineers to minimize the amount of recalculation that needs to be performed, by only performing new calculations for the overpressure scenarios that are dependent on rates; the scenarios that had been identified as rate-independent are believed to be static and are left unchanged.

A generalized expression for pressure relief device sizing is given in API Std 520 Part I (9th ed.) Annex B, equation B.5, rearranged to solve for required area [1]:

$$A = \frac{W}{G\prod[K]} \qquad \text{(Eq. 1)}$$

Where

G	is the theoretical mass flux through the nozzle
W	is the mass flow through the PRV
A	is the effective discharge area
Π[K]	is the product of all applicable correction factors

The scale-up of the pressure relief system design is typically done using the assumption that the only change is to the flow rates through the unit; variables such as fluid composition, operating pressures and temperatures are assumed to remain constant. This effectively limits the relationship between the throughput rate and the relief system required areas to a linear one; as the fluid mass flux is considered constant, and the applicable correction factors are generally constant (exceptions are always possible based on Reynolds' number and actual backpressure resulting from flow), the values of A and W are directly proportional.

For the evaluation of the pressure relief system design, this has led to the following widespread practice:

1. The user takes the heat and material balance at the previous throughput rate
2. The user obtains a list of calculations that had been identified as rate-dependent
3. The user takes the percentage increase in flow and factors this value into the required relief rates
4. The user re-evaluates the pressure relief device requirements vs. capacity (along with other associated concerns where relevant, e.g. pressure losses if calculated at required relief rates)

This process rarely involves revalidating the process unit simulation to confirm the heat and material balance. Typically this practice has been followed for small changes to throughput rates. However, once the procedure has been implemented, the potential exists for repeated applications to increase the degree of deviation from a validated baseline even further. This poses a danger, as excessive extrapolation and repeated application of the scale-up process without carefully considered heat and material balance revalidation can quickly lead to larger and larger deviations, and the potential for undersizing the pressure relief system.

This paper focuses on rate scale-ups that do not involve any change to the as-built system. The rate scale-ups considered are therefore limited to the maximum throughput that the existing equipment can support. Debottlenecking the equipment in the process unit (rotating equipment, exchanger heat transfer area, and pipe sizing) may be required to accommodate further rate scale-up. As a capital project, this effort typically includes development of a new project baseline heat and material balance along with a review of the pressure relief system design.

2 Proper Identification of Rate-Dependency

2.1 How to Identify Rate-Dependency

Smarter decision-making with regard to rate dependency begins with understanding the ways that the traditional definition of rate-dependency may be invalidated by subsequent modeling choices in the determination of the required relief rate.

A general rule of thumb often passed around among novices in pressure relief system design is that overpressure scenario required relief rates that result from equations are rate-independent, and required relief rates that come from a heat and material balance (HMB) are rate-dependent. However, while it may be generally true for many facilities, there are aspects that are commonly overlooked in this approach, and some engineers may take subsequent steps in the required relief rate calculation that can change the classification. These additional steps are often taken as a means to "sharpen the pencil" when the initial determination of the required relief rate results in an undersized pressure relief device. These may include such actions as:

- Taking credit for an outlet that would not be blocked as a result of the scenario (including secondary causes like controller action and shut off systems)
- Reducing heat transfer into the system due to elevated bubble point temperatures
- Limiting flow rates based on the hydraulic limits of the piping, including valves and fittings
- Limiting flow rates to the maximum capacity of rotating equipment that can deliver the stated pressure at the point of relief

For example, if determining the required relief rate in the event of a blocked liquid outlet on an exchanger by taking the normal feed rate as the required relief rate is found to result in a required relief area greater than the currently installed relief area, the engineer may review the inputs and determine that when discharging against the relief pressure, the capacity of the feed pump is reduced below the normal feed rate, potentially making the existing installed relief area adequate. In this example, the overpressure scenario has changed from rate-dependent to equipment-dependent.

Table 1 shows common sources for determining the required relief rate for an overpressure scenario, and potential additional modeling choices that can change the classification.

Table 1. Examples of Dependency and Factors That May Change Classification†

Overpressure Scenario	Rate Commonly Derived From	Dependency for Common Derivation	Factors that may change Classification
Closed outlets	HMB/Simulation, Exchanger duty	Rate-Dependent	Limited by pump/compressor, valve, or pipe capacity
Reflux failure	HMB/Simulation, Exchanger duty	Rate-Dependent	
Cooling failure	HMB/Simulation, Exchanger duty	Rate-Dependent	
Overfilling	HMB	Rate-Dependent	Limited by pump or valve capacity
Inadvertent opening of manual valve	Line capacity	Equipment-Dependent	Credit taken for volume outflow
Failure open of inlet control valve	Valve capacity	Equipment-Dependent	Credit taken for normal minimum flow or outflow
Check valve failure	Valve or line capacity	Equipment-Dependent	
Abnormal heat input	Valve or exchanger capacity, HMB/Simulation	Rate-Dependent	
Exterior fire	Fluid properties, Vessel area	Equipment-Dependent	
Hydraulic expansion (pipe)	Fluid properties, Pipe area	Equipment-Dependent	
Hydraulic expansion (exchanger)	Fluid properties, Exchanger duty	Equipment-Dependent	Exchanger duty driven by hot side flow rate
Tube rupture	Exchanger properties, fluid properties	Equipment-Dependent	
Tube leak	Exchanger properties, fluid properties	Equipment-Dependent	
Plate weld failure	Exchanger properties, fluid properties	Equipment-Dependent	
Column power failure (Complex)	HMB/Simulation, Exchanger duty	Rate-Dependent	
Column power failure (Simple*)	Fluid properties, reboiler duty	Rate-Dependent	
Loss of Absorbent	HMB/Simulation	Rate-Dependent	

*Simple column failure includes overpressure scenarios where the required relief rate can be determined as a function of the reboiler duty and heat of vaporization of the residing fluid.

†Not a comprehensive list

3 Rate-Dependent Factors Influencing Pressure Relief System Design

Although the common practice is to size the pressure relief system as a linear function of process flow rate, the reality is more complex, as other factors that are themselves influenced by process flow rate can also impact the pressure relief system design, primarily through their impact on pressure relief device sizing. These factors can impact the pressure relief device sizing by causing additional deviations in relief rate calculations for rate-dependent overpressure scenarios (i.e., nonlinear relations), as well as creating changes in required relief rates for overpressure scenarios previously categorized as equipment-dependent.

A quick review of three factors influencing pressure relief device sizing will aid in understanding the limits of scaling up the pressure relief system design basis: composition, operating conditions, and heat exchanger duties. Note that this list is intended for illustrative purposes of the effects and is not an all-inclusive list; there are other factors that can influence the design of the pressure relief system that may change during a process rate scale-up.

3.1 Composition

3.1.1 Composition and Pressure Relief System Design

Fluid composition impacts the pressure relief device sizing in two significant ways.

1. Firstly, it can directly impact the determination of the required relief rate; for example, fire calculations use the latent heat of vaporization for a mixture, and depending on the particular modeling choices selected (e.g. inclusion of sensible heat, vaporization interval selected) may include further dependencies on heat capacity, bubble point temperature, vapor to liquid volume ratio, and other variables. Orifice flow calculations (such as are found in tube rupture calculations) depend on the density or molecular weight, compressibility, and specific heat ratio for the fluid.

2. Secondly, the composition of the fluid mixture directly impacts the relationship between the pressure of the fluid and the specific volume of the fluid, which governs the determination of the critical mass flux across the relief device nozzle (G in equation 1).

3.1.2 How Scaling Up the Process Can Affect Composition

It has been commonly observed that the compositional data in a heat and material balance is rarely adjusted when a scale-up of the unit throughput is performed. This assumption can be invalidated by any of the following effects:

1. Changes in the operating pressure and temperature – Changes in operating temperature and pressure can have an impact on the composition by affecting the separation of the vapor and liquid phases within the process equipment. Increasing the operating pressure can raise the bubble point temperature of the fluid mixture, requiring additional heat to achieve the same separation point. Decreasing the operating pressure can reduce the bubble point temperature of the fluid mixture, possibly resulting in more vapor flashing off, with a higher molecular weight.

2. Limitations on feed sources – In order to maximize the throughput for the specific process unit, feeds from multiple sources may be required. If the process unit has excess capacity but insufficient feed is available from on-site units, feed may be supplemented from offsite or from tankage. These feeds may have variations in compositions, and as the new feed is added, the blend ratio can change. For example, a gas oil hydrotreating unit (GOHT) may have additional capacity, but due to changes in product demand the crude column system may be producing a higher proportion of lighter products. To take advantage of the slack

in the GOHT, additional gas oil may be imported, resulting in a change in composition of the feed.

3.2 Operating Conditions

3.2.1 Operating Conditions and Pressure Relief System Design

The operating conditions within equipment in the process unit can directly impact the pressure relief device sizing primarily through the temperature's effect on the fluid conditions (for overpressure scenarios that model the required relief rate using its operating temperature). Temperature also influences the maximum mass flux across the pressure relief device. Since the relief device is sized at the relief pressure, there is no similar direct effect of operating pressure.

More significantly, changes in the normal operating temperature and pressure can affect the required relief rate calculations for overpressure scenarios that require these values as inputs, such as modeling flow across an orifice, flow through piping or valves, or where separation processes are occurring. For example, inadvertent opening of a valve into a separator may involve flow of the fluid mixture across a valve and series of pipe fittings, followed by a separation where the vapor is expected to overpressure the downstream vessel, while the liquid will slowly fill the vessel. The change in overall vapor quality can both affect the required rate to be relieved while the vessel is filling as well as affect the available time for operator intervention – in some cases making the required relief rate better, and in other cases making it worse. Another example is the potential for changes in composition, temperature, or pressure to change the net positive suction head available (NPSHA) for pumps, requiring adjustments to liquid levels in order to maintain proper pump operation, thereby affecting equipment-dependent overpressure scenarios.

It is even conceivable that such a change may alter the applicability of an overpressure scenario, where the existing operating conditions are already near the threshold of applicability – from either direction. Changing the operating pressure may affect the suction pressure of a pump or compressor, thereby altering the deliverable pressure to a downstream system, whose determination of applicability for a control valve failure or inadvertent opening of a manual valve overpressure scenario may depend on the comparison of this pressure to the downstream system maximum allowable working pressure.

Changes to operating conditions can also affect the specification of the pressure relief device. Operating pressures that were previously at the limit of the acceptable range that may now exceed the operating limits of the pressure relief device may require alternative relief device specification (e.g., if the operating pressure exceeds 90% of the PRV set pressure, this may require replacing a conventional PRV with a pilot operated PRV). Significant changes to operating temperature may also affect the set pressure of the pressure relief device. For a spring loaded valve, a new cold differential set pressure may be required.

3.2.2 How Scaling Up the Process Can Affect Operating Conditions

The operating conditions can be impacted in the following ways during a scale-up:

1. The mechanical characteristics of rotary equipment – For example, for a given installed impeller diameter, as the flow rate is increased, the operating point of the rotating equipment must move out on the pump performance curve, resulting in a lower delivered pressure. (This can be countered by changing out the pump impeller, but that is outside the scope of this paper).
2. Piping hydraulic limitations – As flowrate is increased, the pressure drop is expected to increase. With changes in the pressure drop between equipment, the operating pressure profile may be affected. For flow in the turbulent region, for the same diameter of piping, the pressure drop normally follows a trend of

$$\frac{F_1}{F_2}\frac{v_1}{v_2} = \frac{Q_1}{Q_2} = \sqrt{\frac{\Delta P_1}{\Delta P_2}} \qquad \text{(Eq. 2)}$$

Where

F	is the mass flow rate
v	is the specific volume of the fluid
Q	is the volumetric flow rate of the fluid
ΔP	is the pressure drop through the pipe

3. Heat transfer – As flowrates change, when running up against the limits of the heat transfer equipment, changes in temperature and mass quality may result. For example, applying the same duty to a fluid entering at a higher feed rate will soak up more sensible heat and result in a smaller vapor fraction exiting the equipment.

3.3 Heat Exchanger Duties

3.3.1 Heat Exchanger Duties and Pressure Relief System Design

Heat exchanger duties can potentially cause deviations in the relief device sizing basis if the process duties are used as inputs in determining the required relief rates. Note that some users choose to use the design duties of the exchangers in lieu of the process duties. This may mitigate the impact of scale-up (up to the point where new heat transfer equipment is required). Common examples of overpressure scenarios impacted by this factor are blocked outlets on heat exchangers, thermal expansion, and simplified column calculations (i.e. boil-up).

The other route through which heat exchanger duties can impact the pressure relief system sizing is by impacting the operating conditions of the process (see section 3.2.1).

3.3.2 How Scaling Up the Process Can Affect Heat Exchanger Duties

Scaling up the process flow rate through an exchanger can significantly impact the exchanger heat duty if it is limited by flow instead of limited by heat transfer area and overall heat transfer coefficient ("UA-limited"). For the same heat exchanger, the flow pattern correction factor F_T is expected to be constant. In these cases an increase in process fluid flow can result in increased duty, up to the point where the exchanger UA becomes limiting. Note that for forced convection, since the Nusselt number is a function of the Reynolds number, the value for the overall heat transfer coefficient U is not constant; rather, it is itself a weak function of flow rate (as well as composition and operating conditions). If the exchanger duty is limited by UA, then the exchanger heat transfer duty can be still changed by differences in log mean temperature difference (ΔT_{lm}). The temperatures changes may come about as described in section 3.2.2.

3.4 Extent of Deviations and Overall Impact on Pressure Relief System Design

The above discussions have focused on how rate scale-ups can impact the pressure relief system design in ways beyond the typically assumed linear relationship between mass flow rate and required relief area, as well as impacting scenarios that might otherwise not be classified as rate-dependent. While it can be expected that minor changes in flowrate would have a correspondingly minor impact on composition, temperature and pressure, and heat exchanger duties, and therefore a minor impact on the pressure relief system design, it is important to recognize that the extent of deviation from linear behavior can increase with the extent of the process throughput scale-up. Additionally, repeatedly performing scale-ups on a previously scaled-up process, without revalidating the heat and material balance, can lead to the same result in a less obvious manner.

4 Examples

4.1.1 Examples of Compositional Impact on Pressure Relief System Design

For equipment and systems where relief rate is determined by calculating a rate of vaporization, the relief rate typically increases as the proportion of lighter components within the feed increases. Even where the relief rate is not determined by calculating a rate of vaporization, a lighter feed may not change the required relief rate mass flowrate but might instead decrease the relief device capacity, on a mass basis. For example, crude units can be operated with significant variations in the crude slate. These changes in feed composition could cause different effects for relief device design within the crude unit, even with no change in overall unit throughput.

The sizing scenarios for crude column systems commonly include partial power failures that result in continuing crude feed and heat input from the fired heater, with loss of all overhead and pumparound cooling, among other effects (such as backflow and desalter water carryover). In these scenarios, the relief stream is based on the lighter portion of the crude feed, which typically consists of the offgas, liquefied petroleum gas (LPG), naphthas, kerosene, and/or diesel, etc., depending on the heating capability of the feed heater relative to the feed. Therefore, a change in crude slate that increases the proportion of naphtha cuts in the crude while decreasing the proportion of atmospheric gas oil (AGO) cuts can result in a direct increase on the relief rate of the crude column, whereas a change that only changes the heavier portions of the crude feed may not have a significant effect on the crude column. Note that composition changes that do not affect the crude column may affect another system, such as the vacuum column.

4.1.2 Examples of Changes in Operating Conditions on Pressure Relief System Design

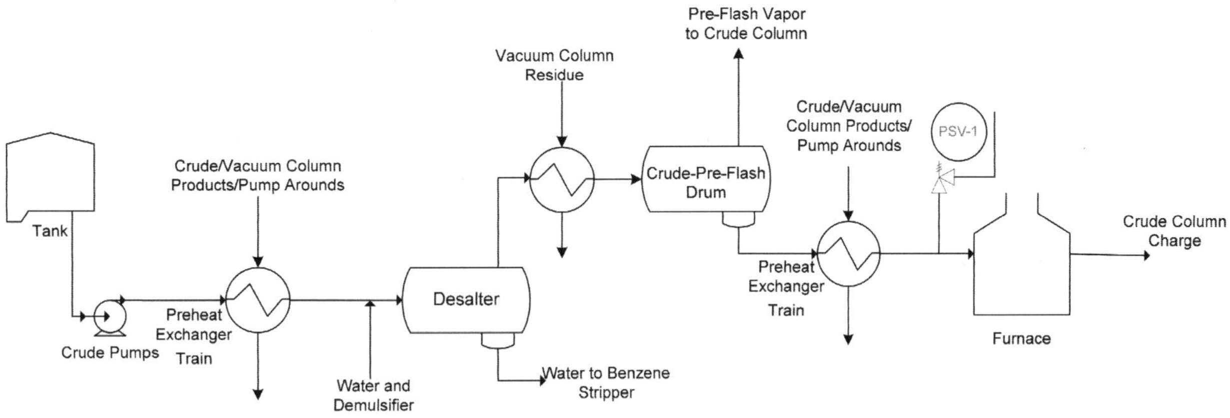

Figure 2: Crude Feed Pre-heating System Process Flow Diagram

The process flow diagram in Figure 2 represents the feed pre-heating section of a Crude Column system. Feed from tankage is pre-heated in a series of exchangers against hot fluid from Crude/Vacuum Column products and pump arounds; the crude feed is then treated for salt content in the Desalter before sending to the Crude Pre-Flash Drum. The Crude Pre-Flash Drum operates significantly lower than the Desalter, allowing the crude to flash. The flashed vapor is directly sent to Crude Column, while the liquid from the drum is sent through another train of exchangers, the furnace, and then to the Crude Column.

a. Effect of Operating Pressure Change on Composition

PSV-1 on the Flashed Crude pre-exchangers is sized for Blocked Outlet, and this case is usually selected as rate-dependent. However, a change in operating pressure of the Crude Pre-Flash Drum, with or without the throughput change, affects the composition of flashed liquid and vapor at the Crude Pre-Flash Drum, because the bubble point of the crude stream components is directly related to the operating pressure. In this case, as the fluid composition is changed, an assumption of a linear relationship between changes in throughput and required area may not be accurate.

b. Effect of Operating Temperature Change on Composition

Similarly, a change in the throughput may result in change in product rates, Crude Column product/pumparound rates and Vacuum Column product/pumparound rates, which could result in change in the heat input at the Crude Preheat exchangers. Similar to pressure, a change in Pre-Flash Drum inlet temperature, may affect the composition of Crude Pre-Flash Drum liquid and vapors, and the rate dependent assumption may no longer apply for Blocked Outlet case for PSV-1.

5 Conclusion

The economic driver for maximizing the production from process units is ever-present. However, it is vitally important to ensure that the techniques used to evaluate the pressure relief system design are appropriate and suitable to the task. Awareness of the assumptions inherent in rate-dependent classifications, and the potential to invalidate this classification with further "pencil-sharpening", is a starting point. Understanding the ways in which process throughput scale-ups can impact other factors in determining the design of the pressure relief system, such as affecting the fluid composition, operating conditions, and heat exchanger duties, and their impact on required relief rate determination or even scenario applicability, helps to illuminate potential problem areas when evaluating a process throughput scale-up, and informs the decision on whether a revalidation of the heat and material balance is appropriate. Altogether, these considerations should enable engineers to make smarter decisions about rate-dependency and ensure safer operation during process throughput scale-ups.

6 References

[1] American Petroleum Institute, *API Standard 520: Sizing, Selection, and Installation of Pressure-relieving Devices Part I – Sizing and Selection*, 9th edition, July 2014.

442298 Learning Lessons from a Big Chemical Disaster in Korea

Abstract:

Learning lessons from a big chemical disaster in Korea

Kwnaghee Lee and Il moon*

Department of Chemical and Biomolecular Engineering, Yonsei University, 50 Yonsei-ro, Seodaemun-gu, Seoul 120-749, Republic of Korea

On September 27, 2012 a massive chemical dispersion disaster occurred at the Gumi National Industrial complex in the southern city of Korea. About 8 tons of hydrogen fluoride gas leaked as the valve of tank lorry was opened due to a worker' mistake. The operators at the factory had never experienced such a big chemical disaster, and the fire station in the area did not have enough equipment at the time of chemical accident, so that an initial counter measurement could not be conducted smoothly. Due to this accident, 5 workers were killed and about 3,600 local residents had to receive a medical treatment at hospitals due to the accident. And the damage proceeded to the second and third stage, causing 17.7 billion KRW of damage to enterprises and spoiling the crops on 323.8ha of land. After the Gu-mi disaster, law and regulations, government organization, and industrial safety system were completely revised and culture have leaded positive effects towards managing and responding to chemical disasters. The governmental action has been led to make more practical plan and manual to control and respond to chemical accident. The governmental reorganization against chemical disaster such as Joint inter-agency Chemical Emergency Preparedness Center which is a specialized agency focused on responding to chemical disasters has also made to increase technicalities of chemical accidents response. In addition, the relevant laws and regulations has clarified the different acts and programs on control, management and assessment of chemicals along as well as its accident. Moreover, private and public enterprises have been also making efforts to systematically manage safety issues and build high-level safety culture. In this paper, we discuss the changes attitudes against chemical accidents from various viewpoints and the successful case in which the relevant agencies responded to major chemical accidents more efficiently after those changes. In order to improve the initial response system against chemical accidents continuously, it is required to strengthen the role of on-site coordinator, establish the Chemical Accident Investigation Committee on a national level. This study covers the big systematic changes in laws, regulations, business culture for the chemical safety management after a big chemical release accidents.

442301 Study on Thermal Radiation of Enclosed Ground Flare

Abstract:

Flare system is an important part of the safety systems in refineries or petrochemical facilities, and on oil and gas platforms. There are generally two types of flares: the elevated flare and the enclosed ground flare. Enclosed ground flares get more and more applications because of its high efficiency, flame invisibility and low nose. During an emergency flaring situation, a portion of the heat produced by the flame will radiate to its immediate surroundings.The amount of radiation heat emitted by flares is important in the design and location of flare systems. But there are no theoretical or empirical methods for predicting the radiation of enclosed ground flares in current literatures. Predicting the thermal radiation becomes one of the greatest challenges in designing a groud flare system. To investigate the radiation of the enclosed ground flare, the flaring process of a series of flare gases were simulated using CFD method.The temperature and radiation field was calculated.Results shows that the radiation flux at ground level were below 1.58kW/m2; at the height above 9m, the 1.58kW/m2 radiation isopleth can reach 60-70 meters radially from the enclosed ground flare axis,so the design of towers or other elevated structures containing operating platforms should consider radiat.

442313 Simple Objects As Indicator of the Severity of Vapour Cloud Explosions

Abstract:

Simple objects such as instrument boxes and oil drums are often found on petro-chemical plant sites. In the event of an accidental vapour cloud explosion (VCE), they serve as reliable indicators of the explosion overpressure at their positions provided there is a good understanding of their behaviour in a VCE.

Such objects were used to great effect in estimating the overpressure magnitude in the Buncefield incident in 2005. In large scale tests conducted since the Buncefield incident (both as part of a joint industry project and subsequently), such objects were located at varying distances from the cloud, as well as some within the cloud. Overpressure measurements were made at or adjacent to the location of these objects and many have been analysed using advanced finite element analysis. The objectives were to (i) gain a better insight into the response of these objects when subjected to different types and magnitudes of pressure loading, (ii) validate finite element models; (iii) assess the suitability of different finite element modelling techniques for predicting the response of these structures and (iv) develop pressure impulse (PI) diagrams which can be used to provide a rapid assessment of explosion magnitude through an examination of the damage level.

This paper presents an overview of the tests performed and the analysis results with comparisons with the actual behaviour of the objects.

Best Practices for Writing Operating Procedures and Trouble-Shooting Guides

Lauren Madden, Process Safety Manager
REC Silicon
Moses Lake, WA 98837
e-mail: Lauren.Madden@recsilicon.com

William G. Bridges, President
PROCESS IMPROVEMENT INSTITUTE, INC. (PII)
1321 Waterside Lane, Knoxville, TN 37922
Phone: (865) 675-3458
Fax: (865) 622-6800
e-mail: wbridges@piii.com

Prepared for Presentation at
12th Global Congress on Process Safety
Houston, TX
April 10-13, 2016

UNPUBLISHED

12TH GLOBAL CONGRESS ON PROCESS SAFETY

Best Practices for Writing Operating Procedures and Trouble-Shooting Guides

Lauren Madden, PSM Manager
REC Silicon

RECSiLICON

William G. Bridges, President
PROCESS IMPROVEMENT INSTITUTE, INC. (PII)

Keywords: Procedures, Human Factors, Risk control, process safety management, PSM

Abstract

There is no complete, best practice guideline or textbook for writing operating procedures and trouble-shooting guides. This paper presents the proven, best approach for developing accurate operating procedures and for ensure the format of the page and steps are optimized to reduce human error rates. This approach and the 35 rules established in this paper are based on the foundations set by Swain and others in the 1970 for control of human error rates, but uses experiences from more than 60 sites where this approach has been successfully followed. The approach and rules for developing procedures for responding to process deviations (such as those needed for Human IPL) are again the best approaches found and have been applied successfully since the early 1990s; these are many times called operational trouble-shooting guides. The paper also discusses how and when to integrate trouble-shooting guides into distributed control systems. Several case studies are provided that show the gains from following this approach. The guidelines in this paper build upon ones presented in 1999 at CCPS.

Background

Operating procedures have always been crucial to the safety, quality, and productivity of process systems. And with the advent of new safety and quality standards such as OSHA's process safety management (PSM) regulation, OSHA's personal protective equipment (PPE) regulation, EPA's risk management program (RMP) regulations, and the ISO 9000 quality standard, many companies are facing the daunting task of developing or upgrading their procedures to satisfy varied and sometimes complex or conflicting requirements.

Although there are overlapping characteristics that the various regulations and standards share, each approaches the procedure-writing process from a somewhat different perspective (e.g., quality or safety, protection of workers or protection of the public/environment), and the required level of detail for documenting the procedures differs greatly between these regulations and standards. For instance, the ISO 9000 quality standard advocates documenting procedures and work instructions that impact quality and suggests keeping the procedures as simple as possible, while the OSHA PSM regulation requires detailed procedures that address all operating modes (startup, shutdown, etc.) and that address operating limits, consequences of deviations, means to avoid hazards, safety and health considerations, and safety systems and their functions. How can companies develop and maintain procedures that ensure productivity and simultaneously satisfy the different regulations and standards? The key is to remember that the ultimate goal of the regulations and standards is consistent: reduce human errors that can impact quality, productivity, and/or safety.

Procedure-related errors are errors that occur because some characteristic of the procedure caused task performance to fail. ***This is currently the most critical human factor at most sites since 90% of accidents have at least one root cause related to mistakes within procedures.*** Reducing these procedure deficiencies can reduce human error rates by a factor of 2 to 10, normally. [1, 2]

The concepts and rules presented in this paper are applicable to all industrial settings. Most importantly, this approach **is effective in satisfying both safety and quality concerns.** With this approach, procedures can and have been developed that provide a sound basis for training and that provide the users and supervisors with an understanding of what can go wrong and how it can go wrong. This approach also provides management with an understanding of the inherent risk (to quality, productivity, safety, etc.) of an activity (i.e., the activity addressed by the procedure) and an appreciation of the safeguards that are being relied upon to mitigate this risk.

An extensive case study is presented by REC Silicon, which illustrates this approach, along with supporting the benefits of the accurate and clear procedures. But the same is true at more than 60 sites where the same approach has been followed.

Fundamentals of Writing Effective Procedures

When developing procedures, or any written material, always remember that the burden of written communication is on the writer, not on the reader. It should never be up to the reader to decipher the meaning of a message. Engineers who write procedure manuals may "tend to include information that is more relevant to them than it is to the operators actually running the unit" [Sutton, 1992].[3] When writing procedures, your guiding principle should always be: if the written procedures are difficult to read, difficult to locate (e.g., if titles are do not have a clear meaning), or inconvenient to use, they will seldom be used [Swain, 1983].[1]

Deficient Procedures are the most prevalent problem in process industries since procedures have not traditionally been developed from the perspective of optimizing human factors; instead, procedures have been traditionally developed to meet a compliance requirement to have written procedures. Examples of procedure deficiencies (inaccuracies) include:

- Incorrect/incomplete/nonexistent (most procedures we have audited have been only 70-75% accurate – the inaccuracies include missing critical steps, steps as written are not what needs to be done, or the steps are out of sequence)
- No/misplaced warnings (for a example, a warning should Never **contain** the action to take; it should instead **emphasize** the action to take)
- Poor format and presentation rules

Figure courtesy of Process Improvement Institute, Inc. (PII), copyright 2003-2016[4]

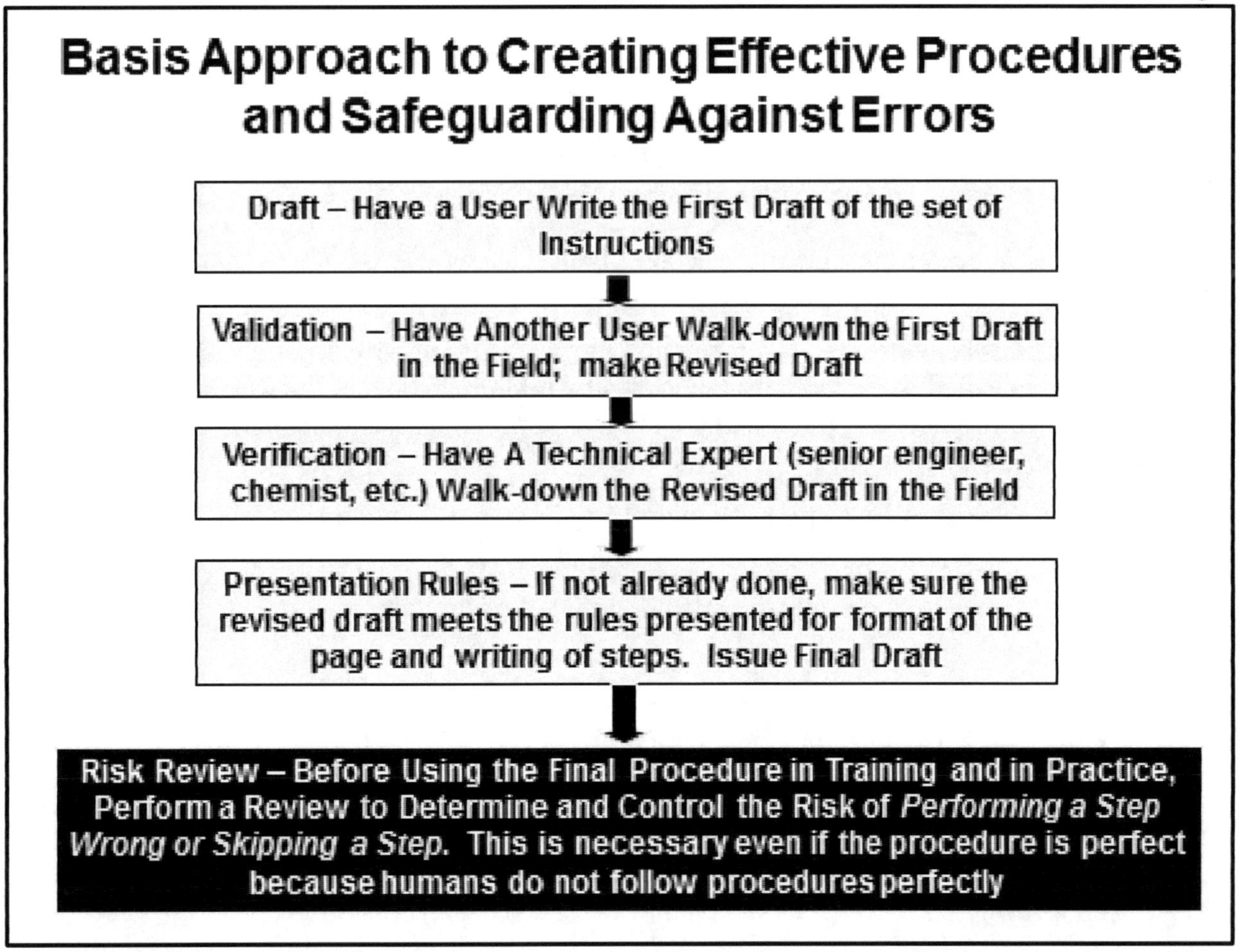

The flowchart above describes the basic sequence of steps for writing effective procedures. Note that the some basic rules are contained in the flowchart, such as "Operators writing operating procedures" and "Maintenance craft-persons write maintenance procedures" and "Lab technicians write lab procedures." Procedures are Not written by engineers or department superintendents. This is a critical rule because it ensures the procedures are written in the common language of the users and is written at the right level of detail.

Universal Purposes of a Procedure[4]

1. A written guide to accomplish initial training of employees
2. A quick reference manual for experienced (trained) employees

The Right Information to Write – Content is KEY[4]

The Content of the procedure controls the human error related to "procedure-based" error – if and content is wrong, the trainee is very likely to learn the step wrong (if the trainer is using the procedure as a basis or when the trainee refers the to the procedure step that is wrong). So:

If the content of the step is wrong it does not matter if the presentation is clear; but if the content is perfect, you can reduce error rates a further 50 to 70% by presenting the information properly.

Ask experienced mechanics or operators to describe a simple task that they perform. You will quickly see that even simple tasks can involve many steps.

You should consider the following questions when writing each procedure step:

- How is the step performed? Does the action involve interacting with a computer terminal, an automatic controller, or simple devices (gauges, valves, micrometers)?
- Can the actions be performed as written and in the sequence written? Can the equipment be operated as specified? Can the steps be physically performed? Does the user have the training or experience to understand and carry out the action using the information available, or is additional information needed?
- Does the user need to be alerted of any potential hazards (Cautions or Warnings) or need any supporting information (Notes) before performing the action?
- Does the user need to know specific operating ranges or limits to:
    - Perform this action
    - Recognize the successful completion of the action
    - Recognize an actual or potential problem
    - Make a decision

For example, does the user have to verify temperature or pressure visually? Is the user waiting for a computer command or verification? How should the process react? Will exceeding a limit result in an accidental chemical release, personal injury, or quality concern? What is the chance of not operating within safe operating limits, and what would be the consequences? If operation of the component is critical and a deviation will probably cause a failure, state the corrective actions. If failure is unlikely, keep going.

- Is needed information found on an instrument, panel, or monitor or is it in the procedure or another source such as a graph, table, drawing, or specification sheet? Should this information be included in the procedure or be referenced?
- What is the next logical step? How is the next step affected by what is performed in the current step?
- What are the results of improper task performance? If an action is critical, spell it out in detail. For example, when starting a special pump that may bind or cavitate if not vented and primed properly, the procedure may require more information than simply stating "start the pump."
- Is the action frequently performed? Is it an action that is easily overlooked? Is this a complex piece of critical equipment that has a low frequency of use? These questions can help determine the level of detail in your procedure.
- Is the action performed so infrequently or is it so complicated that the user is not sure how to do it? Is the action so complicated that nobody is ever certain it's done right the first time?
- Is the decision point clearly defined if a decision is required? Unclear decision points can cause arguments and delays in performing actions.

To address the concerns listed above, a few simple rules are needed for developing content:

- The procedures should be written at the level of someone who has just completed the basic training for that task. Do not write the procedure for someone just hired or for the 10+ veteran.
- Clearly identify ahead of time what activities need procedures to help ensure error rates or controlled low enough, and identify which activities are considered common "skills" of all of the staff.

    Example: Starting a pump is typically considered a "skill." This means a procedure step can simply be "Start the benzene recycle pump (P-119)." The procedure does not need to explain how to start the pump. But, starting a pump is a skill that nonetheless must be learned; it requires doing several sub-steps related to positioning of intake and discharge valves, how the discharge pump is throttled at startup, checking local pressure gage, etc. So, a training module is needed for starting pumps of various types, and the operator needs to learn this necessary skill. Once the skill is learned, it will be applied to so many various cases and so often, that a procedure is no longer needed as a "refresher" on how to do "start the pump."

- The first draft of the procedure must be walked-down in the field by another user. (Simply reading through the procedure does not catch the enough of the mistakes.)
- The revised draft should be reviewed in the field by a technical staff person, such as a process engineer (for operating procedures).

Next:

- The procedure needs to follow the page format and step writing rules described in the next section. This ensures the accurate steps are clearly presented.

The target is to reach an accuracy of 95% or better (so, no more than one wrong or missing step out of 20 steps). The following have been observed in the field by PII staff by direct data collection at more than 110 sites/plants around the world, based on walk-down review of several to dozens of procedures at each unit/plant:

- At 95% accuracy or better and when the same procedures follow 80% of the rules for procedure clarity (presented next), then most users will follow the written procedures and will try to keep the procedures up-to-date.
- At about 85% accuracy or less, about half of the users stop using the procedures
- At about 70% accuracy or less, less than 10% of the users will refer to the procedure or will try to keep it up-to-date. ***So, the written procedures are useless.***
- Unfortunately, the typical Operating procedure walked down by PII staff (accompanied by senior operators, a process engineer, and a shift supervisor) is about 75% accurate (so one step in four is missing or wrong). Usually these inaccurate procedures also follow less than half of the best practices for procedure format, presented next.

PHA/HAZOP of procedures

Just as critical as getting the content accurate and presenting the steps clearly, is performing a PHA/HAZOP of the written steps to determine if you have enough independent protection layers (IPLs) against errors that the "imperfect" operators will eventually make when using the "near perfect" written procedures. This analysis is described in detail in Chapter 9 of the *Guidelines for Hazard Evaluation Procedures*, 3rd edition, 2008[5]. **For an update on the methods for such analysis and for a discussion of the extra emphasis on PHA of procedures by US OSHA, refer to the more recent paper (Bridges, Marshall; 2015).**[6 and 7, 8]

Procedure Clarity

The importance of the accuracy of the procedure steps were discussed above. The clarity of the procedure steps (how they are written and how the page is formatted) is also important. Following best practices for step and page format reduce human errors by a multiplying factor of 3 to 5.[1, 9]

The best practice rules for writing and validating procedures have been published for many years *(see Bridges & Williams, 1997).*[11] These rules include best practices for formatting of the pages and steps; these have been gradually improved over the past decades and now are incorporated into PII's training materials.[4]

The Table on the next is a checklist (in auditing format) based on the current set of best practice rules for developing accurate and clear operating, maintenance, and other work-instructions (procedures).

PROCEDURE QUALITY CHECKLIST *(courtesy PII, copyright 2003-2016)*[4]

#	Issue	Response
	Procedure Content Checklist	
1	Is the procedure drafted by a future user of the written procedure? (Engineers should not author procedures to be used by operators or maintenance staff.)	
2	Is the procedure validated by a walk-down in the field by another future user of the procedures?	
3	Is the procedure reviewed and commented on by technical staff (engineers or vendors)?	
4	Is the procedure checked versus the Page and Step format rules below?	
5	Is a hazard review of step-by-step procedures performed to make sure there are sufficient safeguards (IPLs) against the errors that will occur eventually (when a step is skipped or performed wrong)?	
6	Is the content measured using "newly trained operators" to judge the % of steps that are missing, steps that are confusing or wrong, and steps that are out-of-sequence? (A score of 95% accuracy of content is good; 98% should be the targeted average.)	
	Page Format Checklist	
1	Is the title of the procedure the largest item on the page?	
2	Is the procedure title clear and consistent with other titles, and does it uniquely describes the topic?	
3	Are the document control features the smallest items on the page?	
4	Are temporary procedures clearly identified?	
5	Is white space used effectively? • Is there one blank line between each step? • Does indentation help the user keep their place? • Are the margins large enough to reduce page congestion?	
6	Is type size is 12 pt font or larger?	
7	Is mixed case used for words of steps, with ALL CAPS used only for special cases (such as IF, THEN, AUTO, and WARNING)?	
8	Is the step number very simple (single level of number used)? Only an integer?	
9	Have sections or information not necessary to performing the steps been moved to the back or to another part of the manual or training guide?	
10	Are section titles bold or larger than the text font? Do sections have clear endings?	
11	Is the decision on electronic presentation versus hard copy correct? Are electronic linkages to procedures clear and accurate and easy to use? If paper is chosen, is the color of the paper appropriate?	
12	Is the overall page format (such as Outline format or Two-Column [T-Bar] format) appropriate to the use of the procedure?	
13	Are play script features added for tasks that must be coordinated between two or more users? • Play script is normally used when there are two or more hand-offs of responsibility for steps.	
14	Are rules followed for formatting of Warnings, Cautions, and Notes? (See annotated rules, such as Warnings are for worker safety and Warnings must clearly standout from rest of page.)	
	Step Writing Checklist	
1	Is each step written as a command?	
2	Is the proper level of detail used throughout? This is judged based on: • *Who* will use the procedures • Same level of detail used in similar procedure steps	
3	On average, is there only one implied action per instruction? Best practice is to average 1.2.	
4	Does the procedure indicate when sequence is important? • If sequence matters, each step should be numbered (with an integer or letter) • If sequence does not matter, bullet lists should be used	
5	Are only common words used? Apply "education" level test (5 grade reading level is best)	
6	Do all acronyms, abbreviations, and jargon aid understanding? • Develop a list of such terms for use in procedures *and* communication. • Use terms that users use (within reason)	

7	Is each step specific enough? No room left to guess/interpret: • The meaning of a word or phrase (Check vs. Make sure) • The intent of a step or series of steps • A desired quantity or value • To what equipment the step applies	
8	Is the procedure free of steps that require in-your-head calculations? • Values expressed as ranges rather than targets with error bands • Conversion tables, worksheets, or graphs provided where needed	
9	Are graphics to the user's advantage? • No explanatory paragraphs or lengthy instructions that could be replaced by a picture • No impressive graphics that provide no real advantage	
10	Are references to the user's advantage? • No lengthy explanations or instructions that could be replaced by branching to a reference • No references to a procedure that references still another • No gaps or overlaps between this procedure and a referenced document • If branching, must branch to a procedure, not to a specific step in a procedure	
11	Are rules followed for writing warnings, cautions, and conditional steps? • No more than 2 per page • No actions within a warning or caution (actions must always be numbered steps) • Warnings and Cautions contain descriptions of potential consequences	

Each of these rules is explained in detail in the course *Writing Effective Operating and Maintenance Procedures*, by PII.[4] (Visit www.piii.com for more details.) Each rule is backed by experimental or observed data that indicates there is a clear advantage in error reduction by using the rule. Below are a couple of examples from others on procedure writing rules.

> ***Proper Page Format:*** *In one study, Haney took a procedure presented in narrative format and converted it to column format. He then asked experienced technicians who had been completing the procedure using the narrative format to complete it as presented in the column format. He found that when the technicians used the column format, their errors were reduced by a factor of 3 compared with the narrative format [Haney, 1969].*[10]

> ***Including Supplemental Checklists:*** *Checklists can be effective aids to performance and often should be used to supplement procedures presented in any format. Checklists can be particularly effective in terms of preventing errors of omission. Some of the literature supports using checklists only for simple procedures that do not need to be performed under time or other constraints [Sutton, 1992].*[3] *However, there is also support for using checklists for emergency procedures in which a core set of steps is used to respond to emergency situations, particularly when the actions follow a simple sequence of steps for shutting down a process [Connelly, 1992].*[12] *Overall, experimental data have shown that the probability of errors of omission and commission can be reduced by a factor of 3 to 10 when checklists are used as reminders of key steps in a work instruction [CMA's A Managers Guide to Reducing Human Error, 1990].*[9]

For more data on the history and development of these rules, see Bridges & Williams (1997)[11] and especially refer to originating data and discussion for most of the rules (Swain, 1983).[1] **The procedure on the next page follows the rules**; this is a T-Bar format. Several other page formats are also valid.

PII CHEMICAL COMPANY
Unloading Monomer from Tanker to Storage

Page 1 of 2
Revised 11/4/02, Printed 2/21/16

Unloading Monomer from Tanker to Storage

STEPS	DETAILS
1. Wear standard PPE, plus rubber gloves and full-face organic respirator.	Standard PPE includes hard hat, safety glasses with side shields, and steel-toe shoes.
2. Weigh in the tanker.	Record weight on GROSS line of form.
3. Check bill of lading.	... to verify correct type of material is in tanker.
4. Sign in tanker driver.	Driver must sign in as a visitor and be escorted at all times.
5. Take Certificate of Analysis to QC.	***CAUTION: To avoid contamination, DO NOT unload until you receive approval from QC.***
6. Spot tanker.	A) Direct driver to location. B) Verify brake is set. C) Chock the wheels on at least one side between the two rear axles. D) Ground the tanker (attach grounding strap to an unpainted metal surface).
7. Have the storage area operator make sure the storage tank can hold contents of tanker.	***WARNING:*** *Failure to perform this step can result in an overflow and spill of hazardous Monomer.* Perform **Tank Gauging Procedure (SOP-01-804)** or check the storage tank load cell readout. Storage tank operating limits are stenciled on the control panel.
8. Place 5-psig nitrogen pad on tanker.	A) Connect from local 5-psig nitrogen drop to aft bottom nitrogen connection on tanker. B) Open nitrogen supply valve. C) Then open valve at tanker.
9. Connect Monomer unloading hoses.	A) Verify hoses and gaskets are in good condition. (Hoses are normally stored on hose rack at unloading spot.) B) Connect from tanker to air pump. (Use aft bottom connection on tanker. Remove cap.) C) Connect from air pump to storage tank. (Connect to storage tank line labeled "from TW.")

Trouble-Shooting Guides (Procedures for Responses to Critical Deviation Alarms) [13]

Trouble-shooting guides are a special form of operating procedure; they are written for the actions that we want the operators to take to recover from a process deviation, *before an emergency situation occurs.* They are called guides since rarely can we predict the process conditions at the time the action is required. Trouble-shooting guides (and training and drills of the action) are required for any action that is considered a Human IPL. The Action Limit is what we show as the Min or Max in a Trouble-Shooting Guide. The action limit triggers the demand to use the trouble-shooting guide.

- The response is typically still possible, but it is time dependent. The time available is called the process safety time (PST). The operator must complete the diagnosis, make the necessary change(s), and make sure they are out of harm's way by the end of the Maximum Allowable Response Time (MART). [13]
- We usually set an alarm or a pre-alarm to trigger this action.
- This is usually before the shutdown triggers (ESD occur automatically) or release points (PSV set points) are reached
- The Min and Max shown in the Trouble-shooting Guide table is not the absolute safety limit for a system, but is instead some value that leaves us some time to take action to prevent from reaching the absolute limits.
- There is still time to prevent or avoid the final consequence that could occur if we reach the ultimate limits of the process. Usually, we want the MART to be ½ or less of the PST, and we want MART > 10 minutes for trouble shooting in the field/plant and MART > 5 min for trouble-shooting only from the control room. [13]

If the unit has a good PHA/HAZOP, then it is best to extract information from the HAZOP (or What-If) analysis tables to start the development of each guide. The guide is then finished with input from the process experts.[4]

Examples of Creating a Trouble-Shooting Guide from a HAZOP Table

HAZOP Table Entry	**Trouble-Shooting Guide Entry**
Cause: *Bypass valve is open or passing*	Make sure the bypass is tightly closed
Safeguard: *Isolation valves for the vessel*	Isolate the vessel, if necessary
Safeguard: *Relief valve*	Make sure the relief valve block valves/relief path are open

The key categories of information needed in a trouble-shooting guide are:[4]

- IMMEDIATE ACTION (by system or by operator)
- DECIDE IF ALARM is REAL
- FINDING and FIXING the CAUSE
- FIX or BYPASS PROBLEM

The example below is of a trouble shooting guide for one critical alarm/action.

Trouble-Shooting Guide

Alarm or Indicator:	PAL 4446 – Low Pressure Alarm for Suction of Organic Feed Pump 40-PM-18.445

Action Limit:	5 kPa		
Consequence:	Possible pump seal failure, releasing or spraying organic waste into the berm.		
Process Area:	FB&D Incinerator; Liquid Organic Liquid Feed	**Oper. Mode:**	Normal
Drawing #s:	D-400-PI-013		

IMMEDIATE ACTION (by system or by operator)

- DCS should shut down the organic feed pump (40-PM-18.445).
- From the DCS display, MAKE SURE the organic feed pump is shutdown.
- HAVE the field operator check for leaks near the organic feed pump.
- IF there is a large leak/release, THEN use the ESD switch to shutdown the unit and then follow/complete the shutdown and isolation procedure, OPS-ESD-117.
- IF there is a minor leak or no leak, THEN:
  - COMPLETE the rest of the trouble-shooting,
  - and DECIDE how to contain the leak for now,

DECIDE IF ALARM is REAL

- From the DCS, CHECK the pressure and feed tank level trends. IF the trends indicate the alarm if valid, THEN continue with finding the cause or fixing or bypassing the problem.

FINDING and FIXING the CAUSE

- CHECK valves upstream of the organic feed pump to see if any are closed too far, including checking ESD valves.
- CHECK, by feel with hand, if the heat tracing is on; IF Not, then TURN ON or open heat trace valves
- MAKE SURE nitrogen to the pump seal is at the normal operating pressure.
- CHECK if the line if plugged or frozen (skill)

ABC Chemical Company Prepared by: *Printed copy of this procedure is good for one job task duration.*	OPS-76-TSG-233	Incinerator Unit Revised 8/24/2015 Printed 2/21/16 Page 1 of 2

Trouble-Shooting Guide

Alarm or Indicator:	**PAL 4446 – Low Pressure Alarm for Suction of Organic Feed Pump 40-PM-18.445**

- CHECK if the line if plugged or frozen (skill)
- CHECK if the level is actually low, use the Organic Feed Tank.
- IF the cause is low level in the feed tank, THEN resolve the problem if necessary based on cause that is found (skill)

FIX or BYPASS PROBLEM

- IF necessary, SHUT DOWN the Unit to allow fixing of the problem.
- FOLLOW the proper procedure to resolve problems (repair procedure, line clearing procedure, etc.)
- IF the decision is made to continuing operation without all equipment in normal condition, THEN:
    - FOLLOW Temporary Operating mode, if there is a temporary procedure already written for this possible problem/condition
    - FOLLOW MOC procedures to obtain approval for any non-standard temporary operating procedure or mode

END

Optimal presentation of trouble-shooting information.

Ideally (or ultimately) such trouble-shooting information should be imbedded in the basic process control system (DCS) so that the operators can access the information on demand, with the click of mouse or key.

Comments on Current Industry Consensus Guides for Writing Procedures

None of the consensus textbooks available for procedure writing cover all of the major points of this paper. The best "guidelines" for writing procedures currently exist in the course notebooks used by PII (and likely in the course notebook used by ABS Consulting, as they had similar origins).

Notably, the AIChE/CCPS textbook, *Guidelines for Writing Effective Operating and Maintenance Procedures*, 1996[14]:

- has NO description at all on the development and use and format of Trouble-Shooting Guides. This should have been a major chapter in the textbook.
- mentions only about half of the rules listed in this paper (and the related PII course notebook) for formatting of pages and steps
- emphasizes technical staff or professional procedure writers writing procedures, rather than the Users themselves writing the procedures (it violates the first rule of procedure writing)
- Fails to cover the competency needed by procedure writers.
- recommends and emphasizes having subsections in each work instruction/procedure (between the Title of the task and Step 1) rather than relegating such information either to other sections of the operating manual or rather than turning this information into steps.
- fails to mention the need for a PHA of procedures that focuses on hazard review of deviations from the steps; this is required by best practices and by US OSHA for PHAs under PSM.

This guideline textbook is not worth the investment.

As a better example, but one that nonetheless still falls short of what is needed, the ASM Consortium *Guidelines for Effective Procedural Practices,* 2010[15]:

- has good discussion on the need and thought processes related to handling abnormal situations (trouble-shooting or response to critical deviations), but there is little description on the format and detailed content of the trouble-shooting guides; in fact, the guideline fails to differentiate between routine procedures for startup, shutdown, and online maintenance, and those used for response to critical alarms (trouble-shooting)
- mentions only about half of the rules listed in this paper (and the related PII course notebook) for formatting of pages and steps
- mentions that experience users themselves should write their own procedures, but it does not emphasize this rule and leaves the authorship open to engineers or a committee. And it specifically mentions use of retired senior operators for writing operating procedures. While this is one good way to write the procedures (while also capturing the knowledge from this experienced group is prior staff), we have found that the current users do a better job overall, once they have achieved competency in how to write procedures.
- mentions the competency needed for procedure writers in general.
- mentions the need for a PHA of procedures, but does not reference the method for this in Chapter 9 of *Guidelines for Hazard Evaluation Procedures*, CCPS/AIChE, 3rd edition, 2008,[5] or other related papers on the topic. Also, the description of the PHA is misleading because it focuses on how PHAs should be used in writing procedure content (which is fine for trouble-shooting guides) but fails to emphasize that a step-by-step analysis of deviations from the steps are needed to ensure that the process has enough IPLs against mistakes that can occur when imperfect humans use the procedure.

Though this guideline has deficiencies, it is worth the investment and provides useful insight not found elsewhere. Just be sure to address the deficiencies noted above.

Implementation of this approach

PII staff has helped more than 60 sites implement the approach above. The typical approach for successful implementation is:

1. Perform a gap analysis to determine the average accuracy of procedure content, average clarity (as judged versus the procedure format rules, and identify any examples of excellent procedure writing.
2. Convince management of the problem and get their buy-in on the solution.
3. Train (1.5 to 2 days for operators; 1 day for lab or maintenance staff) and coach the users of procedures (normally for 3 to 8 days, to have them achieve full competency) so they can be effective authors. Normally, 20 to 25% of the users (hands-on operators, hands-on maintenance and inspection staff, lab staff) need to be trained as authors. Not all of those trained will make good writers; the organization needs to judge who does the best jobs. About 15 to 20% of the staff should end up being authors.
4. Get organized, like a project, for tracking and scheduling of the effort. Use master spreadsheets such as those used at REC Silicon (shown later).
5. Use a Traveler form (described later) to make sure each new procedure gets a proper amount of review by other users and by technical staff (such as engineers)
6. Reward procedure writers for their efforts. Especially ensure there are no disincentives for taking the extra effort to write procedures.
7. Track the progress on the effort; fix problems as they arise; enforce schedules (to show management cares).
8. Perhaps use to specialized software make it easy to output the content of procedures into different user-friendly formats such as training tools, training competency records, and field checklists. WordPerfect (formerly from Corel) made this relatively easy with the use of Primary and Secondary merge files; MS Word makes this more difficult. As a result, some specialized software have been developed to help keep formats consistent and to help reformat the content, on demand.
9. Issue procedures through a document control system that does not add extra workload.
10. Of course, ensure there is JSA performed for procedures that have "personal injury" potential and perhaps a PHA of the steps in the procedures whenever there is "process safety" potential.
11. Train all affected workers in the new procedures.
12. Identify the best writers to become the trainers of new authors, over the long run.

Overview of REC Silicon

REC Silicon is one of the world's largest manufacturer and suppliers of

- Silane Gas and specialty silicon gases.
- Granular Polysilicon from Fluid Bed Reactor technology for solar photovoltaic applications
- Ultra-pure float zone Polysilicon for the electronic industry

REC Silicon's Product and Service Offerings

Solar Grade
Polysilicon with various applications for the solar industry

Silane Gas
Product used in production of semiconductors, flat panel displays & thin film technologies

Electronic Grade
Polysilicon delivered to the semiconductor industry as well as other electronic market segments, such as optics & MEMS

Float Zone
High-purity polysilicon delivered to the power electronics industry

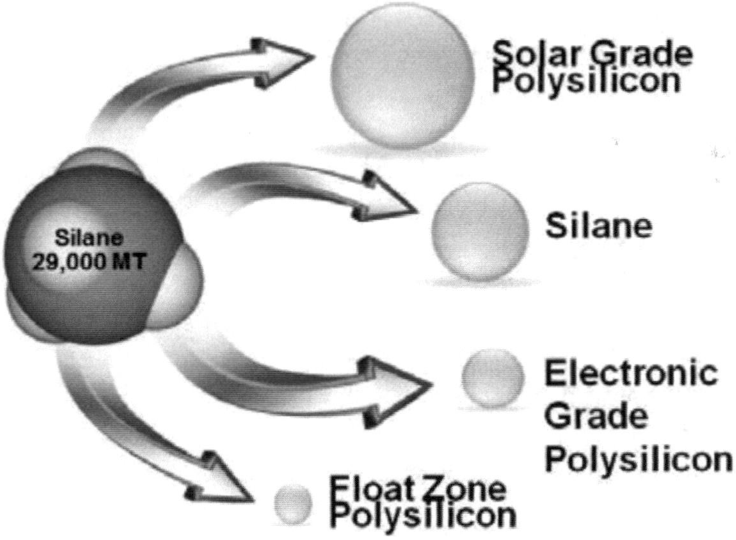

Moses Lake, Washington, USA 396 Employees	Silicon III, IV FBR	Silane gas: 22,000 MT Granular silicon: 16,300 MT
Butte, Montana, USA 255 Employees	Silicon II	Silane gas: 7,000 MT Siemens grade: 3,600 MT

Implementation at REC Silicon

REC production facilities in the USA have been actively implementing process safety for many years, with a renewed interest in 2013 in meeting industry best practices (which goes well beyond US OSHA PSM requirements). To that end, REC commissioned gap analyses versus best practices at both the Butte and Moses Lake production sites. One common finding was that the operating procedures and maintenance procedure for both sites needed improvement with respect to accuracy of the steps as well as clarity (page format and step format). The following were some of the issues with the procedures and related management systems, per the audit results:

- Procedure Inaccuracy (procedure content was less than 80% accurate)
- Technical Staff (such as process engineers) were viewed as owners, not shop floor
- Annual review process ineffective
- Procedure formats were poor; did not follow best practices for reducing human error
- Inconsistent use of Warning, Caution, and Note
- Procedure Naming – not consistent
- Tried to take on all modes of operation in one procedure title
- Procedure accessibility was difficult in the legacy database; multiple searches needed to find tool for job
- Procedure Safety Reviews – no formal methods were applied for procedure safety reviews; i.e., there were no PHA of procedures with process hazards or JSA of tasks with personal hazards

Procedure Author Training – ***the first step in any procedure writing effort***

In early 2014, REC Silicon adopted Process Improvement Institute's (PII's) procedure author training module (*Writing Effective Operating and Maintenance Procedures*[4]) with minor customizations. The training consists of 2 days of classroom lecture and about 6 exercises, followed by practical procedure development and field validation steps in the coaching phase. The classroom course covered all aspects of procedure writing with the focus on optimizing the aspects of the procedures that will lower the likelihood of human error when using the procedures. Topics included:

- human factors overview
- getting the content right
- page format rules and selection
- step format rules
- addressing operating limits (including creating trouble-shooting guides
- nuances of emergency shutdown and emergency operating procedures

- implementation guides and high level procedures for the task of getting all of the work instructions written
- examples/exercises on each topic

The coaching consisted of the instructors from PII, who are expert in organizing procedure writing efforts, helping the procedure writers organize and write operating and maintenance procedures, and trouble-shooting guides. Coaching was for 2+ days. About 20 staff from the Moses Lake and Butte sites attended the initial author training and coaching session.

REC Silicon has revised its management systems to require this competency for all personnel developing or changing procedures and to reach full operator certification. Operators must demonstrate an active role in the procedure MOC process to qualify for recertification every 3 years, otherwise they must repeat the classroom and field training. The unit trainers have taken over leading the procedure author training.

The benefits we have seen from these changes include transfer of ownership of procedures from the technical day staff to the operator since these are the people performing the work in the field. This is necessary to have accurate procedures. **The emphasis on intentional competency development for what is required to write effective operating procedures has greatly improved the quality of our procedures and our operator training and certification program.** The quality of the annual reviews and audits has improved tremendously with the rewritten procedure format and content changes.

Safe Operating Limits and Consequence of Deviation

REC Silicon has developed an engineering standard that governs the content and format for this type of special safety troubleshooting procedure. *CCPS Guidelines for Engineering Design for Process Safety*[16] is the basis for determining the safe operating limits. Production engineers are the owners of these procedures. Engineering competency is required to develop the safe operating limits table based on the available process safety information. A consequence of deviation is then developed by the Process Hazard Analysis team. Finally involvement of certified operators is required to develop steps to correct or avoid exceeding the safe limit. This combined effort produces the safety troubleshooting guide, which complies with the approach listed earlier in this paper.

REC silicon has risk ranked its process areas and is focusing on developing safety troubleshooting guides for high risk process first. To date, for the Moses Lake Facility, thirty six (36) guides have been developed. These have been developed in conjunction with PHA updates to the safe operating limits. We have not yet implemented the formal training and competency testing of these safety procedures. This is planned for the next phase of the initiative.

How REC Silicon approached the re-write effort:

To begin this initiative, each operating department listed all of the existing operating procedures into the procedure development tracking form. Each department also developed a list of "skills" that are used across multiple procedures and that operators or maintenance staff are required to attain before using the written procedures that were developed for more complex tasks (such as starting a pump, or hookup of a chemical transfer hose).

The workers (users) then screened each procedure for safety, equipment damage, or quality risk. This assisted in determining whether 2, 3 or 4 peer walk-downs were required on the validation traveler workflow. Since most of the REC Silicon procedures covered more than one mode of operation, the term "parent" was developed for the original procedure and each procedure was rewritten and broken out into a separate "child" procedures. The procedure re-write assignments were divided up between all four rotating shifts and the superintendent assigned certified authors (e.g., operators trained/coached as procedure writers) from their pool. Each department is managing their pool of qualified authors. For example, our fluidized bed reactor (FBR) unit has 11 qualified authors. Silane, which is a much larger department with 3 production units, has 56 qualified authors. Cross shift validation was utilized to ensure all operating approaches were considered. The assigned author was responsible for consolidating red line comments.

The traveler process introduced by PII was customized to consider information that is marked for removal from the parent procedure. **A Traveler form shown to the right accompanies each new or revised procedure through the writing and editing/review process.** When complete, the traveler form is then kept in the QA file for the procedure to validate long-term how the initial procedure was developed.

Procedure Development Traveler - Moses Lake			
New Document Number:		Area:	
Procedure Title:			
Procedure Type:			
Author:			
Initial Draft Completion Date:			
Author Walkdown Date:			
Author Revision 1 Completion Date:			
Peer 1:		Shift:	
Peer 1 Walkdown Date:			
Optional Peer 2:		Shift:	
Peer 2 Walkdown Date:			
Optional Peer 3:		Shift:	
Peer 3 Walkdown Date:			
Optional Peer 4:		Shift:	
Peer 4 Walkdown Date:			
Author Revision 2 Date:			
Engineer/Verifier:			
Verification Completion Date:			
Author Revision 3 Completion Date:			
QA Person:			
Author Revision 4 Completion Date:			

Images and layouts copyrighted by PII, 2008-2016[4]

A color coding process has been developed to indicate if the content can be permanently deleted, or must be moved into the unit training manual. The unit trainers have been responsible for tracking and reporting weekly progress on the initiative. **See an example of the overall tracking form below.** A total of 184 "parent document" procedures for Moses Lake were tracked through the development, review, and revision process (along with about 200 child documents).

Images and layouts copyrighted by PII, 2008-2016[4]

Procedure Development Tracking Form - Moses Lake

							Initial Development			Validation				
Item	New Doc #	Old Doc #	Area	Procedure Title	Proc Type	Author(s)	Initial Draft	Author Walkdown	Author Rev 1	Peer 1 Walkdown	Opt. Peer 2 Walkdown	Opt. Peer 3 Walkdown	Opt. Peer 4 Walkdown	Author Rev 2
1	FBP.3.xxx	FBP.3.160	FBR	Hydrogen Booster Compressor Cleanup	Cleanup	LT/FG	3/19/14	3/19/14	3/20/14	3/21/14	3/21/14			
2	FBP.3.xxx	FBP.3.160	FBR	Hydrogen Booster Compressor Operation and Op Restrictions	Operation	LT/FG	3/20/14	3/20/14	3/20/14	3/21/14	3/21/14			
3	FBP.3.xxx	FBP.3.160	FBR	Hydrogen Booster Compressor LOTO	LOTO	BCA/OL	3/19/14	3/19/14	3/21/14					
4	FBP.3.xxx	FBP.3.160	FBR	Hydrogen Booster Compressor Standby	Standby	BCA/OL	3/20/14	3/20/14	3/21/14					
5			4.0	STC/TCS from Receiving Tank to STC/TCS Storage Transfer Through Metering Line	Transfer	ME/RR	3/19/14	3/19/14	3/20/14	3/20/14	3/21/14	3/21/14	3/21/14	
6			4.0	STC/TCS from Receiving Tank to STC/TCS Storage Transfer Through Bulk Line	Transfer	ME/RR	3/19/14	3/19/14	3/20/14	3/20/14	3/21/14	3/21/14	3/21/14	
7			4.0	H2 Recycle Compressor Switching to Standby	Standby	ME/RR	3/20/14	3/20/14	3/20/14	3/20/14	3/21/14	3/21/14	3/21/14	
8			4.0	H2 Recycle Compressor Cleanup	Cleanup	ME/RR	3/20/14	3/20/14	3/20/14	3/20/14	3/21/14	3/21/14	3/21/14	

From PII's procedure author training,[4] the Quality and Content checklist was modified and used as quick reference guide to make sure that all the content development and formatting rules are being followed during the re-write process. We used Microsoft Word and made templates for the Tbar, Flat and Consequence of Deviations.

Procedure Management of Change

Through this project REC Silicon has adopted formal safety review methods for procedures. These methods standardize the approach to determine if a procedure is safe to use as written or if revisions are required. These details are outlined in a completely new REC Silicon policy - Procedure Management of Change PSM.30.004.01. The REC Silicon IT department developed a custom tool to execute these requirements in an electronic workflow. The benefits from these changes are formally transferring competency to lead and document job safety analysis to the shop floor supervisors. Prior to this effort, JSA was traditionally a review lead by our safety specialists.

Overall Summary of the procedure writing initiative:

For the Moses Lake facility, 398 procedures (of which 184 were unique and 214 were similar to the 184 unique ones) for normal and routine operations, various maintenance tasks, and trouble-shooting guides have been developed through this project (new and rewrites combined.) Similarly, for the Butte facility, 116 procedures (of which 24 were unique and 92 were similar to the 24 unique ones) for normal and routine operations, various maintenance tasks, and trouble-shooting guides have been developed through this project (new and rewrites combined.) In addition, about 56 existing procedures (that were in "okay" state) were updated and the format improved.

We do not have any data yet for reduction in error rates or increase in production, quality or safety; however we have seen great improvements with the new formats with the ability to conduct job observations in the field. Our job observations assess accuracy of the work being performed, accuracy of the written procedures, and an opportunity for on the job operator training.

This type of project can seem overwhelming for a manufacturing site. It is essential to form a project team with key decision makers from all the correct disciplines and have a project sponsor that can remove barriers. The way REC Silicon was able to accomplish this type of effort was dedicating time to meet each week, setting short term goals, and tracking KPIs to meet those goals. The procedure writing initiative at REC Silicon is still very much "a work in progress". Leaders of the project have to make time to support the staff to keep the momentum going on this initiative and also on the related culture change, until it becomes "the way the work always gets done."

Quotes from workers involved in initiative:

- Operators are receiving better work instructions
- Sequence of steps is well understood
- Much easier to audit operator compliance to procedure
- Standardizes how we do work – protects our equipment from damage, protects our people, protects the environment
- No longer need 20 year operator tribal knowledge and interpretation to complete procedure
- Procedures are not cluttered anymore, they just have the necessary information
- Procedure element now requires involvement of many shop floor workers
- Procedures are really safe to use
- JSA technique is alive on the shop floor
- Procedure can be trusted
- Technical staff no longer doing all procedure work. Freed up to optimize processes and realize other cost saving benefits.

Conclusions

Having clear and accurate procedures is a must. Poor procedures show up as one of the root causes in 90% of the accidents. The approach described in this paper has worked for more than 35 years for one of the authors, and continues to work today. Controlling risk is mostly about controlling human error rates, and having accurate and clear procedures are one key to this control. One other gain from the approach described in this paper that the users become the owners, and therefore they have a vested interested to keep the procedures up to date and clear.

No doubt, writing all of the procedures for a facility is a large task. But, it can be accomplish by site staff, with support from process safety or quality control leadership, after enough users become competent authors; attaining this competency takes less than 2 weeks of work, on average.

Acronyms Used

AIChE– American Institute of Chemical Engineers
ASM – Abnormal Situation Management consortium
CCPS – Center for Chemical Process Safety (a division of AIChE)
CMA – Chemical Manufacturer 's Association, now American Chemical Council (ACC)
FBR – Fluidized Bed Reactor
HAZOP – Hazard and Operability Analysis
IPL - Independent Protection Layer
JSA – Job Safety Analysis
LOPA – Layer of Protection Analysis
MART – Maximum Allowable Response Time
MOC – Management of Change
NUREG – Nuclear Regulation, US Nuclear Regulatory Commission
OSHA – Occupational Safety and Health Administration, US Department of Labor
PHA – Process Hazard Analysis
PII – Process Improvement Institute, Inc.
P&ID – Piping & Instrumentation Diagram
PSI – Process Safety Information
PSM – Process Safety Management
PST – Process Safety Time
SOP – Standard Operating Procedure

References

1. Swain, A. D., Guttmann, H. E., *The Human Reliability Handbook - with Emphasis on Nuclear Power Plant Applications*; NUREG/CR-1278, U.S. Nuclear Regulatory Commission, Division of Facility Operations, 1983.
2. Gertman, D., Blackman, H., Marble, J., Byers, J. and Smith, C., *The SPAR-H Human Reliability Analysis Method,* NUREG/CR-6883, U.S. Nuclear Regulatory Commission, Office of Nuclear Regulatory Research, 2005.
3. Sutton, Ian S., "Writing Operating Procedures," *Proceedings of the International Conference on Hazard Identification and Risk Analysis, Human Factors and Human Reliability in Process Safety*, AIChE, Center for Chemical Process Safety, New York, NY, 1992.
4. *Writing Effective Operating and Maintenance Procedures*, Course Notebook, Process Improvement Institute, Inc., 2003-2016.
5. *Guidelines for Hazard Evaluation Procedures*, 3rd Edition, 2008, CCPS/AIChE.
6. "Necessity of Performing Hazard Evaluations (PHAs) of Non-normal Modes of Operation (Startup, Shutdown, & Online Maintenance)", W. Bridges and Mike Marshall (US OSHA), *18th Annual International Symposium, Mary Kay-O-Connor Process Safety Center, College Station, TX, October 2015.*

7. "How to Efficiently Perform the Hazard Evaluation (PHA) Required for Non-Routine Modes of Operation (Startup, Shutdown, Online Maintenance)," W. Bridges and T. Clark, *7th Global Congress on Process Safety*, Chicago, AIChE, March 2011.

8. Bridges, W.G., et. al., "Addressing Human Error During Process Hazard Analyses," *Chemical Engineering Progress*, May 1994

9. *A Manager's Guide to Reducing Human Error*, 1990, Chemical Manufacturers Association (CMA)

10. Haney, R. W., "The Effects of Instructional Format on Functional Testing Performance," *Human Factors*, 11, 1969, pp. 181-187.

11. "Effective Procedure Writing – A Two-Part Series (Part 1: Reducing Risk of Worker Errors; Part 2: Addressing Process Safety Management Requirements for Operating Procedures)," *Chemical Engineering Progress*, 1997, W. Bridges with T. R. Williams (both of JBFA).

12. Connelly, C. S., "Emergency Operating Procedure Writing for Oil Refinery Process Units," *Proceedings of the International Conference on Hazard Identification and Risk Analysis, Human Factors and Human Reliability in Process Safety*, AIChE, Center for Chemical Process Safety, New York, NY, 1992.

13. *Guidelines for Initiating Events and Independent Protection Layers*, 2015, CCPS/AIChE.

14. *Guidelines for Writing Effective Operating and Maintenance Procedures*, 1996, CCPS/AIChE.

15. *ASM Consortium Guidelines – Effective Procedural Practices*, 2010, ASM Joint R&D Consortium.

16. *Guidelines for Engineering Design and Process Safety*, 2nd Edition, 2012, CCPS/AIChE.

12TH GLOBAL CONGRESS ON PROCESS SAFETY

How Process Safety Information Affects the Quality of a Process Hazard Analysis

Michael Mayo
Smith & Burgess LLC
7600 W. Tidwell Rd, Ste 600
Houston TX 77040
mike.mayo@smithburgess.com

Dick Baum
dick.baum@smithburgess.com

Prepared for Presentation at
American Institute of Chemical Engineers
2016 Spring Meeting
12th Global Congress on Process Safety
Houston, Texas
April 11-23, 2016

How Process Safety Information Affects the Quality of a Process Hazard Analysis

Michael Mayo
Smith & Burgess LLC
7600 W. Tidwell Rd, Ste 600
Houston TX 77040
mike.mayo@smithburgess.com

Dick Baum
dick.baum@smithburgess.com

Keywords: PHA, PSI, MI

Abstract

This paper is a tutorial that shows the importance of how Process Safety Information affects the quality of a Process Hazard Analysis. The Process Hazard Analysis (PHA) is the one requirement of the PSM standard that has the facility stop and review a process in its entirety to ensure safe operation. However, a high quality PHA is dependent on accurate process safety information (PSI) that is consistent with the current equipment state (Inspection History, Changes from original design, operating conditions etc.). A PSI sub-element, relief system design basis documentation, requires significant amounts of other PSI and if organized with forethought, will greatly improve the quality, effectiveness and, potentially, speed of a PHA. If the relief systems documentation were performed in a system that stored all of the other relevant PSI and was cross referenced with current inspection records (MI systems), then this subset of PSI would contain much of the critical information needed for a PHA. The system would have to be designed such that it was accessible to the PHA team and organized to match the PHA workflow. This paper will present a proposed organization of PSI and then show how with a trained PHA facilitator and organized and accessible PSI/MI data, specific items found in PSM audits of PHAs would have been avoided.

1 Introduction

From defining the Causes, developing the Consequence scenarios and ending with Safeguards, accurate PSI information is required to provide the most credible assessment of risk. The best practice of minimizing the assumptions used during a PHA study has two significant benefits;

- Management can be assured (feel good) that the PHA process utilized all relevant and up-to-date PSI information thereby reducing potential inaccurate assumptions and,
- It will reduce time spent for discussions between participants to come to common conclusions and speed up the PHA process while maintaining the integrity of the study.

As learned from many PSM audits, PSI information is typically spread throughout many locations and maintained by numerous groups;

PSV files - Engineering

Mechanical Equipment Data/Inspection reports - Engineering, Maintenance, Inspection Departments

Critical Instrument Logs - Engineering or Controls

In a perfect world, maintaining a common database to store all up-to-date PSI information would be ideal, but the delegation of maintaining and access would be cumbersome. There are many people involved with PSI information and each has their own responsibilities and actions for validating the data. One way to reduce the roadmap for locating PSI data is to maintain relief systems documentation in such a way that it provides most all relevant data in a simplified format and links to other sources of PSI information.

This poster will walk through the PHA process and how readily available PSI information can be located with robust relief systems documentation.

2 Causes

Typically, the detailed cause of a PHA would not reference back to any PSI data. The failure or misalignment of valves are always credible occurrences that are documented. Where PSI information becomes relevant is in the cause category. The example shown in figure 1 indicates the following;

- Inlet PCV-001 on methane supply to V-001 reactor fails open

Cross-referencing to the P&ID indicated a figure 2, it is noted there is a PSV-002 upstream of PCV-001, a vessel, V-001 with its own relief device, PSV-001. A quick check during the PHA study of the relief systems documentation for PSV-002 will show that the set pressure of 400 psig is indeed above the MAWP of V-001, 285psig.

This validates the cause category of high pressure. Had the set pressure of PSV-002 been documented at 250 psig, this cause probably belongs in another category, i.e. more flow.

Processes: 1. V-001 Reactor

Nodes: 1. Feed streams

Deviation	Causes	Consequence	Risk			Safeguard
			S	L	RR	
1. High Pressure	1. Inlet PCV-001 on methane supply to V-001 reactor fails open	1. Potential for high pressure in V-001 reactor, potential **flange leak**, loss of containment, explosion, jet fire, employee injury, damage to equipment and loss of production				1. PSV-001 on V-001 reactor
		2. Potential for high pressure in V-001 reactor, potential **rupture**, loss of containment, explosion, jet fire, employee injury, damage to equipment and loss of production				

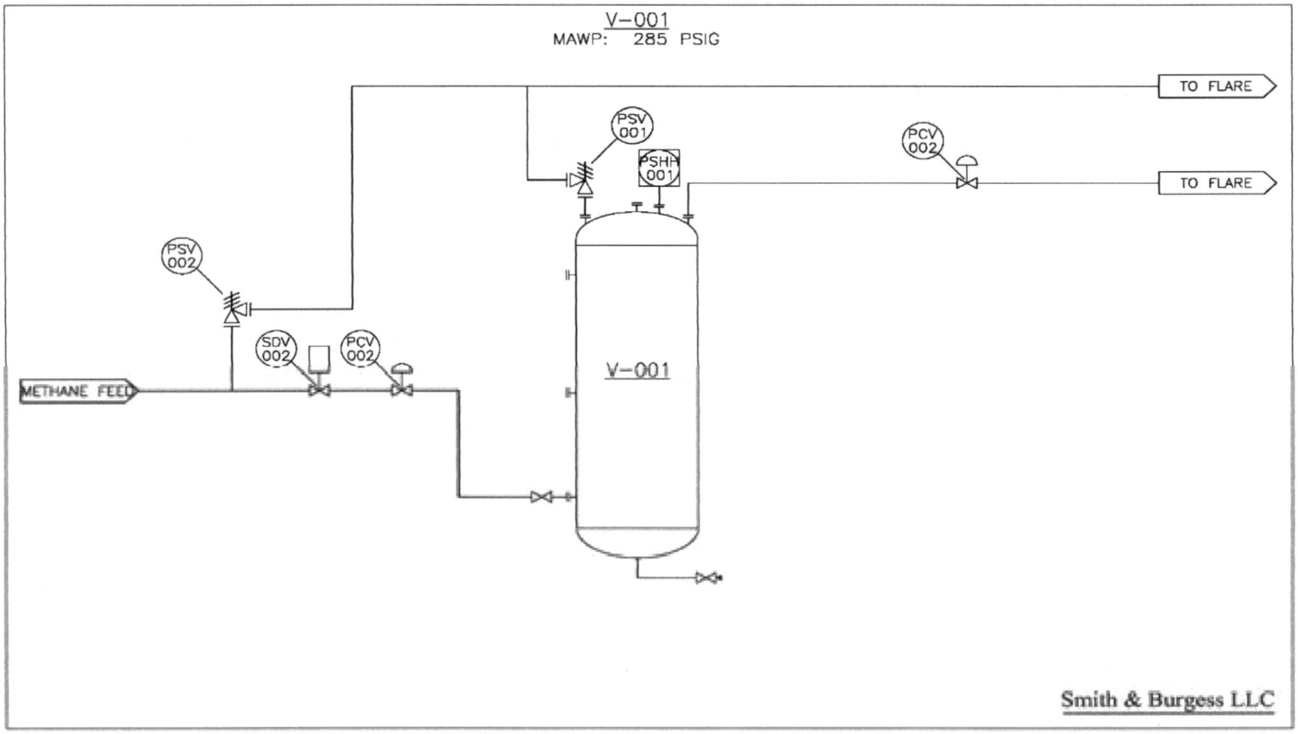

3 Consequences

The consequences developed during the PHA can benefit greatly from accurate and readily accessible PSI information. In the high pressure deviation we can use the PSI information in establishing the overpressure ratio to assist in pinpointing the consequence severity levels. Using the PSI information for PSV-002, a 400 psig supply into vessel V-001 with a MAWP of 285 psig, results in an overpressure ratio of 1.404. Based on your client's guidelines for overpressure and probabilities, this could result in either a flange leak or vessel rupture.

The consequence severity of these two scenarios is quite different but with the PSI data well documented and easily obtained, accurate severities can be obtained during the PHA.

Processes: 1. V-001 Reactor

Nodes: 1. Feed streams

Deviation	Causes	Consequence	Risk			Safeguard
			S	L	RR	
1. High Pressure	1. Inlet PCV-001 on methane supply to V-001 reactor fails open	1 Potential for high pressure in V-001 reactor, potential **<u>flange leak</u>**, loss of containment, explosion, jet fire, employee injury, damage to equipment and loss of production				1. PSV-001 on V-001 reactor
		2 Potential for high pressure in V-001 reactor, potential **<u>rupture</u>**, loss of containment, explosion, jet fire, employee injury, damage to equipment and loss of production				

PSV-002 set pressure can help determine consequence severity

PSI data for V-001 can help determine the consequence severity

4 Safeguards

The safeguards listed during a PHA have the most critical requirement for validation and accuracy during a PHA. They are the protection layers that mitigate or prevent major accidents from occurring and thus leave no room for assumptions or quantitative assessments. PSI information in relief system documentation should be current and accurate for the current operating parameters. A well-documented and arranged system will allow for easy access during the PHA. Information from other PSI data such as P&ID's does not always reflect the purpose of the safeguard. For example, many P&ID's show the governing case of a pressure relief device, which may or may not satisfy the scenario release and could lead to inaccurate assumptions being made during the PHA. (Fire case vs overpressure)

Processes: 1. V-001 Reactor

Nodes: 1. Feed streams

Deviation	Causes	Consequence	Risk			Safeguard
			S	L	RR	
1. High Pressure	1. Inlet PCV-001 on methane supply to V-001 reactor fails open	1. Potential for high pressure in V-001 reactor, potential **flange leak**, loss of containment, explosion, jet fire, employee injury, damage to equipment and loss of production				1. PSV-001 on V-001 reactor set at XXX
		2. Potential for high pressure in V-001 reactor, potential **rupture**, loss of containment, explosion, jet fire, employee injury, damage to equipment and loss of production				PSV-001 set pressure can determine the likelihood

5 Conclusion

Well documented PSI information is critical to the PHA process and provides the validation needed to ensure that the process is being operated and maintained in the best possible safe state. The examples shown indicate how comprehensive pressure relief files can provide much of the needed PSI data or arranged to provide the roadmap for data locations through links to other files. The additional benefit for having accessible PSI information is a smooth transition during the PHA and reduced costs and time for the PHA process.

442025 Experimental Measurements of Flammability at Elevated Conditions Using Thermal Criterion

Abstract:

Accurate measurements of flammability limits for combustible gases and vapors are necessary for the prevention of fire and explosion. The flammability limit, which is the maximum or minimum fuel concentration at which a gas mixture is flammable in a given atmosphere, has been studied for decades. However, due to the measurement conditions and apparatus set difference, the flammability values provided by different experimental methods are inconsistent, especially at elevated conditions. In this study, the flammability limits of methane-air were measured in an 8-L closed cylindrical vessel with upward flame propagation at maximum initial temperature of 300°C and maximum initial pressure of 2 bar. During the experimental measurement, a new thermal criterion was used to define if the gas is flammable and the experimentally determined flammability limits were compared with existing data reported in the literature. Finally, the temperature and pressure effect on flammability was investigated and compared with the modified Burgess-Wheeler Law.

12TH GLOBAL CONGRESS ON PROCESS SAFETY

Use of Process Incident Data for Accident Sequence Precursor Analysis

Erin P. Collins
JENSEN HUGHES
333 E. 46th Street, #11L
New York, NY 10017
ecollins@jensenhughes.com

Bijan Najafi
JENSEN HUGHES
2105 S Bascom Ave, Suite 350
Campbell, CA 95008-3278
bnajafi@jensenhughes.com

Prepared for Presentation at
American Institute of Chemical Engineers
2016 Spring Meeting
12th Global Congress on Process Safety
Houston, Texas
April 11-23, 2016

Use of Process Incident Data for Accident Sequence Precursor Analysis

Erin P. Collins
JENSEN HUGHES
333 E. 46th Street, #11L
New York, NY 10017
ecollins@jensenhughes.com

Bijan Najafi
JENSEN HUGHES

Keywords: accident sequence precursor, data, risk analysis

Abstract

Investigations of recent process incidents have identified multiple process safety management deficiencies that led to the accident. Furthermore, audits of the Safety and Environmental Management System (SEMS) program under the U.S. Department of the Interior's Bureau of Safety and Environmental Enforcement (BSEE) have shown that the "management of risk" elements such as Hazards Analysis and Management of Change (MOC) are not consistently implemented as tools to manage risk. In addition, the requirement to perform hazards analysis is frequently documented and often referenced as a corrective action, rather than as a "system driver". There are several pieces to risk assessment and risk management that allow facility operators to evaluate and measure what is actually occurring in their facilities, including the review of historical event data to identify what went wrong and for near misses, what almost happened. This paper describes techniques that have been used in the commercial nuclear power and other industries to collect and apply historical experience data to improve safety and operability. An approach using accident sequence precursor analysis of incident data and risk analysis techniques is recommended to identify potential accidents before they occur as part of a comprehensive treatment of historical data.

1 Introduction

There are several pieces to risk assessment and risk management that allow facility operators to evaluate and measure what is actually occurring in their facilities, including the review of historical event data to identify what went wrong and for near misses, what almost happened.

For example, the U.S. Chemical Safety Board (CSB) investigation of the February 18, 2015, explosion at the ExxonMobil Refinery in Torrance, California, identified multiple process safety management deficiencies that led to the accident and a serious near miss. In the case of the near miss, "…large pieces of debris from the explosion were thrown into other units of the refinery…One of these pieces of debris hit scaffolding in the refinery's alkylation unit, narrowly missing a tank containing tens of thousands of pounds of modified hydrofluoric acid, or HF. The CSB determined that had the debris struck the tank, a rupture could have been possible, resulting in a potentially catastrophic release of extremely toxic modified HF into the neighboring community." [Ref. 1]

Such near misses are not unique to the chemical process industry, but have been documented in the offshore oil industry as well. As described by EnergyWire [Ref. 2], "Partial failures happen regularly on oil rigs, with a worker catching the errors and correcting them before a major blowout. These are not always reported to regulators, but they are logged carefully and stored as raw data on rigs and onshore in Houston."

The Discussion Paper on Precursor Analysis by the research organization Resources for the Future [Ref. 3] identifies the specific challenge faced by the oil and gas industries of "how to reduce the risk of low-frequency oil spill events, and especially high-consequence events like the Deepwater Horizon accident, when historical experience contains few oil spills of material scale and none approaching the significance of the Deepwater Horizon." Based on a review of successful programs in the nuclear power, aerospace and other industries, the paper recommends the consideration of accident precursor analysis as a means for providing "a risk-informed oversight capability focused on preventing catastrophic oil spills and based on data from ongoing operating experience on the Outer Continental Shelf (OCS)."

The ASP approach is similar to Layer of Protection Analysis (LOPA) [Ref. 4] or Bowtie Analysis in that it postulates accident scenarios and the paths that could lead to a potential accident, but differs in that it uses insights from actual historical near-miss event data to identify the branch points which, if different, could have led to a severe accident.

This paper discusses the collection of historical event data, various programs that undertake such data collection, and proposes steps for use of this data toward risk assessment and management using ASP methods.

2 Collection, Evaluation and Application of Historical Events Data

While HAZOP and FMEA are useful tools for postulating incidents and their consequences, historical event data provides important input to the risk assessment process beyond the development of accident frequencies. Incident data collection and analysis can provide multiple qualitative insights, such as information for developing and refining risk models and a better understanding of system operation and failure (including the human-machine interface and efficacy of procedures and practices).

The CSB investigates incidents at chemical process facilities, produces investigation reports such as the one for the West Fertilizer Company Fire and Explosion in April 2013 [Ref. 5] and conducts public meetings to present and discuss their findings, as with the February 2015 ExxonMobil Refinery explosion [Ref. 6].

Following the Deepwater Horizon accident, the Deepwater Presidential commission report [Ref. 7] found at the time that "Although the raw incident data are online, MMS last produced an analysis of offshore incidents—critical data for promoting the safety of offshore operations—for calendar year 2000."

Probably for this reason, incident analysis was highlighted in the discussion of the Workplace Safety Rule in 2009 [Ref. 8] and one of the key areas where it was expected that the U.S. Department of the Interior's Bureau of Safety and Environmental Enforcement (BSEE) would be focusing their attention during the second Safety and Environmental Management System (SEMS) audit cycle in 2015 was on "Developing representative leading indicators for SEMS performance, while continuing to track industry accepted lagging indicators for evidence that SEMS programs are driving an improvement in safety performance." [Ref. 9] The development of these representative indicators by its nature must come from the review of historical evidence.

The narratives and timelines of previous accidents and near-misses can indicate what triggered the event, the automatic system response and failure, subsequent system failures and their causes, and worker involvement either to cause, exacerbate or mitigate the event. These insights in turn feed back into the data collection process by indicating what trends are of interest to monitor and, therefore, what categories of data should be collected.

The Workplace Safety Rule announcement by MMS in 2009 [Ref. 8] indicated that "As a result of MMS research conducted on accident panel investigations and reports, incident analysis, and [Incidents of Non-Compliance] INCs, it appears that equipment failure is rarely the primary cause of the incident or accident. This is due to technological advances which have provided industry with very efficient and reliable equipment for finding, producing, and transporting offshore oil and gas. However, in most cases, accidents and oil spills can be traced to human error and/or organizational failures." This is consistent with the Deepwater Presidential commission report [Ref. 7], which identified several human error-related issues, including the following:

- Numerous changes to the temporary abandonment procedures for securing the well were made in the two weeks leading up to the operations plan for the worker activities, but "[t]here is no evidence that these changes went through any sort of formal risk assessment or management of change process"

- "The drilling crew and other individuals on the rig also missed critical signs that a kick [unplanned influx of gas or fluids] was occurring. The crew could have prevented the blowout—or at least significantly reduced its impact—if they had reacted in a timely and appropriate manner. What is not now clear is precisely why the crew missed these signals."

- "The crew should have diverted the flow overboard when mud started spewing from the rig floor. While that ultimately may not have prevented an explosion, diverting overboard would have reduced the risk of ignition of the rising gas. Considering the circumstances, the crew also should have activated the blind shear ram to close in the well. Diverting the flow overboard and/or activating the blind shear ram may not have prevented the explosion, but likely could have given the crew more time and perhaps limited the impact of the explosion."

While the human involvement in incidents cannot be denied, the ATHEANA method of human reliability analysis (HRA) developed by the US Nuclear Regulatory Commission [Ref. 10] discusses the situation of "error-forcing contexts" or EFCs in which the plant conditions and so-called "performance shaping factors" (context-specific influences on reliable human performance) conspire to make human error nearly unavoidable. The ATHEANA document provides examples of these EFCs that have occurred in past US nuclear plant accidents, such as mismatches between the actual plant conditions and the operator job aids (e.g., training, experience) during the Three Mile Island event. EFC elements were also identified by ATHEANA for each of the four stages of information processing (i.e., detection, situation assessment, response planning, and response implementation) that describe the range of human activities required to respond to abnormal or emergency conditions. The review of past events and near-misses considering these EFC elements may indicate that organizational factors and gaps in procedures or training may in fact be the root cause.

An independent investigation of the Deepwater Horizon by the UC Berkeley Center for Catastrophic Risk Management stated that "[T]he Macondo system's good record of lost time accidents could not provide effective precursors for the major impending system accident. Efforts to prevent system accidents must address a broader set of organizational issues and hazards associated with all of the elements that comprise a given system. The good record of lost time accidents plausibly led to complacency that contributed to the disaster." [Ref. 11]

A program such as ASP looks at historical events from the perspective of commonality of potential outcomes and identifies the organizational issues that can cause near misses to become severe accidents.

3 Chemical Process and Nuclear Power Industry Experience

Examples of the projects that exemplify the type of incident reviews that factor industry experience into lessons learned are the Accident Sequence Precursor (ASP) program [Refs. 12, 13 and 14], the AIChE CCPS Process Safety Incident Database (PSID) [Ref. 15] and Process Equipment Reliability Database (PERD) [Ref.16], the EPRI Fire Events Database (FEDB) project [Refs. 17 and 18], the EPRI Guidelines for PRA Data Analysis [Ref. 19], the AIChE CCPS Guidelines for Process Equipment Reliability Data [Ref. 20], and the EPRI/NRC-RES Fire HRA Guidelines [Ref. 21]. Jensen-Hughes has been an active participant (lead in some cases) in the development of all of these projects except PERD and PSID.

3.1 Accident Sequence Precursor (ASP) Program

The ASP Program began in the early 1980s under the auspices of the U.S. Nuclear Regulatory Commission in response to the Three Mile Island accident. ASP is concerned with the identification and documentation of operational events that have involved portions of accident sequences and with the estimation of associated frequencies and probabilities. Identification of precursors requires the review of operational events for instances in which plant functions that provide protection against severe accidents (core damage, in the case of nuclear plants) have been challenged or compromised. The intention is to understand how these 'near misses' could have propagated into a severe accident had the circumstances been slightly different and thereby inform the risk modeling, data collection and operational insights.

[NOTE: The following explanation of the US NRC's ASP program is extracted from Reference 11.]

Reports of plant incidents are reviewed to identify those potentially involving failures in systems that provide protective functions against severe accident-related initiating events or the initiating events themselves. Those events selected undergo one- or two-engineer review(s) to determine if the reported event should be examined in greater detail. Events determined to be potentially significant as a result of this initial review are then subjected to a thorough, detailed analysis. This extensive analysis is intended to identify those events considered to be precursors to potential severe accidents, either because of an initiating event or because of failures that could have affected the course of postulated off-normal events or accidents. A variety of other data sources about the same event are also reviewed and compared to thoroughly understand the contributors to each event.

Quantification of accident sequence precursor significance involves the determination of a conditional probability of a subsequent severe accident given the failures observed during an operational event. This probability is estimated by mapping failures observed during the event onto the ASP accident sequence models (event trees and linked fault trees modified to reflect the event), which depict potential paths to the severe accident, and by calculating a conditional probability of the severe accident.

The information gained from the ASP process not only provides quantitative data, but provides insights into the types of events that should be included in risk models. In addition, events that provide insights into unusual failure modes with the potential to compromise continued safety functions but are not considered to be precursors are documented by ASP as "interesting" events.

Building upon this effort, the National Academy of Engineering Program Office undertook the Accident Precursors Project in 2003 to examine the complex issue of accident precursor analysis and management. This seven-month project was designed to document and promote industrial and academic approaches to detecting, analyzing, and benefiting from accident precursors, as well as to understand public-sector and private-sector roles in using precursor information. [Ref. 10]

3.2 AIChE CCPS Process Safety Incident Database (PSID)

The purpose of the CCPS PSID is to pool process safety incident experience among participating companies so they can learn from the experiences of others without suffering the consequences of failures, while minimizing corporate liability. PSID includes process safety incidents with a potentially important lesson to be learned from incidents that did or could have resulted in a fire, explosion, fatality, multiple injuries, significant release of hazardous materials, or any other unique process safety incident (including near-misses).

3.3 AIChE CCPS Process Equipment Reliability Database (PERD)

The equipment reliability database is intended to provide high quality, valid, and useful data to the Hydrocarbon Processing Industries (HPI) and Chemical Process Industries (CPI). This type of data provides input to analyses of availability, reliability and equipment design improvements, maintenance strategies and life cycle cost determination. The database includes Lessons Learned and Contributing Factors, which are data input requirements.

3.4 Fire Events Database (FEDB) Project

The FEDB is a collection of fire events that occurred in nuclear power plants since 1967. These events were reported to the U.S. Nuclear Regulatory Commission or the industry groups such as the Institute for Nuclear Power Operations (INPO) and the Nuclear Energy Institute (NEI). These events were reviewed to extract information on the equipment involved in the fire initiation, response of fire prevention and mitigation systems, response of the fire brigade and the consequences of the fire (degraded incipient or catastrophic). Guidance was also provided for the interpretation and categorization of fire event data. Since its publications the event history has been used, at time in combination with tests and a process called Phenomena Identification and Ranking Table or PIRT, to help define and characterize accident initiation, progression, mitigation and consequences, including:

- Types and characteristics, e.g., timing, of accident initiators,
- Sequence of events, including prevention/mitigation and human interface that helps define accident scenarios (or sequences) with risk implications,
- Severity of the event, e.g., heat release rate of an electrical panel fire or zone of influence from a high energy arc fault in a high voltage switchgear,
- Mitigating systems response, e.g., reliability and effectiveness of the fire brigade response to various types of fires,
- Cable damage and quantification of effects from fire, such as spurious indication of instrumentation and control systems leading to failure of accident prevention/mitigation systems or inadequate/erroneous operator response, and
- Qualitative insights on the types of fires that are most likely to occur, their duration, severity and how the plants address them.

3.5 EPRI Guidelines for Data Analysis for PRA

The EPRI Data Guidelines discusses the nuclear power industry approach to collecting, analyzing and providing summary statistics for several major categories of data types: initiating event data, component reliability data, unavailability (downtime) data and common cause failures. Initiating events are quantified using a combination of historical evidence of event occurrence, both from the specific plant and across the nuclear industry.

3.6 AIChE CCPS Guidelines for Process Equipment Reliability Data

Many of the aspects of data analysis discussed in the EPRI Data Guidelines were identified thirteen years earlier for the chemical process industry in the AIChE CCPS Process Equipment Reliability Data Guidelines. This data book built on the Chemical Process Quantitative Risk Assessment (CPQRA) Guidelines to show the process engineer how to make quantitative estimates of the risk of hazards identified, which can in turn identify the major contributors to risk. This project helped demonstrate the need for process industry data collection efforts like PERD and PSID.

3.7 EPRI/NRC-RES Fire Human Reliability Analysis (HRA) Guidelines

Although its focus is on human reliability, the Fire HRA Guidelines discusses the need to review past incidents experienced at a facility to understand whether the event was human-caused or how the workers' responded to the event as it evolved. These insights not only assist the HRA, but provide an understanding of how workers perceive their role during an accident and how they can influence accident progression, for good or for ill. These findings can then be fed back into the improvement of organizational and safety practices for accident prevention.

4 Considerations for Historical Data Analysis Application in the Chemical Process Industry

The Discussion Paper on Precursor Analysis for Offshore Oil and Gas Drilling [Ref. 3] identifies incident reports required under the Incident Reporting Rule (30 CFR 250.188) for oil, gas and sulphur operations in the outer continental shelf as an example of industry precursor data that would allow a system failure of concern [such as "significant uncontrolled escape of hydrocarbons" (SUEH)] to be evaluated through event trees based on generic commonality of events (such as those used in the US NRC ASP program).

Further, the Discussion Paper recommends an approach for the design of an ASP-enabled incident reporting system, which has been made more general here for application to a wider variety of chemical processes:

a) Define the class of events (such as SUEH) related to the system failure of concern.

b) On the basis of PRA models, design the set of generic event trees that have these events as possible endpoints.

c) Ensure that all nodes in these generic event trees are captured in an incident reporting system, such as PSID or under 30 CFR 250.188.

d) Design a control loop, possibly based on the Potential Incident of Non-Compliance (PINC) checklist inspection system, to ensure coverage of the incident reporting system, with appropriate sanctions for noncompliance.

e) Set up an analysis team to map the key even-related incidents onto generic event trees.

f) Set up a review board comprising industry and government agencies.

g) Design feedback mechanisms for involving industry in efforts to optimize industry-wide performance as measured by the ASP indicators.

h) Design an automation plan to streamline the ASP analysis.

The following additional tasks should be considered for translating these steps into industry guidance:

a) Documentation of insights and development of a draft guidelines for a selected process facility (to be defined) reflecting items a) through h) above.

b) Testing, validation and review of the guidelines. This may be done using one or more well documented industry events.

c) Final guideline for the selected process facility.

d) Adaptation of the guideline to other processes.

5 Conclusion

A recent article in CEP Magazine [Ref. 22] profiles five chemical process industry accidents that, while not as severe as Bhopal or Deepwater Horizon, still had serious results. The article states that "Accidents such as those discussed in this article are more likely to occur. These incidents demonstrate that even small mistakes can have disastrous consequences. Therefore it is imperative that engineers learn from previous incidents to reduce their likelihood of recurring." Near misses, like smaller scale events, are also instances that can provide valuable insights on lapses in process safety management.

As stated in EnergyWire [Ref. 2]. "ASP would use the data from these partial failures, including the failure of rig workers to respond appropriately, to compute the likelihood of a catastrophe. It would identify problems when they occur so regulators can stop them from happening again."

6 References

[1] "U.S. Chemical Safety Board Finds Multiple Safety Deficiencies Led to February 2015 Explosion and Serious Near Miss at the Exxon Mobil Refinery in Torrance, California," News Release, U.S. Chemical Safety and Hazard Investigation Board, Washington, DC, 16 January 2016.

[2] Vaidyanathan, G., "GULF SPILL: Could risk analysis prevent future deepwater disasters?" EnergyWire, Environment & Energy Publishing (E&E), http://www.eenews.net/stories/1059963200, 20 April 2012.

[3] Cooke, R.M., Ross, H.L. and Stern, A., "Discussion Paper - Precursor Analysis for Offshore Oil and Gas Drilling, From Prescriptive to Risk-Informed Regulation", RFF DP 10-61, Resources for the Future, Washington, DC, January 2011.

[4] Guidelines for Initiating Events and Independent Protection Layers in Layer of Protection Analysis, American Institute of Chemical Engineers (AIChE), Center for Chemical Process Safety (CCPS), New York, NY, 2015.

[5] "West Fertilizer Company Fire and Explosion", Final Investigation Report 2013-02-I-TX, U.S. Chemical Safety and Hazard Investigation Board, Washington, DC, January 2016.

[6] Presentation slides from CSB Public Meeting on ExxonMobil Refinery Explosion, U.S. Chemical Safety and Hazard Investigation Board, Washington, DC, 13 January 2016.

[7] Deep Water: The Gulf Oil Disaster and the Future of Offshore Drilling, Report to the President, National Commission on the BP Deepwater Horizon Oil Spill and Offshore Drilling, January 2011.

[8] The Bureau of Ocean Energy Management, Regulation and Enforcement, Office of Public Affairs, Fact Sheet, The Workplace Safety Rule on Safety and Environmental Management Systems (SEMS).

[9] Memorandum from Doug Morris, Chief, Office of Offshore Regulatory Programs to Brian Salerno, Director BSEE, SEMS Program Summary—First Audit Cycle (2011-2013), July 23, 2014. http://www.bsee.gov/uploadedFiles/BSEE/Regulations_and_Guidance/Safety_and_Environmental_Management_Systems_-_SEMS/SEMS%20Program%20Summary%208132014.pdf

[10] Technical Basis and Implementation Guidelines for A Technique for Human Event Analysis (ATHEANA), NUREG-1624, Rev. 1, U.S. Nuclear Regulatory Commission, Washington, DC, May 2000.

[11] Final Report on the Investigation of the Macondo Well Blowout, Deepwater Horizon Study Group (DHSG), Center for Catastrophic Risk Management (CCRM), University of California – Berkeley, Berkeley, CA, March 2011.

[12] Minarick, J. W. and Kukielka, C. A., Precursors to Potential Severe Core Damage Accidents: 1969-1979, A Status Report, NUREG/CR-2497, U.S. Nuclear Regulatory Commission, Washington, DC, June 1982.

[13] Belles, R.J. et al., Precursors to Potential Severe Core Damage Accidents: 1998, A Status Report, NUREG/CR-4674, US Nuclear Regulatory Commission, Washington, DC, July 2000.

[14] Phimister, J.R., Bier, V.M., and Kunreuther, H.C. (Eds.), Accident Precursor Analysis and Management: Reducing Technological Risk Through Diligence, The National Academies Press, Washington, DC, 2004.

[15] Presentation slides, Overview: Process Safety Incident Database (PSID), American Institute of Chemical Engineers (AIChE), Center for Chemical Process Safety (CCPS)

http://www.aiche.org/ccps/resources/psid-process-safety-incident-database

[16] American Institute of Chemical Engineers (AIChE), website "About PERD"

http://www.aiche.org/process-equipment-reliability-database-perd/about-perd

[17] EPRI 1000894, Fire Events Database for U.S. Nuclear Power Plants: Update Through 1999, Electric Power Research Institute (EPRI), Palo Alto, CA, October 2000.

[18] EPRI 1025284, The Updated Fire Events Database: Description of Content and Fire Event Classification Guidance, Electric Power Research Institute (EPRI), Palo Alto, CA, July 2013.

[19] EPRI 3002000774, EPRI Guidelines for PRA Data Analysis, Electric Power Research Institute (EPRI), Palo Alto, CA, December 2013.

[20] Guidelines for Process Equipment Reliability Data with Data Tables, American Institute of Chemical Engineers (AIChE), Center for Chemical Process Safety (CCPS), New York, NY, 1989.

[21] EPRI 1023001/ NUREG-1921, EPRI/NRC-RES Fire HRA Guidelines, Electric Power Research Institute, Palo Alto, CA and U.S. Nuclear Regulatory Commission, Washington, DC, July 2012.

[22] A. Ness, "Lessons Learned from Recent Process Safety Incidents," CEP Magazine, American Institute of Chemical Engineers (AIChE), New York, NY, March 2015.

442384 Quantitative Risk Analysis of the Expansion of the Chlorine Room of the Industrial Water Plant of the Petrochemical Complex Jose Antonio Anzoategui of Petroquimica Venezuela Pequiven S.a

Abstract:

Quantitative risk analysis of the expansion of the chlorine room of the industrial water plant of the petrochemical complex Jose Antonio Anzoategui of Petroquimica venezuela PEQUIVEN S.A.

Venezuela Petrochemical PEQUIVEN SA has a management responsible for risk control, which makes this quantitative risk analysis to estimate the risk to employees and third parties for process safety events that may occur as a result of the loss of containment in the facilities and thus indicate recommendations aimed at reducing the level of risk associated with such events. Quantitative risk analyzes are made from the standard rules of Petroleos de Venezuela , PDVSA as IR-S-02: criteria for quantitative risk analysis and other international standards such as Center of Chemical Process Safety Guidelines for Chemical Process Quantitative Risk Analysis and the netherlands organization Yellow book: Methods for calculation of physical effects, through the computational tool PHAST RISK 6.7 . Developed by DNV Company.

PEQUIVEN has three complexes nationwide, in which fertilizers for the food industry, olefins and plastics for residential construction, among other products are produced. The Jose Antonio Anzoategui complex will make an expansion of the chlorine room of the industrial water plant and because of the dangers associated with chlorination process, a quantitative risk analysis has requested to engineering management risks.

The process of a quantitative risk analysis (QRA) started setting goals, objectives and depth of the study, once these are established, and in conjunction with materials, inventory, environmental conditions and the description of the process, one proceeds to identify scenarios. In this case, the HAZOP tool was applied, which gives the place a list of possible scenarios, of which those that could cause the worst consequences. With the previously selected scenarios, the model building on which the estimated frequency and consequence will be held is started. The frequency is estimated based on a review of probability theory study according to Chemical Process Safety Fundamentals with Applications Standard, through a theoretical data extracted from the norm Guidelines for Process Equipment Reliability Data, a probability of occurrence of the scenarios is established.

The estimation of Consequence answer a series of complex mathematical models that simulate the behavior of these events based on the data of the others that occurred already trough PHAST RISK software. Once the simulation is made, the software generates a series of results of contours risks in function of exposure to chlorine gas to the personnel working in this complex and surrounding facilities, allowing emit a series of observations than are aimed to decrease level of Risk associated with the expansion of the chlorine room.

Assessing Cognitive Human Errors for the Process Industries

Erin P. Collins
JENSEN HUGHES
333 E. 46th Street, #11L
New York, NY 10017
ecollins@jensenhughes.com

Bijan Najafi
JENSEN HUGHES
2105 S Bascom Ave, Suite 350
Campbell, CA 95008-3278
bnajafi@jensenhughes.com

Prepared for Presentation at
American Institute of Chemical Engineers
2016 Spring Meeting
12th Global Congress on Process Safety
Houston, Texas
April 11-23, 2016

Assessing Cognitive Human Errors for the Process Industries

Erin P. Collins
JENSEN HUGHES
333 E. 46th Street, #11L
New York, NY 10017
ecollins@jensenhughes.com

Bijan Najafi
JENSEN HUGHES

Keywords: cognitive, human error, human reliability analysis

Abstract

Through the analysis of actual incidents and the understanding of the underlying human errors that have contributed to them, it is apparent that 'first generation' HRA methods are appropriate for some instances but have their limitations. For example, failure to recognize the significance of high fluid returns during the negative pressure test and failure to perceive the importance of drill pipe pressure during well monitoring activities were key contributors to the Deepwater Horizon incident. These diagnosis and decision-making errors are not well addressed by first generation HRA methods and require particular timing, error mode and performance shaping factor considerations. This paper provides an introduction to HRA methods and then discusses those available for evaluating cognitive errors and their applicability to the chemical process industry.

1 Introduction

Layer of Protection Analysis (LOPA) provides a simplified risk analysis technique for a select set of postulated incident scenarios, evaluating an initiating event (IE) and subsequent independent protection layers (IPLs) as a series of probabilities of failure on demand (PFDs). Human error is assessed in LOPA as either an IE or IPL. Techniques that have been recommended for human reliability assessment (HRA) by the CCPS Guidelines for IEs and IPLs in LOPA [Ref. 1] are generally considered in other industries as 'first generation' HRA methods, meaning that they are appropriate for some instances but have their limitations. These limitations have become more apparent through the analysis of actual incidents and the understanding of the underlying human errors that have contributed to them. For example, failure to recognize the significance of high fluid returns during the negative pressure test and failure to perceive the importance of drill pipe pressure during well monitoring activities were key contributors to the Deepwater Horizon incident. These diagnosis and decision-making errors are not well addressed by first generation HRA methods and require particular timing, error mode and performance shaping factor (PSF) considerations. Several HRA quantification methods are available to assess the human error probability associated with these so-called "cognitive" aspects of a task, as well as the equipment manipulation.

While many HRA approaches had their genesis in the nuclear power industry, they have been successfully applied to other industries, such as chemical weapons destruction plants and nuclear fuel transport and storage, and their results have been accepted during peer and regulatory agency reviews. In each instance where HRA is applied to a new industry, however, an evaluation must be made regarding the degree of applicability of the standard error types, PSFs and quantification methods. Since the error types and PSFs have an underpinning in human factors, psychology and cognitive science, they generally apply regardless of the industry. However, it is important to evaluate the applicability of HRA tools and techniques to ensure that industry-specific issues are being appropriately addressed.

This paper provides an overview of HRA principles and methods, then discusses those available for evaluating cognitive errors and their applicability to the chemical process industry.

2 HRA Principles and Methods

HRA provides a framework to evaluate, prioritize, quantify and improve the human interface with complex systems. HRA tools and techniques were pioneered by the commercial nuclear power industry, but are based on fundamental precepts of human reliability and data collection and analysis that are not technology specific. They have been applied to an increasing number of facilities across various different industries, such as aerospace [Ref. 2], chemical weapons destruction [Ref. 3], HVAC system optimization, nuclear fuel disposal facility design risk evaluation [Ref. 4] and detection and suppression activities in response to fire events in the military [Ref. 5]. This experience and repetition have allowed lessons learned to be identified and fed back into safe practices, the development of a safety culture, and the ability to meet standards [Ref. 6] and regulatory goals (which are quite stringent in the nuclear power industry).

This history and associated lessons learned indicates that wider application of HRA in the chemical process industry could provide greater understanding of risks (both accident and economic), how to control them (by improving procedures, focusing maintenance and surveillance efforts on the actual errors), and how to develop a safety culture that focuses on human error prevention.

The steps of the HRA process discussed below are adapted from NUREG-1921, the EPRI/NRC-RES Fire Human Reliability Analysis Guidelines [Ref. 7]. The process begins with an identification and definition task to identify actions performed by various workers and to define the actions in terms of human failure events or HFEs. These actions may include both failure to perform a required task or undesired actions with deleterious impact on system performance and safety. For the chemical process industry, this identification and definition step could be expected to be very closely tied with the subsequent step of qualitative analysis because the same information gathering that helps define the events also feeds into the development of an "operational story" or HFE narrative that assists in further understanding the events.

A sound qualitative analysis allows the HRA to provide feedback to the facility operator on the factors contributing to both the success and failure of a worker action. This involves the evaluation of performance shaping factors (PSFs), which are situation-specific influences on human error probabilities (HEPs) and include:

- Cues and indications that indicate a system problem
- Timing of the task
- Procedures and training
- Complexity of the task
- Workload, pressure and stress
- Human-Machine Interface (HMI)
- Environmental factors, such as heat or confined spaces
- Special equipment, such as tools and protective gear

These PSFs are used to flesh out the HFE narrative and identify the key elements of the HFEs that require further understanding through the performance of subject matter expert interviews.

Finally, quantification or quantitative analysis is performed by applying the information gained from identification, definition and qualitative analysis of HFEs toward the estimation of HEPs. Quantification can be done at varying degrees of detail and rigor. The three levels of analysis performed in HRA are: Screening, Scoping and Detailed.

Screening analysis involves the assignment of a preliminary and conservative expert judgment value as a means for "screening" out the risk-significant HFEs from the non-contributing ones.

The Scoping approach, discussed in NUREG-1921, was developed to provide additional rigor to a semi-screening process, but still results in conservative values for initial quantification.

Detailed analysis is performed by evaluating the action steps, timing and PSFs in detail and assessing the probability of errors in the cognitive aspects, or diagnosis and decision-making process for the correct action to take, as well as the execution aspects, such as the equipment manipulation steps.

Several HRA quantification methods are available and it has been found to be most effective if the analyst evaluates the strengths and weaknesses of each method against the nature of the human error being evaluated to make the best match. Having a large HRA method toolkit to draw from allows the analyst the flexibility to assess the driving failure mechanism behind the error rather than simply filling out a checklist.

The following HRA quantification methods are most widely used in the nuclear power industry, and are contained in Electric Power Research Institute's (EPRI's) HRA Calculator software [Ref. 8]:

- Human Cognitive Reliability/Operator Reliability Experiment (HCR/ORE) [Ref. 9], which focuses on the cognitive or decision making errors that are constrained by time;
- Cause-Based Decision Tree Method (CBDTM) [Ref. 9] that addresses cognitive errors as well but from the standpoint of the error-causing mechanisms, such as unclear instrumentation or lack of procedural direction;
- The Technique for Human Error Rate Prediction (THERP) [Ref. 10], which has been used for many years for the evaluation of task performance failures (also called execution errors) that can occur either before or after an accident;
- The Accident Sequence Evaluation Program (ASEP) [Ref. 11] that provides a simplified version of THERP for human errors that contribute to but occur prior to the accident; and
- The SPAR-H method [Ref. 12], which only permits highly simplified assessments of cognitive and action errors (and has not been used in the latest generations of nuclear plant risk assessments since it does not allow the depth of analysis to support safety improvements).

For HRA applications other than nuclear power, such as chemical weapons destruction plants and nuclear fuel transport and storage, other quantification methods have been successfully applied and their results accepted during peer and regulatory agency reviews. These include the Cognitive Reliability Error Analysis Method (CREAM) [Ref. 13], and the Human Error Assessment and Reduction Technique (HEART) [Ref. 14]. As the name implies, CREAM focuses on the decision-making errors by providing HEPs for a standard set of task types, but also includes adjustment factors to evaluate facility and scenario-specific influences. HEART contains a list of generic task types and their HEP estimates that are modified using factors associated with 'error producing conditions'.

While the application of a method and generation of HEP values is important, the primary output of the HRA is the identification and prioritization of the human error issues of greatest risk significance and the development of risk insights. The benefit of conducting more detailed HRA is in gaining a better perspective on what influences human failure and therefore what can be done to improve human performance.

The decomposition of human error into cognitive vs. execution tasks and a detailed evaluation of these tasks has proven to be particularly important since the approaches to their improvement are different. For example, more instrumented processes attempt to provide information to the operator that allow him/her to diagnose the situation and take the appropriate action. In other cases, however, the process may not include such instrumentation or it may be not be available due to failure, so diagnosis becomes a critical issue. Even when there is no instrumentation or early warning system provided for an action, the HRA can provide an understanding of the cues required for diagnosis and allow these to be included in training and procedures/manuals. These cues may even be sensory in nature, such as the sound, smell or sight of a process that is going awry. This does not mean that the HRA will recommend instrumenting everything. If correctly done, the HRA instead allows the most critical information used by the worker to be identified and suggests how it should be presented or emphasized to optimize performance.

The reliability of execution tasks is generally improved by providing a formal set of procedures and training that are continually reviewed and refined. However there may be instances where procedures and training are not yet formalized, such as when the facility is a new design or is in the pre-operational testing phase, or because of historical precedent or preference. In these instances, the worker action is evaluated in HRA as a so-called 'recovery' action, where the worker utilizes previous experience with tasks they consider to be similar and applies them to a new situation. These insights are important to evaluating the reliability of task performance and can even be subsequently captured in procedure modifications and training enhancements. The risk benefit assessment might even recommend conducting a design change to the human-machine interface to improve the execution task reliability.

In summary, HRA is

- Widely used
- Applied across a variety of industries, which demonstrates that it provides value
- Allows the user to quantify and prioritize the human influence on safety
- Provides the means for improving process safety and reliability in the short term and a safety culture to reduce future errors over the longer term.

3 Cognitive Errors in the Process Industries

The Deepwater Presidential commission report [Ref. 15] identified several human error-related issues that led to the severity of the incident, including the following:

- Numerous changes to the temporary abandonment procedures for securing the well were made in the two weeks leading up to the operations plan for the worker activities, but "[t]here is no evidence that these changes went through any sort of formal risk assessment or management of change process"
- "The drilling crew and other individuals on the rig also missed critical signs that a kick [unplanned influx of gas or fluids] was occurring. The crew could have prevented the blowout—or at least significantly reduced its impact—if they had reacted in a timely and appropriate manner. What is not now clear is precisely why the crew missed these signals."
- "The crew should have diverted the flow overboard when mud started spewing from the rig floor. While that ultimately may not have prevented an explosion, diverting overboard would have reduced the risk of ignition of the rising gas. Considering the circumstances, the crew also should have activated the blind shear ram to close in the well. Diverting

the flow overboard and/or activating the blind shear ram may not have prevented the explosion, but likely could have given the crew more time and perhaps limited the impact of the explosion."

These issues can be considered as relating to cognitive human errors since they concern the lack of procedural guidance that assists with diagnosis and the decision-making on the part of the crew that appears to have been ill advised or ill informed. Hubbard and Embrey [Ref. 16] assessed these issues from a human factors standpoint and they raised the following concerns related to the cognitive errors during the incident:

- Why were procedures not used – is this typical?
- What was the effect on decision making and on practice of the pressure to get the well tested and capped?
- Why did the crew use the mud gas separator rather than pipework that would have discharged the mud more quickly?
- What emergency training did the team have? Did they have training or job support to help them identify and respond to escalating situations?

Questions such as these are key to the evaluation and assessment of cognitive errors.

A recent article by Ness in the CEP Magazine [Ref. 17] outlined lessons learned from several process accidents, many of which involved cognitive human errors:

- ARCO Channelview wastewater tank explosion, 1990 – "The wastewater tank was not considered part of the operating plant. Hence, the management and workers did not understand that a chemical reaction was taking place in the tank, generating oxygen. The lack of understanding enabled a series of poor decisions, such as discontinuing the nitrogen purge…no management of change (MOC) review of these decisions, and no pre-startup safety review."
- Terra Industries ammonium nitrate explosion, 1994 – "…the lack of procedures for shutdown and monitoring the equipment during shutdown led operators to perform actions that sensitized the AN solution and provided energy to the decomposition reaction."
- Formosa Plastics vinyl chloride release, 2004 - "The operator overrode an interlock, which led to a release of hot, pressurized VCM. Formosa Plastics did not have comprehensive written standards…for managing interlocks on the vessels. Employees were unprepared for a major accident at the facility."

These incidents demonstrate faulty understanding of the process and lack of procedural direction, which in turn led to improper decision-making and catastrophic consequences.

4 Treatment of Cognitive Human Error in Process Guidelines and Studies

The CCPS *Guidelines for Preventing Human Error in Process Safety* [Ref. 18] mentions cognitive failure assessment methods such as CBDTM, CREAM and HEART but does not provide much guidance to the user for implementing them. Furthermore, in the ten years since that guidance was issued, the Deepwater Horizon incident has occurred and much has been learned in various industries on the analysis and mitigation of human error using HRA.

The recently issued CCPS *Guidelines for Initiating Events and Independent Protection Layers in Layer of Protection Analysis* [Ref. 1] states that "developing PFD values for human IPLs is not straightforward." Note that PFDs as defined in these guidelines include not only the HEP, but the PFD of the cue for response, namely the "instrumentation or equipment used to detect the process deviation and/or take action to return the process to a safe state."

The IPLs for human actions that are cited in these guidelines are:

- Human response to an abnormal condition
- Human response to an abnormal condition with multiple sensors and indications and the operator has >24 hours to accomplish the required response action
- Adjustable movement-limiting device – this is actually equipment that is intended to prevent operator misoperation of equipment beyond a limit that would initiate a scenario of concern
- Personal Protective Equipment (PPE) – safety gear that acts as a barrier between a person and a hazard.

In an HRA context, the first two IPLs would be considered Human Failure Events (HFEs), but the specification of sensors/indications (cues) and time available for response would be evaluated in detail for both cases, in addition to other scenario-specific PSFs, such as procedures and training. Most importantly, the time available would be balanced against the time *required* for the action to assess the time margin and the HEP.

Note that the *nature* of the human response is not specified for these IPLs, so there is no differentiation between cognitive and execution errors, which have different aspects and methods for evaluation.

The adjustable movement-limiting device IPL would not itself considered to be a human failure, but would be considered as a mechanical form of interlock that would need to be defeated (either through another human action to leave it out of position after maintenance or through equipment failure) to allow the human error to occur. A Quantitative Risk Assessment (QRA) model allows such equipment and human failures to be functionally linked.

PPE is addressed in HRA not as an HFE but as a performance shaping factor on an existing human response action since it is considered to cause sight, sound and manipulation limitations on the human action, and can even lead to increased stress.

Papers presented in recent years discuss application of HRA to the petroleum industry through the use of the SPAR-H quantification method [Refs. 19 & 20]. Gould, Ringstad and Van De Merwe state that "The main advantages of SPAR-H is that it provides clear guidance on assigning numerical weights to the performance-shaping factors, is not very resource-demanding, and includes worksheets that promote analyst consistency." While SPAR-H provides a useful screening analysis of factors related to cognitive error, such as Available Time, Procedures and Ergonomics, it would seem to be best considered as the first step toward identifying risk-significant cognitive human errors, which would then need to be evaluated in greater detail. The benefit of detailed HRA lies in identifying the driving mechanisms behind the error that can then be applied to steps for error likelihood reduction.

Another current study [Ref. 21] has successfully applied the CREAM cognitive error evaluation technique to the cargo loading process of LPG tankers "to systematically predict human error potentials for designated tasks and to determine the required safety control levels on-board LPG ships." The human cognition and action context are determined in accordance with Common Performance Conditions (CPCs) which provide a well-structured and comprehensive basis for characterizing the circumstances and to define sets of possible error modes and error causes.

5 Information Processing Model for Cognitive Activities

Assessment of HFEs can be guided by a model of higher-level cognitive activities, such as an information processing model. The model that has been used by the authors is based on the discussion in Chapter 4 of the ATHEANA method [Ref. 22] and consists of the following elements:

- Monitoring and detection - Both of these activities are involved with extracting information from the environment. Also, both are influenced by the characteristics of the environment and the person's knowledge and expectations. Monitoring that is driven by the characteristics of the environment is called data-driven monitoring. Monitoring initiated by a person's knowledge or expectations is called knowledge-driven monitoring. Detection can be defined as the onset of realization by operators that an abnormal event is happening.

- Situation awareness - This term is defined as the process by which operators construct an explanation to account for their observations. The result of this process is a mental model, called a situation model that represents the operator's understanding of the present situation and their expectations for future conditions and consequences.

- Response planning - This term is defined as the process by which operators decide on a course of action, given their awareness of a particular situation. Often (but not always) these actions are specified in procedures.

- Response implementation - This term is defined as the activities involved with physically carrying out the actions identified in response planning.

The final bullet concerns the execution of the response, but is predicated upon the successful cognitive assessment of the course of action.

By assessing these features of a human action, the analyst can then select from the available HRA models to evaluate the HEP and moreover, identify the driving issues leading to error that can be factored into error prevention.

6 Conclusions

Recent studies are showing that existing HRA methods can be applied to assess cognitive human errors in the process industries, but these insights have yet to be incorporated into industry guidelines. While initial assessments can be made with simplified tools, the key to addressing and preventing major incidents is the identification of underlying human error mechanisms so that risk-mitigating solutions that truly address the diagnosis and decision-making process can be developed.

7 References

[1] *Guidelines for Initiating Events and Independent Protection Layers in Layer of Protection Analysis*, Center for Chemical Process Safety, American Institute of Chemical Engineers, New York, NY, 2015.

[2] Collins, E.P. and Fragola, J.R., "Advances in Human Error Evaluation", 1995 Reliability, Availability, Maintainability Symposium (RAMS), Washington, DC, January 1995.

[3] Umatilla Chemical Agent Disposal Facility Quantitative Risk Assessment, Report No. SAIC-00/2641. Volume I, Science Applications International Corporation (SAIC), Abingdon, MD, 2002.

[4] Yucca Mountain Project Canister Receipt and Closure Facility Reliability and Event Sequence Categorization Analysis, 060-PSA-CR00-00200-000-00B, Bechtel SAIC Company, LLC (BSC), November 2008.

[5] Garvey, M. (Dept. of the Navy), Joglar, F. and Collins, E.P. (JENSEN HUGHES, Inc.), "HRA for Detection and Suppression Activities in Response to Fire Events", 2014 Annual Reliability and Maintainability Symposium (RAMS), Colorado Springs, CO, January 2014.

[6] ASME/ANS RA-Sa-2009, Addenda to ASME/ANS RA-S-2008, "Standard for Level 1/Large Early Release Frequency Probabilistic Risk Assessment for Nuclear Power Plant Applications", The American Society of Mechanical Engineers (ASME), New York, New York, 2009.

[7] Electric Power Research Institute and U.S. Nuclear Regulatory Commission, EPRI/NRC-RES Fire HRA Guidelines, EPRI 1023001/ NUREG-1921, July 2012.

[8] Electric Power Research Institute, The EPRI HRA Calculator® Software Manual, EPRI 3002004030. Palo Alto, CA, May 2014.

[9] EPRI TR-100259, An Approach to the Analysis of Operator Actions in Probabilistic Risk Assessment, Electric Power Research Institute (EPRI), Palo Alto, CA, June 1992.

[10] Swain, A. D. and Guttman, H., Handbook of Human Reliability Analysis with Emphasis on Nuclear Power Plant Applications, NUREG/CR-1278, Washington, DC: U.S. Nuclear Regulatory Commission, August 1983.

[11] Swain, A., Accident Sequence Evaluation Program Human Reliability Analysis Procedure, NUREG/CR-4772, U.S. Nuclear Regulatory Commission, Washington, DC, February 1987.

[12] The SPAR-H Human Reliability Analysis Method, NUREG/CR-6883, Idaho National Engineering Laboratory for U.S. Nuclear Regulatory Commission, Washington, DC, August 2005.

[13] Hollnagel, E., *Cognitive Reliability and Error Analysis Method – CREAM*, Elsevier, New York, NY, 1998.

[14] Williams, J., HEART - A Proposed Method for Assessing and Reducing Human Error, 9th Advances in Reliability Technology Symposium, National Centre of Systems Reliability, University of Bradford, Bradford, UK, 1986.

[15] Deep Water: The Gulf Oil Disaster and the Future of Offshore Drilling, Report to the President, National Commission on the BP Deepwater Horizon Oil Spill and Offshore Drilling, January 2011.

[16] Hubbard, A. and Embrey, D., "Deepwater Horizon – Summary of Critical Events, Human Factors Issues and Implications," Human Reliability Associates, Ltd., 23 September 2010.

[17] Ness, A., "Lessons Learned from Recent Process Safety Incidents," CEP Magazine, American Institute of Chemical Engineers (AIChE), New York, NY, March 2015.

[18] *Guidelines for Preventing Human Error in Process Safety*, American Institute of Chemical Engineers (AIChE), Center for Chemical Process Safety (CCPS), August 2004.

[19] Gould, K.S. and Ringstad, A.J (StatOil) and Van De Merwe, K. (Det Norske Veritas), "Human Reliability Analysis in Major Accident Risk Analyses in the Norwegian Petroleum Industry", Proceedings of the Human Factors and Ergonomics Society 56th Annual Meeting, San Diego, CA, October 2012.

[20] Boring, R.L (Idaho National Laboratory) and Øien, K. (SINTEF), "Defining Human Failure Events for Petroleum Risk Analysis", Probabilistic Safety Assessment and Management PSAM 12, Honolulu, HI, June 2014.

[21] Akyuz, E. and Celik, M., Application of CREAM human reliability model to cargo loading process of LPG tankers, Journal of Loss Prevention in the Process Industries, Vol. 34, pgs. 39-48, Elsevier Ltd., 2015.

[22] Technical Basis and Implementation Guidelines for A Technique for Human Event Analysis (ATHEANA), NUREG-1624, Rev. 1, U.S. Nuclear Regulatory Commission, Washington, DC, May 2000.

442486 Critical Considerations in the Vacuum Protection of Butane Spheres

Abstract:

Vacuum protection of pressure vessels, including spheres, is not well defined in the ASME Boiler and Pressure Vessel Code (BPVC) Section VIII, Division I or API Std. 520. Vacuum protection of pressure vessels usually follows API Std. 2000 on low pressure tanks for guidance.

Vacuum can occur on butane storage spheres at temperatures of 25 to 30 F due to butane condensing. This paper addresses the challenges in vacuum protection of butane storage spheres, including both the positive pressure rating of vacuum devices and sizing of vacuum vents for a butane condensing scenairo.

442493 Assessing Risk in the Supply Chain Using the Bowtie Method

Abstract:

After a number of recent incidents involving the transportation and storage of hazardous materials, many industry end-users wish to understand their level of risk exposure during the supply chain. This exposure can either be via actual ownership or when the product can be associated with the concerned party. Ultimately, this may mean that an end-user's procurement group will want to see that their suppliers of products and services are performing as needed in the control of these risks.

A qualitative risk assessment is a technique by which these risks can be identified, understood, and evaluated. This paper will describe the application of one such method, where the bowtie barrier analysis is central to the means by which this assessment is completed. The barriers identified are those necessary to prevent and mitigate undesired events that could result in a fatality or a catastrophic incident to the environment, public, or a company's reputation.

The overall method employed in this process includes the following steps:

1. Hazard identification
2. Bowtie diagram development and barrier identification
3. Barrier screening to identify safety critical elements - those that provide a credible reduction in the risk of fatalities or other catastrophe
4. Performance standard development for the safety critical elements

A quantitative risk assessment can also be completed to provide additional risk knowledge to aid in prioritization and decision-making. The performance standards are then enforced appropriately throughout the supply chain. This will ensure that the transportation and handling of hazardous materials meet the purchasing company's expectations with regards to proper risk management.

442386 Reference Framework for the Application of Quantitative Risk Analysis for Hydrocarbon Pipelines, Coupled with Uncertainty Treatment Methods: Techniques for Analysis and Modelling Methods

Abstract:

Risk assessment is imperative to reduce consequences of accidents generated by leaks on pipelines. After a failure in an urban pipeline, the contents can be released rapidly forming a spreading pool in the surface or a vapor cloud that can lead to different events such as fires and explosions, affecting the population in the area. Generally, deterministic assessments are carried out where certain assumptions are made in the models used to calculate failure and event probabilities, event consequences and risk values. This causes the existence of uncertainty that is not considered and therefore it generates results only for "mean" or "worst-case" scenarios. Not taking uncertainty into account can lead to decisions with an unknown degree of conservatism: overestimate results could increase cost of decisions and underestimate results can lead to poor risk management. There is a lack of a clear directive to use methods of uncertainty assessment in risk analysis and for that reason, a framework is needed for carrying out quantitative risk assessment coupled with uncertainty treatment methods in order to handle, reduce, and properly inform uncertainty to the decision makers.

Uncertainty appears when we try to represent a system using different analysis techniques and therefore, depends on how accurately this process can be done. This in turn relies on how complex is the system under analysis and which techniques can be used. As part of the framework mentioned above, a classification is suggested to adequately select methods in probabilistic risk analysis to take into account different types of uncertainty in different kinds of representation and system analysis techniques. The classification is first based on the complexity of the system to be analyzed and then in the complexity of the modeling technique that is used to deal with system complexity. System complexity depends mainly on three variables: 1) understanding about the system (variables and its interactions) 2) the number of elements in the system 3) complexity of system interactions. Depending on system complexity, different techniques to represent it and analyze these systems in risk analysis are proposed and then, methods to propagate uncertainty are mentioned. A comparison of methodologies is presented to provide the engineer with tools to use adequately these methods in different situations and provide guidelines to deal with uncertainty in risk assessment.

Dynamics Operational Risk Management in Organizational Design, the Challenge for Sustainability

Salvador Ávila Filho
Universidade Federal da Bahia, UFBA
Rua Aristides Novis, Federação 02, Salvador Bahia
avilasalva@gmail.com

José Filipe Michel Gagliano Ferreira
Universidade Federal da Bahia, UFBA
Rua Aristides Novis, Federação 02, Salvador Bahia
gagliano.fm@gmail.com

Carlos Roberto de Oliveira Sousa
Universidade Federal da Bahia, UFBA
Rua Aristides Novis, Federação 02, Salvador Bahia
carlososousa.cs@gmail.com

Ricardo Araújo Kalid
Universidade Federal do Sul da Bahia, Itabuna
Ricardo.kalid@gmail.com

Prepared for Presentation at
American Institute of Chemical Engineers
2016 Spring Meeting
12th Global Congress on Process Safety
Houston, Texas
April 11-23, 2016

Dynamics Operational Risk Management in Organizational Design, the Challenge for Sustainability

Salvador Ávila Filho
Universidade Federal da Bahia, UFBA
Rua Aristides Novis, Federação 02, Salvador Bahia
avilasalva@gmail.com

José Filipe Michel Gagliano Ferreira
Universidade Federal da Bahia, UFBA
Rua Aristides Novis, Federação 02, Salvador Bahia
gagliano.fm@gmail.com

Carlos Roberto de Oliveira Sousa
Universidade Federal da Bahia, UFBA
Rua Aristides Novis, Federação 02, Salvador Bahia
carlososousa.cs@gmail.com

Ricardo Araujo Kalid
Universidade Federal do Sul da Bahia
Itabuna
ricardo.kalid@gmail.com

Keywords: business risk and corporative, human factors and organizational, tools, sustainability

Abstract

The maintenance of the organizational sustainability requires acknowledgment that business risks are dynamic and, therefore, depend on the balanced analysis of issues relating to product and process with people and culture. The analysis of the dynamic risk occurs in three steps as described as follows. The first analyzes the balance between the human and social factors with process and product factors, generating preliminary recommendations from probable scenarios through techniques such as LOPA and PRA with recommendations to avoid accidents and technical failures. Risks related to the environmental and human aspects are analyzed with PRA-SH to anticipate future events that may cause technical failure or evoke a stressor environment in the workplace. The second step requires knowledge of the organization in its aspects of management, technical and non-technical, inputs and productivity benchmark to define the assumptions and criteria used to project performance of operations, including the skills required by the business. The result of this study is the definition of technical, human, organizational and natural factors for the sustainable operation of production system. The techniques applied in this second stage are: HAZOP and HAZOP-Social; requirements for the task (GMTA, PADOP); events cause and consequence analysis; and analysis of barriers for events (BOWTIE). In the third stage of dynamic risk analysis rescues the criteria and assumptions established in the previous step by defining the management model and detailing the technology used. In this phase, it will be handled operational routines, asset control, control of failures and possible crisis. Thus, it will designed the systems of technical area (equipment, processes, products and control modes), interface (task, management, and human-machine interface), and social area (culture and biases, cognitive decision, stress analysis and behavior). The factors that make up the control of operations are analyzed in the standard procedure way and in the failure analysis using the following techniques: equipment (FMEA, FMEAH); product and process (monitoring and statistical control); task in complex mode (GMTA, CARMAN, PADOP); man-machine interface and control room (PRA-SH and Regional Criteria for control room design); culture (human reliability committee for organizational change and archetypes analysis); cognitive decision (study of cognitive failure); stress and decision (LOPA); and, stress and occupational health (LESHA). These techniques for analysis of operations in a dynamic way have metrics based on the operator's speech from the results of routine by reflecting the occupational consequences (accident), environmental (pollution and waste), climate (CO2 emissions) and economic (competitiveness losses based in the rates of consumption and productivity). The result of these measurements allows the routine adjustments in the design of operations, the organization's project settings and business project. A deployment exercise will be proposed for the chemical industry. Diagram will be presented reflecting the dynamic in the management of operational risks.

Glossary

ANP - National Petroleum Agency
FMEA – Failure Mode and Effect Analysis
FMEAH – Human Failure Mode and Effect Analysis
FPSO – Floating Production Storage and Offloading
FTA – Fault Tree Analysis
GMTA – Goals-Means Task Analysis
HAZOP – Hazard and Operability Study
HAZOP-Social – Social Hazard and Operability Study
HSE – Health, Safety and Environment
LESHA – MultiLayer Stress Effect on Health & behavior Assessment
LODA – MultiLayer Operation Decision Assessment on Progressive Stress
LOPA – MultiLayer Operation Protective Assessment
MDA – Methyl Diphenyl Amine
PADOP – Patterns and Procedures in the Operation
PHR – Preliminary Hazard Analysis
PRA – Preliminary Risk Analysis
PRA-SH – Socio-Human Preliminary Hazard Analysis
THERP – Technique for Human Error Rate Prediction

1 A discussion on reliability and risk – an introduction

1.1 Industrial risk analysis

The initial thought on the issue of risk and subsequently his management was primarily driven by the financial sector, specifically post-World War II [1][2]. However, in parallel with accelerated development of industry and processes increasingly complex - which implies exposure to workers, society and the environment to hazards ever growing - this issue gained notoriety in the industrial environment.

The occurrence of major industrial accidents in Brazil, and around the world, became the subject Risk Analysis and Management more popular to meet the demands of society and governments about the safety of industrial processes. It is in this context [3] that is needed better management and mitigates the impacts of accidents. One reason for the importance of developing risk management was the change in constraints for operating license after the Seveso accident in Italy (1976).

Schneider and Martins [4] define risk as a result of the frequency of occurrence and the consequences generated by an unwanted event, so that risk reduction is by reducing the frequency of undesirable events and the mitigation of the consequences. The risks that can be quantified are risks of injury to the person, damage to property and environment, and risk of loss of reputation. Risk management depends on the level that is being analyzed: equipment, process unit, facilities, and other levels.

According to Bharatiya [5], an industrial hazard example is the handling of flammable products that can be released causing fire and explosion and resulting impacts on workers and third parties, as well as economic losses such as: profit loss caused by shutdown plant; damage to equipment; penalties for release of toxic material, etc. In this regard, it is necessary application of qualitative and quantitative techniques of risk analysis and its review to ensure a safe working environment for employees and neighborhood.

Qualitative techniques of risk analysis such as PHA/PRA (Preliminary Hazard Analysis / Risk) and HAZOP (Hazard and Operability Studies) allow identifying and assessing risks and hazards associated with industrial processes, and identify operational problems through scenarios identification. According to Fleming and Garcia [6] PHA/PRA technique performs a risk assessment for each identified scenario, develops mitigation measures and specific studies to calculate the probability and analyzes the effects through the vulnerability of these scenarios. The HAZOP allows analyzing risk scenarios using a procedure that generates questions and answers in a structured way by guiding words that enable the identification of possible deviations of the operation. The review team examines the links between the causes and consequences of these deviations.

Regarding the semi-quantitative techniques for risk assessment, there is a tool for layers of protective analysis (LOPA) that is used in the safety integrity level setting against unwanted events in safety instrumented system [7]. His concept was first treated by Center for Chemical Process Safety [8].

In quantitative analysis, the various methods are distributed from the analysis of consequences, which is measured the extent of damage caused by adverse events, to quantify the frequencies and probabilities of each event causing accidents [4]. Among the most widely used techniques are Vulnerability, Fault Tree Analysis (FTA) and LOPA. The techniques can be combined in a study of quantitative risk analysis.

The development of quantitative criteria can follow internationally recognized parameters. In general, the aim should be to develop criteria also based on the existing company culture and the most effective process in this direction would be a hybrid of internal and external internationally accepted criteria [9].

1.2 Risk analysis services

Brazil has been placing on a world stage major events that generate therefore great responsibilities. Mega-events such as road races, international shows, World Cup soccer (2014) and the Olympics (2016) demand new ways to manage the risks. Experiences with carnivals indicate the need for organization in public management to meet the security needs of the population, quality in the health area, mobility, and social life.

For each event has a positive impact on society and economy, it is important to have a control of risks in the event of scenarios. According to Correia [10], the factors that pose major difficulties in quality control and risks in the event derive from the fact of being subjective experiences and difficult to measure.

Each event is unique and presents its own security risks. Major events bring large crowds with factors to be analyzed for risk management [11]: demand complex planning with different stages in the design and operation; have different participating organizations that serving people of different customs; It works with many volunteers and inexperienced staff; It has national and international media projection; demands less than the time required for the complexity of the services; It has regional, state and national economic impact; and requires careful consideration for risk management.

For large events are ephemeral, not admit failures, delays, and its facilities and spaces are in most provisional, at the same time these experiences are characterized by subjectivity, It is more efficient to analyze the risk qualitatively [10]. In this way the analysis generates a more tangible and simple approach, easy to study and implementation by the organizing companies.

1.3 Integrating the reliabilities: equipment, operation and human

The first steps in the study of reliability were been conducted in the early twentieth century (the end of the World War I), but the concept of reliability as it is treated today only began to take shape during the World War II in an attempt to improve the reliability of missiles [12]. For this classic model of reliability, Kardec and Nascif [13] define reliability as "the probability of the system or device to perform the function successfully required, under specified conditions of operation and a certain time".

The development process of reliability engineering was important in product development and process improvements. In the latter case your application has a strong relationship with the evolution of maintenance management acting not only in the transition between corrective, preventive and predictive models, but above all, the definition of maintenance strategies where each model has a degree of importance that depends on the results of investigations into the system failure patterns. That is, even if the corrective, preventive and predictive models represent, respectively, the 1st, 2nd and 3rd phase of maintenance management, any one of them may be applicable depending on the operational context of the organization and its failure patterns. The concept of reliability then goes to lead the efforts for efficient maintenance.

To meet the demands of analyzing the nature of the failure and its consequences several tools were developed to improve the reliability of equipment. The most used are: Failure Mode and Effect Analysis (FMEA), and Root Cause Failure Analysis, Method of Analysis and Troubleshooting and Reliability-Centered Maintenance.

However, while many companies design the best systems based on this reliability model, taking into account the great difficulty in maintaining the availability and reliability of equipment and machinery at satisfactory levels and the cases of accidents with environmental impacts and fatalities, it is clear that such practices are not sufficient to maximize system reliability and/or guarantee the effectiveness of crash prevention barriers. Therefore, in order to fill this gap, many studies have been conducted to include the role of human in this process, what we call human reliability analysis. These studies began to gain strength from the sixth annual meeting of the Human Factors Society which had as its theme the quantification of human error. So was born the first formal method for human reliability analysis – the THERP technique [14].

According to Pallerosi *et al.* [15], human reliability can be defined as "the probability that a person does not fail to comply with a task in a given period of time, in appropriate environmental conditions and available resources to do it". The author also classified two types of procedures to quantify the human failures. One is called the prediction methods and includes tools as TESEO, HEART, SPAR-H, THERP, ASEP, MPPS and others. The other method is classified as evaluation methods and use theories and probability distributions as Gamma, Weibull, Exponential, Binomial and other distributions applied to each type of task analyzed.

The elements or factors that contribute to human error happen are related to the cognitive and operational processes. These elements of the cognitive process of decision-making are: perception, memory, abstraction, emotions, stress, fatigue, knowledge, skills, communication, tasks, procedures, organizational culture etc. Ávila [16] discusses these factors in the operator's performance in the control of socio-technical processes in task execution and learning stage. These factors can be measured by individual results in the task accomplishing, from the quantitative approach presented in verification methodology of learning parameterized (RGA - Reliability Growth Analyses).

The human factors analysis allows the projection and the calculation of risk and also the calculation of human reliability. Among the analytical techniques are: HAZOP SOCIAL, PRA-SH, FMEAH, LODA, etc. In the organizational dimension, the API 770 standard [17] aims to assess the company culture and the impact on human and operational reliability diagnosing aspects of policy, procedures, man-machine interface, and issues relating to the roles and task settings. The latter issues are present in David Embrey (2000), Frank Lees (1996), Rasmussen (1997) and Avila studies (2009).

The understanding of cognitive processes is essential to increase the human and operational reliability in organizations. Study the social, economic and emotional utility [16], cognitive function [18][19] and models for decision-making [20] are required to analyze the commitment and quality of the decision.

Therefore, approaches to socio-technical areas of reliability should not be an isolated study for each functional part of the industry [21]. The reliability of each part is interconnected with each other and their level of integration depends on the level of automation, the type of process, the manager's management style and the environment promoted by the organization. By understanding about the relationship between the functions of these systems makes it possible to study the integrated reliability. This knowledge enables the formatting of factors that increase the performance of staff and equipment leading to greater organizational efficiency.

1.4 Routine management and crisis management

The failure analysis and reliability project should be carried out at various stages of the project lifecycle - from conceptual design (detailing and assembly), through operation (control of operations and routine) to crisis management. The study of fault begins with the perception of signals from the process, people, equipment and tasks and advances in the analysis of incidents and / or accidents that cause financial loss (or image). These events, if not treated properly, can lead to loss of control in a chain reaction and It requires contingency actions not always known by the leaders, characterizing a crisis situation.

According Constantinides [22], crisis management is a systematic attempt to identify and detect potential crisis scenarios and take action to prevent them, contain its effects or interruption, and finally the recovery of the system normality. Muller [23] highlights the need for early warning systems projects of undesirable events and resolve them before they turn into a real crisis. He added that in case of failure of preventive measures is essential to take an immediate action.

2 Performance of GRODin and publications

The Group of Dynamics and Operational Risks (Grodin) located at the Federal University of Bahia intends to conduct diagnostic systems from applied research and works to develop solutions to complex problems in industrial sectors and services over the integration of academy, industry, and Public sector. Its constitution allows multidisciplinary making the investigative process dynamic starting in the subjective transformations of human and social behavior until the objective accomplishing of the task in industrial facilities with the search of metrics for decision making in organizations.

Currently (2015) the group consists of 25 masters, 5 doctoral and 10 graduate students with graduate completion forecast for masters and doctors around 10 per year. Large areas of concentration of Grodin's publications are: (a) Process Safety; (b) Human Factors; (c) Reliability and Risk analysis in socio-technical systems; (d) Energy Efficiency and Sustainability.

(a) Process safety

Publications performed in process safety seek always discuss the inclusion of human and organizational factors among the aspects considered in the accident process. Thus, several articles were discussed and presented at conferences as described: (a1) risk analysis of the factors that define human performance for the tasks in equipment and processes classified as critical [24]; (a2) the analysis of the performing task environment, the relationship between man and technology that result in failures and the discussion about leadership and the moment of decision [25]; (a3) analysis of underlying causes (human error) which are involved in the accident process (economic bias, risk management, conflicts between policy and practice, cultural traits, task planning, contracted management) [26]; (a4) analysis of decisions of the operation under progressive stress on the sinister situation [27]; (a5) analysis of high frequency and lower impact accidents considering the inclusion of human factors and complexity of the task [28]; (a6) analysis of social relations in oil platform and the implications for an enabling environment to an accident [29]; (a7) analysis of organizational change, drafting checklist, impact analysis of possible process safety events and the need for a more detailed discussion of the human factors [30]; and (a8) risk analysis of public services in cultural mega events (carnival) [31] and sporting events (World Cup) [32].

(b) Human factors

The publications performed on Human Factors always tries to discuss the analysis of the intrinsic variability of the man in an organizational and cultural environment and that brings consequences as gains or losses for the society and industry. Thus, several articles were discussed and presented at conferences as described: (b1) usability of procedures based in operators opinion of the fertilizer industry [33]; (b2) complex task evaluation taking into account the requirements of criticity [34]; (b3) cognitive deficit analysis during the performance of executive function in performing tasks [35]; (b4) social assumptions that change cognitive decision for the review of ergonomics standards in the control room [36]; (b5) analysis of archetypes and their derivatives to avoid improper behavior on safety of industrial plants [37]; (b6) Operator behavior analysis and leadership in emergency situations, high stress caused by chain reactions [27]; (b7) trace analysis of the Brazilian and global culture, biases, and bad habits that hinder the establishment of operational discipline [38]; (b8) stress analysis in industry, health care and diving and its consequences on the task, behavior and health [39]; (b9) changing operator behavior based on social dynamic culture, difficult to detect changes before impacting events happening like accidents with economic, human and environmental losses [40].

(c) Reliability and risk in socio-technical systems – complex processes

Publications performed in Reliability and Risk Analysis in complex processes discusses how to detect and calculate the risks and failures always considering the presence of human performance factors. Thus, several articles were discussed and presented at conferences as described: (c1) relationship between loss of industrial processes and connection to the cognitive processes with impacts measurement [24]; (c2) the operational failure exergy relating the energy loss in cognitive failure with the loss of energy in the industrial process heat systems [41]; (c3) evaluation of failure in complex processes, complex tasks and complex social relations in the chemical industry [34]; (c4) dynamic risk analysis for industrial sustainability [42]; (c5) human factors and equipment failures analysis using tools such as SPAR-H and Weibull ++ in the nuclear and chicken industries [43][44]; (c6) Analysis of socio-technical reliability in the refining industry [21]; (c7) risk analysis of the pre-salt considering: phenomenology, reliability model and risk factors [45][46].

(d) Energy efficiency and sustainability

Publications performed in energy efficiency and sustainability seek to discuss what are the opportunities to reduce consumption of energy and water and ensuring the sustainability of business. Thus, several articles were discussed and presented at conferences as described: (d1) mass balance and energy in cooling towers [24]; (d2) audit in organic effluent and rainwater to prevent process losses; (d3) audit of environmental deviations in industrial routine [47]; (d4) contamination control program at source by mapping effluent [48]; (d5) application of mass balance techniques and energy to identify opportunities to reduce energy and water losses in the industry; (d6) a discussion of systems and cooling towers, challenges to avoid energy losses and water; (d7) identification and calculation of thermal performance and energy efficiency in equipment and industrial processes [49]; (d8) Analysis of loss of condensate and steam heating systems and energy recovery [50].

3 A vision of sustainable organization based on risk

Maintaining organizational sustainability requires an understanding that the business risks are dynamic and the operational risks depend on a balanced analysis of issues related to product and process with the people and culture issues. So, to start the discussion on sustainability in high-risk industry a management model will be presented.

An efficient risk management of business develops premises to organizational project considering social-technical aspects. This risk management draws an industrial operation format including multidisciplinary analyzes of different levels of normality (default to failure and the crisis). It is intended with this model to measure the effects and impacts on society and then review the criteria and procedures adopted by project operations.

3.1 Inputs in the Risk Management and Organizational Design

Important inputs for the analysis of dynamic risk are: entrepreneur demand (vision and mission), technological conceptual design (product and process), analysis of the availability of resources and production factors (energy, materials and people), and external demands represented by the law and behavior of the stakeholders in the existence of this business. The risk analysis stage to address the dynamic risk intent to insert a discussion about the process and product taking into account the type of task (or the procedure), the cognitive models for decision (local or multicultural) and the social behavior of internal and external community of the organization.

3.2 Step one: Risk Management - socio-technical risk analysis

The first step of socio-technical risk analysis balances human-social factors with process and product factors generating preliminary recommendations. In the case of technical risks are evaluated the product, process and the likely scenarios through LOPA and PRA and getting as result recommendations to prevent accidents and technical failures. For non-technical risks such as environmental and human aspects is performed an PRA-SH to anticipate future events related to human errors that can cause technical failure or generate an stressor environment at the workplace. At this time are considered the risks of events that involving cognitive faults and failure on the task resulting from social processes.

3.3 Step two: organizational project - detailing of technical and non-technical project

The second step requires knowledge of the detailed organization design in the technical and non-technical aspects to define the assumptions and criteria used for the design of operations. Thus, the organizational project aims to open the product technology, process, service, and control to discuss which the necessary demands on leadership, behavior, attitude, and skills.

The intrinsic variability policies, humanities, social and variability of nature (climate, biota, etc.) are scored as important for maintaining the environment in which the industry is installed and intends to stay in a sustainable manner. The organizational design analysis requires detailed information of technology, knowledge in integrated management system, consumption analysis and energy benchmark, water, materials and people (productivity) to establish performance indicators. The result of these studies is the definition of technical, human, organizational, natural factors for the sustainable operation of the production system. The techniques applied in this second phase are: HAZOP, HAZOP-Social; requirements for Task (GMTA, PADOP); cause and effect analysis of the events; barriers analysis of events (BOWTIE).

PADOP is a method that starts with the analysis of the requirements for feeding the following phases: environment, planning and execution of the task. Others steps such as task control and conformity analysis under review are performed in operational routine (see item 3.4).

3.4 Step three: design and control of operations

In the third stage of dynamic risk assessment is intended to rescue the criteria and assumptions set out in the previous step, organizational design, where technology is detailed and an integrated management mode is defined. In the project of operations, it will be handled operational routines, asset control, and control of failures and possible crisis. So, it will be designed systems for operation in the technical area (equipment, processes, products and control modes), the interface (task management, and human-machine interface) up to non-technical area (culture and biases, cognitive decision and stress and behavior analysis). The results of these measurements allow adjustments on the routine of the design of operations and business and organization projects.

These factors and elements that make up the control of the operations are analyzed on standard procedure mode and failure analysis in the following areas and using the following techniques: equipment (FMEA, FMEAH); product and process (monitoring and statistical control); task in a complex way (GMTA, CARMAN, PADOP); man-machine interface and control room (PRA-SH and regional criteria for control room projects); culture (human reliability committee for organizational change and archetypes analysis); cognitive decision (study of cognitive failure); stress and decision (LOPA); and stress and worker health (LESHA). These techniques for analysis of dynamically operations present measurements and indicators that are based on operator's speech. This speech is announced from the routine results and reflects the occupational consequences (accident), environmental (pollution and waste), climate (CO_2 emissions) and economic (loss of competitiveness based on consumption rates and productivity).

In this case, the PAPOD technique intent to compare the standard designed for task control rates with measurement of the process after the task performed. It is worth noting that these standards include variables of integrated management (safety, quality, environment, occupational health, etc) as well as socio-humans aspects (quality of communications). In case of deviations it will be realized preventive corrections with necessary revision to the task or written procedure.

4 Application of exercise

An implantation exercise is proposed based on the application of different technical and management systems in research in various chemical process industries. Activities and analyzed processes involve energy products (oil and LPG), chemical (sulfuric acid, methyl diphenyl amine – MDA, and explosives), petrochemicals (basic petrochemicals and fertilizers), polymers (polycarbonates) and plastic (plastic bottles).

These various research opportunities involve aspects of routine at the work place and characteristics of management that promote the company's results close to normal or fault situations. The experiences are gathered in an integrated way toward sustainability considering that risks are dynamic and that the systems are socio-technical.

Some of these techniques are applied with quantitative results indicating the intensity of the problem and the volume of investment to be taken and others indicate the existence of the failures and suggest the construction of safeguards, but in a qualitative way.

This exercise will only treat of techniques relating to the socio-environmental-human risk and it will quote only within the management system proposed and which evaluations (of process and product) should be interconnected.

4.1 Inputs in the risk management and organizational design

For this model that considers the dynamic risk, the entrepreneur discusses the social variability in the possibility of failure and practices a vision and mission more sustainable, which admits the existence of a multidisciplinary approach to the definition of the operating pattern. This discussion is part of the control of the operational routine. So, inputs to this management system from experienced cases it is suggested.

<u>Entrepreneur demand</u>

Case 1: the entrepreneur demand to identify the routine deviation is a requirement to draw up barriers that prevent systemic failure. To discuss the causes of failures that lead the operation to a mode of low productivity, an entrepreneur of the LPG industry recognizes the need to analyze the routine. The operator's speech feeds the construction of the fault tree for routine events and together with the task analysis identify the GAPs of competence that lead to failure [51].

Case 2: the perception of the manager must be attentive to behavioral deviations that can lead to accidents. In this case, the corporate HSE manager of the explosive industry believes that the study of human typology is necessary to change the criteria in risk management. This belief is based on the possible change of behavior, which caused two consecutive accidents not expected in different countries, after a history of 20 years without accident. This entrepreneur corroborates the following statement: "cultural changes alter the behavior of the workers and there is a need to study new criteria for dynamic risk management".

<u>Analysis of the availability of resources and production factors (water, energy and people)</u>

Production factor: energy. The water crisis in the state of São Paulo [52] and the reduction of energy quotas for petrochemicals due to its unavailability caused by climate change indicates that the analysis of availability is critical input in the organizational design to address the variability of nature.

Production factor: water. CEOs of Brazilian refineries are being instructed by National Agency of Water (for now voluntarily) for an actual 20% reduction in water consumption. This highlights the urgent need to change behavior and decisions in this organizational design.

Production factor: people. The opening speech of the Brazilian Association of Risk and Reliability (ABRISCO) indicates that the Brazilian engineer does not know and does not quantify industrial risks indicating the unavailability of appropriate skills to quantify investments in safeguards to prevent accidents.

<u>External demands legislation for new behaviors of stakeholders</u>

Stakeholder – National Petroleum Agency (ANP). After the FPSO production ship accident in the city of São Mateus, an investigation realized by the ANP [53] indicates the need to discuss the issues of contract management, omissions, commitment, crisis management and decision in stress. These issues are part of social- human variabilities.

Stakeholder – University which reflects demand for workers. In Risk Management is important to define the requirements necessary from technology and local culture for planning the task (PADOP). From the master's degree studies on the task analysis it was found that most of the 450 operators of a fertilizer industry accepted the importance of establishing social and human premises to carry out the critical task.

Stakeholder – Multinational oil companies. From international meetings with reliability consulting groups that serve the oil industry, it was found out the need to discuss regionalities to change the ergonomic criteria in control room. This indicates that the CEOs of major companies (Chevron, Statoil, BG) are reflecting on the question of risk review to include in their analysis the dynamic and social aspects (cultures and people).

4.2 Step one: risk and opportunity management - socio-technical risk analysis

After the hazard analysis and technical risks by conventional application of the techniques APP and APR (to business, processes and products) it is natural to use the vulnerability analysis in the most critical scenarios to indicate the level of impact on human and nature. Similarly to analyze the social and human risk seeks to detail the various performance shaping factors [54] that indicate the level of consequence in case of human errors.

Fault analysis (in 20 cases) in the chemical process industry indicates that elements such as organizational climate, leadership, skills, knowledge, and decision-making processes are related to human error in performing the task.

Several cases involving deviations, failures, incidents and accidents were analyzed and allowed the construction of the anatomy of failure in complex mode relating human and organizational factors [26]. Among the cases are: director's behavior in the plant starting management (polycarbonates); deviations in mitigation of waste generation reaction (MDA); operational control to avoid stoichiometric reaction stop (MDA); skill in the perception of vacuum loss in sulfuric acid unit; failures caused by improper turn change (MDA); investigation of effluent contamination source (MDA, sulfuric acid, lab, fertilizers); diagnosis to map the gaps in the process (MDA, polycarbonates, LPG); negotiation of new operating parameters without causing downtime (MDA); improper use of individual protection equipment and omission by fear of guilt (MDA); oring diameter change at startup leading to failures with solvent losses (polycarbonate); and others.

Based on the investigation of cases the initiators factors to human error are [26]: design criteria, risk management, economic bias, cultural traits, task planning, and conflict between policy and practice. It is also important to identify the social dangers to avoid accidents before assembling the installation and make the hires to operate the industrial site.

A specific case was applied at the stage of design of an oil exploration platform in northeastern Brazil during the graduate student research in the area of security and human factors analysis in 2013 [55]. After seven risk analysis sessions for human and social factors, the group of six experts (socio-technical) listed the following factors that can cause human errors in this installation that come into operation in 2015: difficulties in routine leadership; non-commitment to the safety result; low perception to failure; and improper internal communication.

It is worth listing the technique steps: identification of social and human dangers, social and human risk analysis, risk classification, interaction between risks (interaction matrix to correct risks), the final risk map, and treatment of social and human risk.

An item that is repeated in the workplace discussion refers to the effect of stress on the effectiveness of the task [56] and occupational health [57]. Thus, it interconnects the worker performance factors to the result of task with economic results for the industry. The causal relationship between work environment, cognitive fault and failure in the task are motivating ongoing research that result in techniques and applications.

Technique 1: analysis of social relations in the workplace. Application: oil platform where the rituals of loading and unloading, the distance and isolation are analyzed to list critical topics that can cause stress and human error [58] [29]. Technique 2: protocol for the analysis of stressor environment. Application: industry [56], nursing services [39] and diving services in deep waters. Technique 3: task analysis [33]. Application: chemical process industry. Technique 4: analysis of cognitive impairment [35]. Application: chemical industry.

4.3 Step two: organizational project - detailing of technical and non-technical project

Step 1 of this management model for sustainability is related to the fit project of product, process, and business, in a discussion of technology and economic viability. The analysis of social and human risk in the task is performed along with the APR technique of this future operation.

Step 2 is the organizational design that includes the adaptation of the technology for the future operation that includes product and process aspects, and non-technical issues related to teamwork in performing the task (culture and people). At this stage (detailing phase) are carried out studies of the conceptual design (made in step 1) and details the risks of failure to cause economic loss and / or the company's image.

<u>Application of HAZOP-Social</u>

The process deviation analysis during operation indicates the possibility of danger containment loss related to the industrial activities. The Social HAZOP tool works similarly to the conventional technique [16]. In its application the cognitive decision-making is tested through the deviations and recommendations are announced to prevent failure of attention, memory, competence, managerial decisions and others. Social HAZOP was applied to a draining operation of a decanter vessel in the refining industry as having maximum result the accident with fatality and with the critical lack of attention, centralized management profile and commitment to the task.

Diagnostic of skills gap

Identification of fault root cause in the chemical industry that has characteristics of processes and complex tasks [59] [34] requires the study of process variables and production beyond the qualitative and quantitative analysis of operator's speech [60][61]. This discourse analysis allows the organization to verify the main failure modes in socio-technical systems by locating the critical variables in the tasks, equipment, skills and operator behavior. When comparing the type of fault carried out (based on competence offered) with the demand for knowledge and skills in technology (competence demanded) have been related what are the gaps in competence.

This technique was applied in PhD research [51] of the gas compression unit (LPG) in northeastern Brazil. In this case, the database was collected from one year of statistical analysis of process and production variables, and four months of shift book analysis in low performance periods. Diagnosis of gaps indicates the lack of skills in the following areas [25][62]: Task control (abstraction and logical); concepts (thermal and material exchange); failure (rotary, motorized valve and burner); skills (thermal cycles, pressurized system and logistics); process control and safety (supervisory and risk in the routine).

The gap of skills already identified in the LPG plant may result from lack of commitment of the people, the team cooperation and inappropriate profile of leadership. A classification model of commitment, the leader's behavior and aggregation of the team is suggested by Avila [51][25][62].

Research on behavioral intrinsic variability

Demand for leaders in chemical industry and oil organizations during the GCPSs events indicates the importance of analyzing the changing worker behavior as a result of Dynamic Culture. A methodology is proposed to find the biases of global and regional culture that promote behavioral changes towards the accident [40]. This method was structured to identify potential changes in teams of explosives chemical company with unusual situations accident in facilities overseas. The suggested steps with the respective models are: (1) analysis of social culture; (2) analysis of the social and human types to work; (3) analysis of the dynamic factors of culture or affective socio-economic utility; and (4) evaluation of the result in the worker's behavior performing the critical task.

Archetypes and regionalities analysis

Entrepreneurs from different areas (chemical and oil) have discussed the ergonomic criteria for control rooms and raise the possibility that tacit aspects influenced by behaviors are not being considered. This hypothesis is due to continuity failures in critical operations such as plant startup and shutdown. These questions led the researchers to analyze the control room construction criteria based on the ergonomics requirements in the EEMUA and ISO standards [36]. The result of the research is the mapping of critical situations that can result in accidents.

In the case of a LNG industry control room were related aspects of local and global culture that impact on the cognitive processing leading to failure in decision making. This tool depends on the study of product risks and social risks (APR and APRSH), the analysis of the major failures and relations with human error in similar facilities, and brings results in a cross-matrix, which relates the social influences, cognitive functions with interfaces and control room devices.

The regional or global behavior that can turn into social risk (eg, default and inadequate leadership) may be due to cultural biases from primary or secondary archetypes. A case of industrial gases plant is analyzed and actions are suggested in the form of ritual to neutralize or deflect improper behavior [37].

Variability of climate change

The urgency of actions to reduce GHG emissions, fuel consumption and water motivate researchers in techniques seek to adjust the demand for water and energy in relation to confirmation of the offer. The demand adjustment actions involving routine management, knowledge of techniques for thermal performance analysis and new technologies such as process intensification or modularization of projects. The research conducted (LAB, Olefins, Urea) indicate that the equipment involved in research and routine adjustment are: cooling towers, boilers, furnaces, condensate systems and unit operations with their reflux [49][50].

4.4 Step three: design and control of operations

The Control of Operational Routine depends on the knowledge of appropriate standards for the maintenance of a stable process with the appropriate cost. It is increasingly announced by managers of high-risk companies that the economic result is guaranteed if the image is positive - that is, the non-occurrence of events related to accidents.

The definition of operating standard to routine management is part of a policy initiated by the management of technical and non-technical risks. So, the foreknowledge keeps the assets in normal mode. The occurrence of deviations, failures and incidents in the industrial operation can cause a crisis situation that requires a new leadership profile (crisis and contingency management).

Cases of task analysis

In LPG industry there are high-risk activities that can cause fire or explosion. Purification of LPG avoids failure through moisture removal. The moisture-removing columns have a saturation point and are removed from operation to be regenerated. This, in turn, requires the moisture expulsion using heated LPG itself resulting from passage through a furnace.

During the investigation concerning the study of root cause analysis in LPG process [51] was made the analysis of the operation in the task, startup and breakdown in regeneration furnaces using techniques such as GMTA[20], CARMAN[63] and PADOP[26][33]. These techniques used for fault analysis in the task involves the logic and the nature of the failure, the chronology of events, connectivity between factors beside the point where failure goes out of the social mode (eg. communication) to the physical mode in GLP [51] and Polycarbonate [60].

The task control actions include the procedure review (requirements, targets, or steps) with training. When the risk, complexity of processes, the complexity of social relationships and stress levels are high (in part or in full), the training actions has no effectiveness demanding action on culture [56][33] or to the redesign of equipment and processes.

Cases of failure in equipment and process

The benzene hydrogenation is an intermediate step in the production of caprolactam. The recycle gas is maintained at constant pressure to control the reaction through a centrifugal turbo-compressor of the brand Demag. Repeated interruptions on Monday (after the weekend) raised the possibility of having human error in the operation and/or maintenance which did not allow the effectiveness of technical actions.

A FMEAH study was performed to identify the failure mode which problems of perception, attention and expertise were occurring in routine production. The method resembles the traditional FMEA, but includes analysis of human error in the equipment failure mode [64].

The degradation systems, changes in technology, organizational changes [30] and climate change may cause variations in the process which results in the modification of the possible event scenario, which in turn increases the risk of accidents. In this regard, through the signals emitted in routine, seeks to review the product risk, process and operations through the application of new techniques such as APR, APRSH, HAZOP, Social HAZOP.

Human performance case in state of stress

A startup case of sulfuric acid purification discussed the influence of stressful environment. The elements that make up the work environment to achieve a state of stress that affects human performance are: task risk level; health impacts, the cost of business and social environment; function of the task in the process; and analysis of the psychological and organizational types. The protocol indicated that the level of stress was 4.8 exceeding the stress level manageable (under 4).

An accident analysis in progressive stress

An accident involving LPG furnace, as discussed in this paper, was used to build a fault tree of technical and non-technical events in different levels of stress.

The application of LODA technique [27] from gradually more dangerous events indicates that the decision of the shift varies according to the level of stress. The progressive stress is promoted by the need for decision-making by building possible increasingly serious consequences scenarios at the following levels: deviation from the routine, lack of process, lack of equipment and task, accident and sinister.

Unfortunately, the psychological unpreparedness of staff in dealing with contingencies, especially the leaderships, makes the human error probability goes from 10%, in steps of process deviations and lack of control of process, to 50% in sinister steps.

Cognitive failure cases and vices in the industry operators behavior

In a startup of a chemical industry reactor [35] are analyzed psychological functions in the executive work of the task. From social relationship models and the executive function analysis it was identified the following types of failure on the team with priority level 1: lack of knowledge and difficulties in memory recovery.

In the sulfuric acid industry [40] were analyzed behavioral addictions through fault trees that indicated the need for vaccines for cognitive diseases, among which: "frozen" perception, low motivation to cooperate, not enabled memory, commitment corrupted, communication and leadership inadequate, and very simple mental map. This finding allows the construction of behavior adjustment programs.

In industrial gases and chemical industries were found two types of behaviors in the control room which allowed the analysis of two different groups archetypes. Some archetypes favor a robust safety culture and others hinder their establishment [37].

Root-cause analysis in complex systems

The operator speech analysis is a tool that links events occurred in the shift such as: problems with equipment, product quality, double-shifts, missing and other facts. The form of communication, the statistics of events, the possibility of omission and team cooperation are some of the aspects included in this speech.

On several occasions in the chemical process industries [60] were applied root cause analysis techniques in complex systems. In the production of alkyl amines it was noted that the delays were the result of errors in the setup of the campaign. In the production of methyl diphenylamine it was observed that the interruptions were strongly influenced by the formation of groups that are not communicated with each other and by the lack of knowledge of the importance of recycle line for the quality in the process. In polycarbonates production sought to reduce loss of methylene chloride. In this case the operator's speech has enabled to review: procedures (from the task analysis) and manager behavior in the plant startup in control room.

In LPG industry in which the concern is the leakage events, fire or explosion, the operator's speech helped to identifying critical areas, gap of competence for decision-making, and team behavior in emergency situations [51].

5 Conclusion

The model of operational risk management requires technical and social aspects are correlated at different levels of normality / abnormality of the processes with the consequences and impacts that produce in their own business, on environment and society. By multidisciplinary involved in the analysis of these risks, the definition of management strategies should take into account the dynamics of the risks in business.

In some cases analyzed were used the operator' speech as a tool for building the event fault tree in the routine that jointly with task analysis identified the gaps of competence that lead to failure where the criteria and procedures adopted can be reviewed in the design of operations.

In another case (explosives factory) was found to be necessary to investigate the human type to change the criteria in risk management. In this case, the entrepreneur recorded accidents with serious injuries to the worker after a history of 20 years without accident (plants located in different countries) and admitted that cultural changes alter behavior which highlights the need to consider new criteria for dynamic risk management.

From the perspective of production factors such as water and energy, is also recorded vulnerability in the sustainability of the business in major change situations in the availability of these resources requiring that teams know how to prioritize investments to overcome such difficulties.

The applications of various techniques mentioned in this work are structured in a sustainable model of risk management. This model requires generation of multidisciplinary expertise (socio-technical aspects) to properly handle the dynamics of risk.

6 Annex

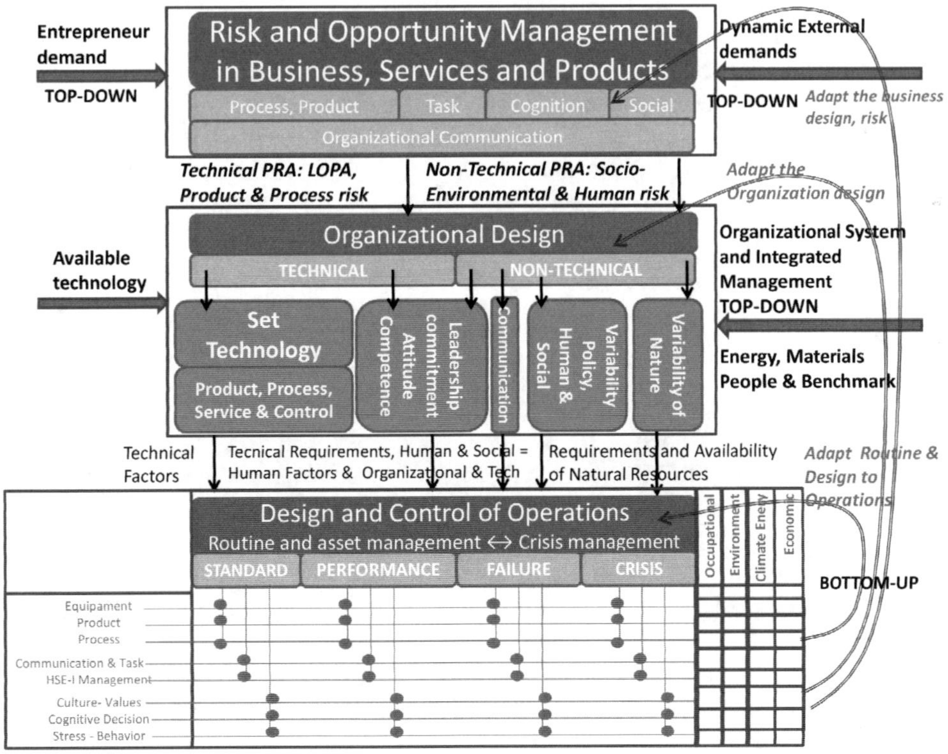

Figure 1 – Risk Management Model for Sustainable Organization

7 References

[1] G. Dionne Risk Management: history, definition, and critique. Risk Management and Insurance Review, 16 (2013), 147-166.
[2] J. L. Vesper. *Risk Assessment and Risk Management in the Pharmaceutical Industry: Clear and Simple*. Parenteral Drug Association, New Vernon, 2006.
[3] M. R. Martins *et al.* Undesirable events Consequences analysis during the transport, loading and unloading of Liquefied Natural Gas (LNG) on ships. Proceedings of Brazilian Association of Risk - ABRISCO. Rio de Janeino, 2013.
[4] A. M. Schleder and M. R Martins. Methodology for quantitative analysis of the consequence of flammable leaks and / or toxic environments with barriers or confined. Proceedings of Brazilian Association of Risk - ABRISCO. Rio de Janeino, 2013.
[5] B. K. Bharatiya. Projet Hazard Analysis In Hydrocarbon Processing Industry. Proceedings of 4th Latin American Conference on Process Safety - CCPS. Rio de Janeino, 2012.
[6] P. V. Fleming and C. B. Garcia. Industrial and Environmental Risk Assessment With Preliminary Hazard Analysis (PHA) and Fuzzy Logic. Proceedings of Brazilian Association of Production Engineering - ENEGEP. 1999.
[7] R. Freeman. Quantifying Lopa Uncertainty. Process Saf Prog 31(2012), 240-247.
[8] CCPS. *Guidelines for Safe Automation of Chemical Processes*. Center for Chemical Process Safety, American Institute of Chemical Engineers, New York, NY, 1993.
[9] J. N. Shan. Understanding and Developing Quantitative Risk Criteria. Proceedings of 8th Global Congress on Process Safety, Houston, 2012.
[10] A. H. L. Correia. Strategic Marketing of Sporting Event. International Seminar of Sports Event Management. Ministério da Juventude e do Desporto. Centro de estudos e formação desportiva. Lisboa, 2001.
[11] Worksafe Victoria. Advice for Managing Major Events Safely. 1st Ed. Austrália. 2006. Available at: https://www.worksafe.vic.gov.au/__data/assets/pdf_file/0010/12502/major_events_organisers_advice.pdf. Accessed on January 3, 2015.
[12] F. S. Flogiatto and J. L. D. Ribeiro. *Reliability and Industrial Maintenance*. Elsevier: ABEPRO, Rio de Janeiro, 2009.
[13] A. Kardec and J. Nascif. *Maintenance – Strategic Function*. 4th Ed. Qualitymark Editora, Rio de Janeiro, 2013.
[14] R. L. Boring. Fifty Years of THERP and Human Reliability Analysis. U.S. Department of Energy National Laboratory. U.S.A., 2012.
[15] C. A. Pallerosi; B. P. M. Mazzolini and L. R. Mazzolini. *Human reliability - Concepts, Analysis, Evaluation and Challenges*. AllPrint Editora, São Paulo, 2011.
[16] ÁVILA FILHO S.; PESSOA F.L.P. and ANDRADE J.C.S. Social HAZOP at an Oil Refinery. Process Saf Prog - 32 (2013), 17-21.
[17] D. K. Lorenzo. API770 – A manager's guide to reducing human errors, improving human in the process industries. API Publishing Services, Washington, 2001.
[18] J. Reason. *Human Error*. Cambridge University Press, New York, 2003.
[19] R. Stenberg Junior. *Cognitive Psychology*. 3rd Ed. Artmed, Porto Alegre, 2008.
[20] E. Hollnagel. *Human Reliability Analysis Context And Control - Computers And People Series*. Academic Press Inc, San Diego, 1993.
[21] S. Ávila Filho. Reliability analysis for socio-technical system, case propene pumping. Engineering Failure Analysis, 56 (2015), 177-184.

[22] P. Constatinides. The Failure of Foresight In Crisis Management: a Secondary Analysis of the Mari Disaster. Technological Forecasting & Social Change, 80 (2013), 1657–1673
[23] R. Muller. Corporate Crisis Management. Long Range Planning, 18 (1985), 38-48.
[24] S. Ávila Filho *et al.* Risk Management and Prevention Program to Avoid Loss of Process (LP3) based on Human Reliability, Practical Application and Measurement. Proceedings of 7th Global Congress on Process Safety – GCPS. Chicago, 2011.
[25] S. Ávila Filho and A. C. Carvalho. Human Factors Assessment to Avoid Accident at LPG Industry. Proceedings of 6th Latin American Conference on Process Safety – CCPS. Rio de Janeiro, 2012.
[26] S. Ávila Filho. Assessment of Human Elements to Avoid Accidents and Failures in Task Perform, Cognitive And Intuitive Schemes. Proceedings of 7th Global Congress on Process Safety – GCPS. Chicago, 2011.
[27] S. Ávila Filho. Dependent Layer of Operation decision analyzes (LODA) to calculate Human Factor, a simulated case with PLG event. Proceedings of 7th Global Congress on Process Safety – GCPS. Chicago, 2011.
[28] S. Ávila Filho. Review of Risk Analysis and Accident on the Routine Operations in the Oil Industry. Proceedings of 5th Latin American Conference on Process Safety – CCPS. Cartagena de Indias, 2013a.
[29] S. Ávila Filho. Risks in the Platform: Social and Human Guidelines Avoid Accidents and Alter Standards in Oil Production. Revista Proteção, 2013b.
[30] S. Ávila Filho; E. T. Dantas and J. Duarte. The Requirements & Tools to Treat Process Safety Risk as Result of Organizational Change. Proceedings of 9th Global Congress on Process Safety – GCPS. Austin, 2015a.
[31] S. Ávila Filho, *et al.* Analysis of risk scenarios and mapping of communication processes to avoid incidents in Salvador Carnival in 2015/16. Proceedings of Brazilian Association of Risk – ABRISCO/PSAM. Rio de Janeiro, 2015a.
[32] S. Ávila Filho, *et al.* Risk Analysis in Sports Mega-events, UFBA Project in the World Cup. Proceedings of 34th Brazilian Association of Production Engineering - ENEGEP. Curitiba, 2014.
[33] S. Ávila Filho; C. R. Sousa and A. C. Carvalho. Assessment of Complexity in the Task to Define Safeguards Against Dynamic Risks. Proceedings of 5th International Conference on Applied Human Factors and Ergonomics – AHFE. Las Vegas, 2015.
[34] S. Ávila Filho. Failure Analysis in Complex Processes. Proceedings of 19th Brazilian Chemical Engineering Congress – COBEQ. Búzios, 2012.
[35] S. Ávila Filho and C. Costa. Analysis of Cognitive Deficit in Routine Task: as a Strategy to Reduce Accidents and Increase Industrial Production. Proceedings of European Safety and Reliability Conference – ESREL. Zurich, 2015.
[36] S. Ávila Filho and F, L. P. Pessoa. Proposition of review in EEMUA 201 & ISO Standard 11064 Based on Cultural Aspects in Labor Team, LNG Case. JofPM, 3 (2015), 6101 – 6108.
[37] S. Ávila Filho, M.L.A. Menezes. Influence of Local Archetypes on the Operability & Usability of Instruments in Control Rooms. Industrial Production. Proceedings of European Safety and Reliability Conference – ESREL. Zurich, 2015.
[38] S. Ávila Filho, *et al.* Influence of Brazilian Culture in the Workers' Behavior and Operational Discipline. Proceedings of Brazilian Association of Risk – ABRISCO/PSAM. Rio de Janeiro, 2015b.

[39] K. R. V. Nunes, *et al.* Working Hours Excessive and its Influence on Accidents, Productivity and Decision Making of Nursing Technicians in an ICU. Proceedings of Brazilian Association of Risk – ABRISCO/PSAM. Rio de Janeiro, 2015b.
[40] S. Ávila Filho; C. C. Silva and C. R. O. Sousa. Worker Behavior Changing by Dynamic Culture. Proceedings of 9th Global Congress on Process Safety – GCPS. San Antonio, 2013.
[41] S. Ávila Filho. The Exergy of Operational Failure, a Philosophical Discussion with Basis Practices. Proceedings of 19th Brazilian Chemical Engineering Congress – COBEQ. Búzios, 2012.
[42] S. Ávila Filho, J. R. N. Lopes and C. R. O. Sousa. Strategies to Industry Sustainability. Annuals of 20th World Renewable Energy Forum – WREF. Denver, 2012.
[43] A. C. Alcoforado and S. Ávila Filho. SPAR-H Application in the Crushing Operation of Uranium Ore. Proceedings of Brazilian Association of Risk – ABRISCO/PSAM. Rio de Janeiro, 2015.
[44] S. S. Dórea and S. Ávila Filho. Failure Investigation in Operations of Chicken Industry with Based Human and Technical Factors. Proceedings of Brazilian Association of Risk – ABRISCO/PSAM. Rio de Janeino, 2015.
[45] M. N. G Abreu and S. Ávila Filho. Assessment of Risk in the Exploration and Production of Oil at Pre-Salt Layer – Brazil, a Discussion about Social-Technical Factors. Proceedings of Rio Pipeline Conference & exposition. Rio de Janeiro, 2015.
[46] M. N. G Abreu and S. Ávila Filho. Assessment of Reliability and Risk at Dynamic Systems: a Discussion of Social-Technical Factors in the Exploration and Production of Oil at Pre-Salt Layer – Brazil. Proceedings of Global Congress on Process Safety – GCPS. Austin, 2015.
[47] S. Ávila Filho, *et al.* Environmental Deviation Assessment, Record, Failure, Probability and Investment. Chemical Eng Trans, 26 (2012a), 441-446.
[48] S. Ávila Filho, *et al.* Program to Avoid Product Leakage at Source, Fertilizer Case. Chemical Eng Trans, 26 (2012b), 447-452.
[49] S. Ávila Filho, *et al.* Iniciatives of Energy Efficiency Toward Sustainability, Opportunities & Challenges In: Rio Oil Gas Conference & exposition. Rio de Janeiro, 2014.
[50] S. Ávila Filho, *et al.* Avoiding Loss of Energy in a Petrochemical Industry, Operation and Design. Proceedings of World Renewable Energy Congress – WREC. Linköping, 2011.
[51] S. Ávila Filho. Etiology of Operational Abnormalities in Industry: a Model for Learning. Doctoral thesis - Chemistry School. Federal University of Rio de Janeiro. Rio de Janeiro, 2010.
[52] ANP. Investigation Report of an Explosion Incident in FPSO Ship in São Mateus Town on November 02, 2015. National Agency of Petroleum, Natural Gas and Biofuels. Available at: http://anp.gov.br/?pg=78836&m=&t1=&t2=&t3=&t4=&ar=&ps=&1452081674560. Accessed on January 15, 2016.
[53] R. Bonalume Neto. 30% Loss of Water During Distribution Contributes to Crisis in São Paulo. Sustainability Forum. Folha de São Paulo. São Paulo, 2014.
[54] D. K Lorenzo. API770 – A manager's guide to reducing human errors, improving human in the process industries. API Publishing Services, Washington, 2001.
[55] M. P. Barroso. The influence of human factors in APP and APR risk analysis techniques: a study case of an oil drilling platform in the northeast. Master's Dissertation – Polytechnic School. Federal University of Bahia. Salvador, 2013.
[56] G. G. Gonçalves, *et al.* Protocol to evaluate the influence of stress on the task involving human types, behavior and health: An exercise proposal. ESREL 2015. Zurick. 2015.

[57] S. Ávila Filho. LESHA – Multi-Layer progressive stress & impact assessment on health & behavior. Proceedings of 9th Global Congress on Process Safety. San Antonio, 2013.

[58] S. Ávila Filho. Human Errors Assessment, Social Environment at Oil Platforms. Proceedings of 4th Latin America Conference on Process Safety. Rio de Janeiro. 2012.

[59] C. Perrow. *Normal accidents: living with high-risk technologies.* NY: Basic Books, 1984. 453 p.

[60] S. Ávila Filho. Methodology to Minimize Generation of Waste at Source from the Investigation of Operational Abnormalities. Master's Dissertation – Polytechnic School. Federal University of Bahia. Salvador, 2004.

[61] E.S. Drigo, S. Ávila Filho and C. A. O Sousa. Operator discourse analysis as a tool for risk management. Proceedings of European Safety and Reliability Conference – ESREL. Zurich, 2015.

[62] S. Ávila Filho. Diagnosis in Socio-Technical Systems Aiming to Increase Human Reliability, a LPG Case. Proceedings of International Applied Reliability Symposium – ARS. Fortaleza, 2011.

[63] D. Embrey. Preventing human error: developing a consensus led safety culture based on best practice. London: Human Reliability Associates Ltd. 2000.

[64] S. Ávila Filho, *et al.* Human Factors Analysis in Turbocharger Equipment Failure FMEAH in Chemical Plant. Proceedings of 27th Brazilian Maintenance Congress. Rio de Janeiro, 2012.

Water and Energy availability analysis, a proposal for management for chemical industry

Salvador Ávila Filho
Federal University of Bahia – UFBA
Salvador – Brazil
avilasalva@gmail.com

José Rafael Nascimento Lopes
National Service of Industrial Learning – SENAI
Salvador – Brazil
jrafael.lopes@gmail.com

Jean marcel Prazeres Silva
Federal University of Bahia – UFBA
Salvador – Brazil
jmarcel_7@yahoo.com.br

Prepared for Presentation at
American Institute of Chemical Engineers
2016 Spring Meeting
12th Global Congress on Process Safety
Houston, Texas
April 11-23, 2016

Water and Energy availability analysis, a proposal for management for chemical industry

Salvador Ávila Filho
Federal University of Bahia – UFBA
Salvador – Brazil
avilasalva@gmail.com

José Rafael Nascimento Lopes
National Service of Industrial Learning – SENAI

Jean Marcel Prazeres Silva
Federal University of Bahia – UFBA

Keywords: water availability, energy analysis, sustainability

Abstract

The evidences that the chemical industry is not ready to deal with interruption of operations resulting from climate change and the complexity of decisions is increasingly apparent. In the state of São Paulo / Brazil, petrochemical industries are using served water from sewage treatment plants to supply the cooling tower makeup water. However, industries are not striving to analyze the availability of water and energy to suit the demand to supply.

The setting of premises to build scenarios of water availability, energy and skills is required. This will allow to make strategic plans, control programs, techniques and methods of investigation that aspects of products and processes are considered as well as the characteristics of the culture and behavior of managers and employees.

The water supply scenario and its use in the agriculture, livestock and cities indicate that there will be reduced availability of water for industry. As these consumer sectors are disorganized and there are losses in the distribution of water in the order of 30%, it is up to industry to worry about meeting their demand while reducing your water consumption.

The aim of this paper is to present an analysis model of availability of water and energy in the chemical industry, along with a behavioral adjustment model to enhance a large reduction in water consumption and energy supply to meet the expectations of these. The reduction in water consumption and energy depends on: management actions; switching procedures; behavior changes in the short term; application reuse techniques and reprocessing wastewater and energy in the medium term; and development and installation of technologies to increase the supply of water and energy and to reduce demand for water and energy in the long run.

Initially normal trend of events without intervention in the provision of management or demand management is considered. If occur climate change altering the supply of water and energy, on the other is the population and industrial growth by increasing demand for water and energy. In this case the industry will become hostage to a situation of lack of water and energy when looking at demand in relation to supply. In our case this will occur in 2022.

The analysis of availability compensates for the reduced supply with even greater reduction in demand. The analysis method of availability is divided into the following steps considering the comparison between the supply of water and energy with climate change with the demand for water and energy with preventive actions. Should perform prediction of what should happen in the future to adjust the present with the routine programs. This methodology is divided into the following steps: (1) Definition of current and future supply scenarios based on IPCC; (2) Identification of current demand scenario and future to compare Offer; (3) Analysis of the availability of water, energy and competence; (4) Culture Analysis and People for behavioral change; (5) Analysis of Routine and Integrated Management; and (6) Compliance Analysis of consumption of water and energy. A case of the chemical fertilizer industry are tested.

1 Introduction: Context and Background

Water is essential for life and it is the base of almost all human activities. In this context, water is used in power generation processes, domestic use, agriculture, cattle raising, and industrial sector among others. Provide access to clear water in quantity and quality represent one of the major contemporary challenges [1].

According to the UN World Report, the demand for water is expected to grow by 55% worldwide by 2050 due to the fast urbanization mainly in developing countries. The report also states that the world's water supply will suffer a deficit by 40% until 2030 if the water management do not improve. The agricultural sector will produce about 60% more food, and the industry sector will increase water demand up to 400% until 2050. Developing countries such as Brazil, Russia, India, China and South Africa will drive this [2].

The concern about water scarcity increased in Brazil after the water crisis in the state of São Paulo the most populated area located in the southeast of the country. This region from Brazil instead of the Northwest was an area that there is not this kind of problems. The water crisis didn't harm only the life of the local population, but also the industrial sector which account for 32% of the Brazilian GDP [3].

The industry needs to be alert and prepared to deal with this kind of event, because this issue affect the running and growth of production [4]. Moreover, scarcity and water pollution have amplified some planetary boundaries known by scientists as safe operating space for humanity. Be in the planetary boundaries is a precondition for sustainable development. According to Johan Rockstrom *et al.* freshwater problems is influencing and affecting some planetary boundaries as biodiversity, food security, population health, and the ecological functioning, such as climate regulation and carbon sequestration [5].

According to ANEEL (National Electric Energy Agency), approximately 75% of Brazilian power generation come from the use of renewable sources mainly hydropower with 62% [6]. Water is also a driving force in the generation of electricity in thermal power generation (steam generation), and necessary to chemical and petrochemical industries for heating (thermal energy) or cooling, among others services [7]. Thus, the availability of water is associated to the availability of energy in socio-economic activities in Brazil and the world.

The lack of structure industrial sector to deal with the interruption of operations resulting from non-availability of water and energy (effects of climate change) is increasingly evident. The structure is related to technological aspects (e.g., automation) and of complex decisions. Many decisions involve multiple factors such as climatic and behavioral aspects.

Although industries are trying to develop projects to reuse of water from sewage treatment plants to supply cooling tower makeup water as was done in the state of São Paulo during the water crises, it is not a tendency. The reuse of wastewater in the industrial sector has been considered synonymous with economic gains, in addition to the positive image transmitted. Nevertheless, the concern is the concentration of pollutants and the possibility of increased losses by corrosion. Although reuse of water from sewage treatment demonstrate that management procedures are related to the basic principles of preventing water shortages, energy and environmental impacts, it isn't clear a commitment to analyze the availability of water and energy in order to adequate demand over supply considering scenarios for 2035 [4].

2 Method

The method intend to compare the supply of water and energy to their demand in the present time and a projection for the future - forecasting. This future scenario permit to take place the necessary actions to maintain the water and energy availability over time – backcasting [8]. Furthermore, prediction scenarios of greenhouse will help adjust the current emissions through routine management programs and new technologies.

Thus, short and long-term actions can be preventive as reducing the emission of greenhouse gases (replacement of energy by renewable or decrease their use or reduction of fuel burning) or corrective as the reduction of water consumption.

The analysis of availability method (shown in Figure 1) is divided into the following steps. (1) Definition of current and future supply scenarios based on the IPCC. (2) Identification of the current demand scenario and future to compare with the supply. (3) Analysis of the availability of water, energy and competence. (4) Culture Analysis and People for behavioral change. (5) Analysis of Routine and Integrated Management. (6) Compliance Analysis of the consumption of water and energy.

2.1 *Definition of current supply scenario and future based on IPCC*

The scenarios of use and supply of water indicate further reductions in the availability of water for industry. Water users (agriculture, livestock and population) are messy and work with distribution losses that can reach up to 34% [9]. Thus, the industrial sector need to care about increasing the supply of water and reduce their consumption at the same time through demand management and local source control programs.

The discussion about the energy consumption is linked to the greenhouse emission. The rise of GHG into the atmosphere depends on the burning of nonrenewable fuels or the lower conversion efficiency producing CO. The environmental impacts result in climate change and consequently alterations in water availability for society and for the economy as well as impact on biodiversity.

According the annual report about the inventory of CO_2 in the atmosphere, the amount of emission should be minimal to reverse climate change. Thus, the planet earth must send less and less CO_2 into the atmosphere, gradually, even with the growth of economies. The value target is 14 GTY in 2050 [10] and [11]. The forecast on the growth of the economy, with no GHG emission reduction action is to increase the current 50 GTY to above 62 GTY in 2050 (value stated in 2009) [10] - indicating the irreversibility of the greenhouse effect.

This paper aims to demonstrate how an analysis model on the availability of water and energy in the chemical industry and a behavior adjustment model may cause a large reduction in water consumption and energy in order to meet the expectations mapped by future supply of these inputs. The reduction in water consumption and energy depends on: management decisions; switching procedures; behavior changes in the short term; application reuse techniques and reprocessing of waste water and energy in the medium term; and development and deployment of technologies to increase the supply of water and energy and to reduce demand for water and energy in the long run.

2.2 *Identification of the actual and future demand scenario to compare with the offer*

A analysis of production and shutdown reports allow to build, in a preliminary way, the knowledge about the cause of losses and the behavior type to deal with these losses mainly of time, materials water, and energy. A rapid assessment of water and energy restrictions can give an idea of the scenario in the company. The production growth rate indicates if the demand will increase with the view to forecast the water consumption in the future, without including the preventive and corrective actions to demand management. When the supply and demand are compared, we can estimate the gap and which information is required to improve the analyses.

Water is classified in industry in various levels to meet the demand of the production plant and the administration. Large volumes of water are used in operations such as washing, transportation, ion exchange and others. One of the largest industrial water uses are in the cooling towers that meets the warm currents of the separation unit operations in the chemical process industry. Clarified water is the type of water used in cooling towers and represent 70% of the raw water inlet to the industry [12].

The measurement of water consumption is a legal obligation in order to inform the government environmental agency who granted a license with a utilization limit. This grant lets to use a quantity of surface or underground raw water and returns it (in effluent format) a quality better than the shot. Thus, it is necessary knowing the use of water, qualities (restrictions of equipment and processes), disposal and the behavior of the team and the manager regarding this precious resource to reduce the current water consumption and future need.

Regarding to energy consumption, classifying the type of energy used by industry is the first step to establish a demand profile. The electrical or thermal energy are primary energy and they can be converted into chemical energy and work. Thus, the main inputs in chemical industry are electric power and fuel. The measurement of consumption allows calculating the use of energy per unit of product giving an idea of the efficiency of its use. When a baseline and target (benchmark) is established and these are compared with the requested reduction of emission of GHG it is possible to establish the programs and investments in medium and long term.

2.3 Analysis of the availability of water, energy and competence

The starting point of the availability analysis is building the graph indicating water consumption and how much in terms of grants is authorized consume (immediately). In the case of supply, it is necessary to build an algorithm to increase the supply due to new sources or the reduction of losses. Another algorithm must be built when the source of water decreases due to climate change, increasing urban population, rise of irrigation area to agriculture, and so on. However, demand for water could consider two possibilities: (1) any action to reduce demand during the time (2016-2035) indicating only increase in the forecast increase in production or (2) actions and investments in process design, control routine and team behavior to reduce water consumption.

When the annual supply is projected for the future (including economic changes in society and the climate) and compared to future demand without considering corrective action, there is an indication of interruption of industrial operations for lack of water. Thus, actions to reduce water demand is indicated.

The availability of energy analysis actually compares the possible amount of CO_2 emission (which changes with time due to more restrictive situations in the future) from estimates by the International Energy Agency, with the current and future amount of CO_2 emission in different situations as the action taken or not to prevent irreversible climate change.

The initial approach of water/energy supply and demand needs to be confirm by mapping the thermal systems, equipment controls, water and energy saving programs, and analysis of sources and water and energy uses. The demand mapping is confirmed and compared with the preliminary data, they are made of the demand adjustment exercises in the timeline considering new best-case scenarios or more critical every five years.

Thus, the master plan for water and energy is builds up with the demand adjustment regarding the supply by technologies, programs, and technologies to be installed gradually. It is considered to predict supply by 2035 criteria related to industry growth, population, issues related to agriculture and irrigation, rainfall forecast, and the probable faults keeping the same demand.

This initial analysis should be reviewed after a comparison with external benchmark and improve the quality of information from direct and not secondary measurements.

Among the technologies are: reverse osmosis, geothermal, flexible cooling towers for reuse, process intensification, and others. Among the techniques are: water balance, energy balance, pinch, reuse of wastewater, Exergy analysis, analysis of material losses and energy, and others.

2.4 Analysis of Culture and People for behavioral change

A series of procedures and attitudes in the levels of implementation, management and strategy are based on increasingly complex decisions [13] and [14]. It is necessary to know how these decisions occur based on local cultural characteristics [15] and likely rules that influence not reduce water and energy consumption. It is important to analyze the influence of equipment failure, processes and operations of the direct losses and indirect materials to the reprocessing [16]. The decision to adjust the quality of the product can impact on the increase in the consumption of water, energy and the rework of workers.

An analysis of people and cultures will be able to identify indicators that evidence the inappropriate behavior that enhance water loss and energy. This allows building a program to control losses from these resources. With this identification, it is possible setting a program of training or behavioral adjustment to be implemented in the company.

The cultural influence is discussed for building a robust safety culture and is based on the dynamic behavior of the staff. This behavior is the result of a field of cultural forces: organizational desire, local culture and features of global culture. Some trends should be strengthened and other inhibited and this action are based on archetypes analysis [17].

The human factors analysis allows to establish programs to increase human reliability. Human factors can be [18]: conflict between policies and practices, the conflict between production and safety priorities, control facilities, managerial aspects, skills analysis, and preparation procedures are part of human factors that should be analyzed and monitored.

The moment of decision is then influenced by biases of culture and human factors causing cognitive impairment [19] occurs following human errors mechanisms as the default [20]. In addition to the analysis of cognitive impairment seeks to reduce the risk of accident through the joint analysis of technical factors and not technical. This strategy allows us to discuss what are the aspects that can impact the security processes in an organizational change [21].

2.5 Analysis of Routine and Integrated Management

After identifying the behavioral constraints or characteristics that lead the employee and the manager to start faults that cause loss of energy and water, it is important to analyze the routine. Check the risks of system stop on a daily basis allows you to work in a review of equipment and critical tasks. The knowledge of water systems, heat exchangers, energy integration, wastewater allows for maximum use of energy by increasing the energy efficiency. The water and waste energy reuse will rationalize the consumption reducing dependence on bulk supply facilitating sustainability of the company.

2.6 Analysis of the availability of water, energy and competence

The establishment of comparison criteria for use of water, energy and behavior that avoids the technical and human error adjusts techniques, technologies, decisions, behaviors and knowledge. So this step enables the continuous improvement of the risk management program in Energy and Water. (ISO 50001) [22].

3 Demand management

Some global factors have to force the industry to make efforts to increase productivity and reduce the consumption of basic materials as raw materials, water and energy. The water is already being alert reason for society and for the industrial sector, where specific events of lack of water can turn into long periods of shortage impacting on the economy and the population.

Energy is a key input to ensure the economic and social development and the rationalization of its use is an essential strategy to avoid losses during periods of restrictions resulting from the emission of CO_2 and the scarcity of natural resources. Energy rationalization might be to change procedures, behavior change, control techniques and technologies to enhance the use of energy and thus improve efficiency and improved thermal performance of the equipment and systems.

The water consumption in the industry is directly linked with energy demand, for the use and distribution of water in a production process, the energy has a fundamental role to keep the equipment operating. Similarly, the electrical and thermal energy coming from a unit of basic inputs need the aid of water for its existence.

Industrial activities that involve the consumption of basic inputs has suffered deployments due to significant consumption and existing waste. Cycles of concentration in cooling towers, water use outside of specified boilers, mass loss and energy in steam distribution systems, heat losses in electric power generators, discarded wastewater without prior analysis, are some examples that show opportunities for deployment techniques or technologies that contribute to the reduction of utilities, waste and therefore environmental impact.

However, to reduce the consumption of water and energy are proposed a number of solutions management, various techniques used by the STAFF during routine operations and technologies.

3.1 Management

The management plays a key role in the implementation of programs to control routine related to the behavior of the staff and control of deviations. Each technique has suggested goal, steps and impacts are described below.

The control program at the source [23] is to identify equipment and pathways of contamination to the final disposal, liquid (effluent), solid (waste inventory) and gas (emissions or losses fugitive emissions). Understanding on which the carrier fluid and how it is possible to establish measures to avoid losses. The additional technical material balance loss is the mass calculation of losses and the impact on fixed costs, variable and billing.

Control of critical task, named as POS, Procedures and Operational Standards [24], aims to identify, in this case of water and energy availability, which tasks are that they can produce losses and analyze the environment, planning, execution, control and the failed task.

The analysis of environmental deviations [25] want based on the fault tree method and field observation, identify deviations that lead to contamination of everyday life. When this technique is well understood in the organization it enables a behavior change operation.

This research is in development and is about identifying what the proper conduct of the operation team to prevent energy loss. Follows the line of CCPS in Operational Discipline [26] and includes aspects of the company culture. It takes into account that energy is intangible and less noticeable to control losses.

The calculation of the reliability of socio-technical systems [27] depends not only on engineering and includes the social and managerial area of the company. For this reason it will be led by the manager. When the reliability of each functional part of the system (process, equipment, operation and human) is identified, after the process of the complexity of the classification, the calculus of the system reliability level is possible. The system reliability is then corrected through social attractiveness that indicates the strength of culture, the level of cooperation (communication) and adaptive leadership to new realities. Through this analysis it becomes possible to more policy action on production controls.

By calculating the reliability of systems and identified the impact of human reliability begins research on each performance factor indicated by SPARH [28] and discussed by API770 [18]. Thus, it becomes possible to define critical areas for action by the manager. These analyzes made by the management has focused on reducing energy losses!

3.2 Engineering

The engineering staff has the role to develop adaptations to current processes and equipment and provide new equipment to avoid or mitigate energy losses. These solutions require equipment yield calculation, energy efficiency and thermal performance.

Initially the operational control equipment and thermal systems adjusts the best operating point through procedures and control systems. It is therefore important that the process engineering and equipment to track and control the income and performance of furnaces, boilers, cooling towers, electric motors, power contracts, distribution of steam, condensate distribution, reflux in the distillation, and balance energy in the process equipment and utility in general.

This control can identify the need to revise the performance calculation methodology to be open systems, such as cooling towers. In furnaces and boilers may be a need to review the maintenance procedures (installation of refractory) or operating as purging. In collecting and recovering the condensate, on the other hand, several factors may impact on live steam loss to the atmosphere, it depends on the thermal load of the plant, steam quality, operation of steam traps, and quality of thermal insulation.

The effluent mapping to identify the loads of contaminants requires the establishment of networks of sources and uses of water by continuous or intermittent processes procedures. It is needed when discussing energy sources and uses, in this case, less visible, and thus requiring additional technical field, such as using pyrometer surface of hot equipment.

The mass balance and energy are more detailed technical and require methodologies for utilities and process equipment. simplifications may be needed, where process streams with low heat capacity variation are classified to 80% balance of the impact, reducing the number of streams.

For combustion systems are used to heat the lower and higher power. For open cooling tower systems we use the useful heat exchanged. In the case of energy balance that includes the axis of work it is important to check the income and losses of energy through heat. The energy balance is important to compare the fuel consumption with the issue of greenhouse gases.

3.3 Technology

In terms of technology we will discuss those involving new criteria for design of industrial plants and new equipment that reduce energy losses and water or increase its offer. Among the technologies involving new standards for process design and its control: modularity, intensification, integration and exergy. Among the technologies involving new equipment to reduce losses and increase in supply: desalination, heat pump, cogeneration, trigeneration and combustion bomb.

4 Application in the Chemical Industry

Applying this analysis methodology on the availability of water and energy it is based on the experiences in research projects to rationalize water use in the chemical industry of linear alkyl benzene (LAB) and the fertilizer industry (UREA) [29]. Other research has enabled the discussion of opportunities to reduce energy losses in the petrochemical industry from energy balances in processes and utilities [30] AND [31].

The experiences in the LPG industry and own fertilizer industry [24] allowed the present discussion of cultures and people. And finally, the argument based on ISO31000 and ISO50001 enabled compliance analysis.

The implementation of these steps during the period of two years will have as a result the establishment of control programs in source and strategic plans for water, energy and skills. Every five years is necessary to readjust the supply condition and analyze the indicators that demand occurred in accordance with the master plan. Organizational changes can alter the demand for water, energy and time.

The intended application for two years will be addressed through the following topics: contextualization, supply management, demand management, strategic analysis of the availability, culture and people analysis, routine management and compliance analysis.

4.1 Contextualization

Audits for use and water sources that used the effluent mapping technique and performance analysis in cooling towers indicated the high water consumption in the industry LAB mainly to feed the cooling towers [32]. It was noted also that occurs excessive contamination of waste water with several points of discharge of effluents and products. The LAB production process is complex, with the possibility of reprocessing with handling high toxicity products such as hydrofluoric acid, but a well-structured staff to decisions simply to achieve better control, access to appropriate techniques for controlling consumption energy and water.

The energy and water balance techniques, due to the complexity of petrochemical processes, indicates that more rigorous methods require mathematics reconciliation data. In other cases seeking the simplification based on the impact of the amount of energy involved, where the reduction of line processes is necessary, within the volume control for mass balance.

In the petrochemical industry of basic inputs where the type of naphtha defines differences in production costs, the equipment, the process parameters and the quantity to be produced of each product (olefins or BTX), it is important to review the energy balance. Flow rates in these processes are very high, so errors in the way of operating equipment can cause the plant to have high energy losses, indicating a need for high fuel consumption and high or make-up water for cooling towers.

The demand by up to 30% water cuts and agreements of pressure climates, and briefly change the intensifying industries in energy use. The lack of assertive decisions or appropriate behaviors have removed the goals in relation to the consumption of energy and water in the routine. This indicates that the transformation requires greater care in the industrial segment!

4.2 Supply management: control and planning for 2035

Petrochemical industry of cases and fertilizer supply of real data and water demand and actual data supply and energy demand are discussed.

<u>4.2.1 Water supply management: premises for the analysis of supply management in the fertilizer industry in northeastern Brazil</u>

Assumptions: (a) average climate change scenarios A1B with rainfall reduction impacts in Bahia (2070-2100) in about 60% on the coast and 35% in semi-arid with 2-5ºC temperature increase [33]. (b) increase in supply due to improvements in the operation of river basins and due to climate change by 2018; (c) reduction due to climate change estimated at 28% to 60% of the current offer in surface and groundwater respectively, for 2035; (d) reduction of reliability in service of the Paraguaçu River Basin [34] for 2011 to 2040, mainly for irrigation, industry and power generation; (e) Increase in the number of months during which there were failures in care:

the Savior supply (1% to 5%) and industrial demand (1% to 9%) with reduced power generation by 66% in Dam of Horse Stone [35].

Supply: the water supply to the industrial sector in the basin of Paraguacu and Salvador city between 2012 and 2034. **Offer 2012:** Surface water = 20.9 m^3/s, groundwater = 3.9 m^3/s, total = 24,8 m^3/s. **Offer 2018:** Climate change with increased rainfall: 24,8 to 78,8 m^3/s (3.2 x). **Offer 2034:** Reduced rainfall = 25% in 2018, 19.7 m^3/s.

Fertilizer unit: in the specific case of fertilizers unit [29] is considered that the initial flow supplied (2012) is equal to the demanded flow rate, that is, 700 m^3/h. **The Offer in 2018:** since part of the overall offering for the industry shifts to supply the city (an increase of 2 X instead of 3.2 X), that is, 1400 m^3 / h. **Offer 2034:** decrease due to climate change by 33%, that is, 462 m^3/h. The demand for water in the fertilizer unit should be less than this flow.

4.2.2 Power supply management: premises for the analysis of supply management in the petrochemical industry in northeastern Brazil

Considers that energy supply is limited by the capacity of the atmosphere of the planet receive GHG to prevent climate change. Thus, the supply limit is based on climate studies by the IPCC. On the other hand, the demand for energy in industry depends on the activities carried out by thermal and electrical systems. Thus, the balance between environmental restriction is sought (supply) and the actual greenhouse gas emissions by industry (demand).

The power availability analysis will be handled by the comparison between supply and demand. Supply management will have an upper limit maximum environmental capacity to receive GHG emissions to prevent climate change (14 GTY in 2050), this is the blue scenario. In this scenario emissions are monitored since 2005 with 28 GTY emissions and tend to grow up to the value of 62 GTY when the climate change process becomes irreversible. This emissions growth occurs if no action is taken towards sustainable economic growth of BRICS and developed countries [10].

Global targets need to be resized to establish regional or industrial sector or by enhancing industry in energy consumption targets. This schedule allows you to define local action strategy seeking to guarantee the reduction of GHG emissions.

In the case of a petrochemical in Camaçari (Brazil): since 1000 there are on planet earth petrochemical same capacity and the global industry emits 100 times the amount of CO_2 (based on fuel consumption) of the global petrochemical segment. The proportion of the global emission reduction for local equivalent to a reduction of 10-5 Industrial overall goal [10] for this local industrial. In the case of switching to renewable matrix, the overall goal is 9 GTY after the local design amounts to 0.09 MTY for this petrochemical industry.

Offer in 2050: (a) replacement by renewable energy → 10 gigatons year of global emissions or 0.1 mega ton year local issue; (b) reduction in CO_2 consumption by effective use of thermal energy resulting from the burning → 11,5 gigatons year of global emissions or 0.115 mega ton year local issue [10].

4.3 Demand Management: control and planning for 2035

Petrochemical industry of cases and fertilizer supply of real data and water demand and actual data supply and energy demand are analyzed.

4.3.1 Water demand management: assumptions for demand management analysis on the fertilizer industry in northeastern Brazil

Assumptions: demand management analysis in a fertilizer industry in the Northeast of Brazil: (a) average growth of 2,5% per year of urban demands of the cities of the RMS (increased population projection and reduction of per capita consumption) [36] and (b) growth in consumption by industrial expansion and irrigation.

Urban and irrigation demands will reflect the reduced availability of water for industry.

Demand: the demand for water for industrial thread in the basin of Paraguaçu and nearby cities including Salvador city between 2012 and 2034 considering that there is no demand tuning activities. **Demand in 2012:** 15,5 m^3/s city and 6,2 m^3/s for industry, total = 21,7. **Demand in 2034:** 40% of consumption by the population increase and reduction of losses of approximately 25% due to investments in the operation of water supply equivalent to 5% increase in the city, 16,3 m^3/h. New industry and new units with 15% increase indicating 7,2 m^3/s for the industry. Total 23,5 m^3/s. Indicating that the offer will be lower (19,7) than the demand (23,5 m^3/s). That is, if no action is taken to adjust to demand, the industry will be at high risk of interruption.

Fertilizer unit: in the specific case of fertilizers unit [29] is considered that the initial flow supplied (2012) is equal to the demanded flow rate, that is, 700 m^3/h. **Demand by 2034 without setting activities,** will increase by 15% in water consumption = 805 m^3/h. This unit of fertilizer also have problems which need to consume less than 462 and will be consuming 805 m^3/h.

4.3.2 Energy demand management: assumptions for demand management analysis in the petrochemical industry in northeastern Brazil

The suggestion by the International Energy Agency (IEA) is that there are different ways to reduce GHG emissions: energy efficiency (electrical and thermal), renewable energy (wind, solar, biomass), nuclear energy and carbon capture. The IEA also shows that the main actors for this reduction are: processing industry, power industry, transport and buildings.

Avila *et al.,* 2011, showed that a petrochemical site:

(a) replacing 20% of energy consumption based on fuel oil with biomass energy. Following the assumption that this biomass, which compose a forest previously cut off from the atmosphere by photosynthesis, the same amount of CO^2 emitted in their use. This petrochemical be taking 20% of CO^2 emissions initially issued.
(b) 50% reduces the loss of live steam (estimated at 30%) in the condensing system, steam systems, boilers and furnaces.
(c) recover 50% of losses due to non-use of cooling water recirculation flow in heat exchange in cooling towers (ideal thermal performance (dti) = 95%; dti real = 75%; 20% difference). Thus, an original temperature difference of 10°C, with higher

energy efficiency reaches 11°C, a reduction of 1°C in cold water → bringing as positive impact improving the profile of the separation columns and reduction of fuel demand thereof magnitude and CO^2 emissions as a result, 10%.

Note that:

1. The amount of biomass used (petrochemical-based local → global petrochemical → global industry) inhibits the emission of CO^2 with the change of 20% for renewable matrix 8,45 GTY, compatible with the demands of the International Energy Agency.

2. Increased energy efficiency in thermal systems (petrochemical-based local → global petrochemical → global industry) reduces burn and CO_2 emissions by 12,5 GTY, compatible with the demands of the International Energy Agency.

4.4 Analysis of availability: strategic plan

Exercises for analysis of the availability of water, energy and skills

The program to control wastewater at the source and master plan provided operational activities and new projects to achieve the demand reduction of 750 to 310 m3/h, so below the 462 m3/h.

Scheduled actions are [29]: (2013-2018) (1) pattern of change in the final product plant shutdown; (2) stability in the operation of purification and reuse; (3) reprocessing the waste water treatment, reuse in cooling tower; (4) increase in operational reliability; (5) reuse of disposal with a high level of contaminant; (5.1) purifier contamination reduction - reduction of product load; (6) condensate polishing removal of organics. (2019-2020) (7) energy conservation project, condensate (I); (8) control work in the steam generation (II); (9) condensate return for demineralized water; (10) the entry of new plants with improved design pattern. (2020-2024) (11) of rainwater use project; (12) improvements in the use of cooling towers; (13) Water reuse for fire and utilities; (14) improvements in water treatment plant; (15) rationalize water use in the area and the administrative use. (2025 and 2028) (18) effluent reuse in cooling towers; (19) advanced technologies to reduce vapor loss - cogeneration and heat pumps. (2029 and 2034) (20) new techniques for heating and cooling: intensification; (21) new cooling technology projects: geothermal energy and application of the products.

The availability of energy and competence [8] are discussed from Figures 1 and 2. Figure 1 is considered the emission of CO2 in a temporal analysis which compares the issue (with and without adjustment of shares) with the offer is the maximum capacity of the atmosphere to process GHGs without climate change. In Figure 2 we analyze the defendant competence for technology with the expertise offered by people with and without adjustments regarding the offer.

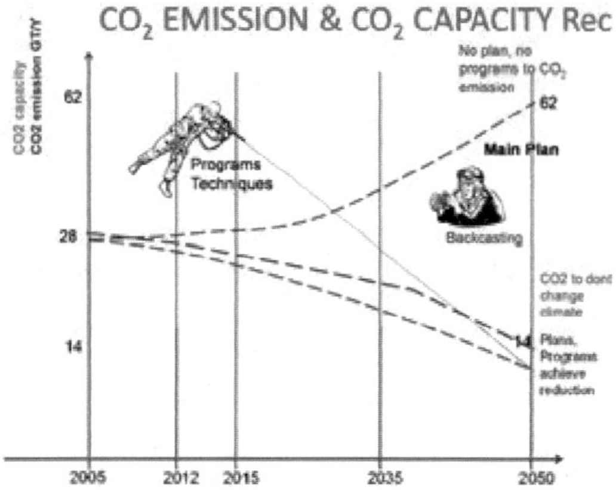

Figure 1 - Supply and demand GHG [8] Figure 2 - Supply and demand Competence [8]

<u>Master Plan for Water and Energy [8]</u>

The Master Plan is a strategic tool that enables meet the challenges of climate change and its consequences for the industry. Planning for adjustment and availability of water & energy for the industry is complemented with technical and management tools for the implementation of the Master Plan.

The Master Plan discusses the legal and institutional aspects of the laws dealing with water, energy and environmental impacts. supply management depends on legal issues that allow water and energy management in industry. The current and future availability is performed by comparing supply of water resources in local basins with the demand of these resources to the industry seeking the necessary adjustments. This discussion should also is made to address the constraints of the Brazilian energy matrix with the conditions for water use and CO2 emissions.

Analysis of water and energy classes and specifications of equipment and processes enables us to diagnose the possibility of water and energy reuse. The identification of classes used in the industry indicates their effective use to have qualities in different sources. The water treatment and effluent depends on the water balance enabling reuse projects.

Concepts and applications are detailed for adaptive management of supply and demand and which proposals for long-term strategic adjustments as credit demand reduction, industrial symbiosis concepts and opportunities, the importance of monitoring indicators, and the risk of lack of water for 10 and 40 years.

The ways to manage supply take into account: availability of water sources, climate change, hydro-geological perspective, costs and energy content. A calculation procedure is suggested for the adjustment of supply relative to the projected demand. In case of power must calculate the CO2 emission target for the specific case of the installation. The management of future demand includes the control program in the source material balance losses and others.

In the Annex of this plan are presented technical and managerial and operational methods. The intention is to keep demand and controlled supply. The main technologies to maintain demand and controlled supply for Industry are classified and detailed.

Control Equipment and Thermal Systems

The thermal control equipment and systems, process control and reliability of the systems are extremely important to regulate the demand for water and energy. The demand adjustment requires lower value to water supply (based on state water policy) and energy (based on local and global constraints of GHG emissions).

The equipment that has high impact on fuel consumption and the emission of greenhouse gases are: (a) cooling towers; (B) heat exchangers that utilize utilities; (C) generating steam boiler; (D) heating stoves or pyrolysis reaction; (E) steam distribution and condensate energy recovery (including traps); (F) current of high enthalpy process; (G) operating point and balance of separations such as distillation; and (h) electrical systems in general.

At this point will be discussed energy efficiency in some of these systems: (a) Deten-water research project in 2003, Ávila noted the importance of working with clean cooling water circuit to avoid blockages in the upper pools and highlighted the potential losses if maintaining the cooling tower structure (and water quality) is poorly made, with losses that may cause heating of water up to 1 ° C. By maintaining one should prevent the cooling water flow paths follow without heat and mass transfer with the dry air. (b) This project was highlighted the importance of working with cycle highest concentration in the towers to reduce water consumption. This control increases the risk of corrosion in order to be header exchangers and hinders water reuse in cooling towers [37] [38].

(c) There are several opportunities to recover waste heat in the boilers that generate steam in enhancing industry in energy consumption [30] [31] [38]. Since preheating of the water, control of stoichiometric combustion, the specification and maintenance of refractories, the utilization of heat of flue gas (new boiler, flue pump). (e) the design assumptions for steam distribution, steam traps installation and condensate energy recovery (including traps) should be based on the expansion possibilities of production and operating limits of the systems [30] [31] [38] . Thus, due to the limitations of operating with traps and thermal systems, search is the best point and with new needs, install new production modules.

Water balances, effluent, fuel, GHG and energy

In research projects to rationalize water and energy were used mathematical techniques mass balance (FAFEN-WATER, DETEN-WATER) and enthalpy (Ecobraskem) with multiplier weights varying according to quality information and then closing balance by data reconciliation. This technique was used for LAB production plants, Olefins, urea and petroleum [30] [31] [38]. As for the balance of utilities in equipment such as cooling towers, boiler and furnace energy balance it was based on ICPs and the formulas simplified the treaters water that is based on the wet bulb temperature, the flow and the temperature difference [38].

Sustainable projects: new techniques and technologies

The need to reduce energy consumption requires changes in equipment design and process. The main reason for these changes is to reduce the gaps, allowing a reduction in the consumption of utilities through direct heat transfer without the need for thermal fluid design adjustments considering the external environmental changes and changes in thermal load.

The spare capacity of operations and equipment is adjusted to increase the capacity of industrial plants. In industrial processes it is possible to carry out this adjustment due conservative design factors that enhance the capacity of the equipment, but in the case of utilities is not always possible. Thus, you can use the spare capacity of reactors, pumps, but it is not possible to trap and electric motors. With restrictions on the use of energy and knowledge in human reliability, it is intended to modularize the projects through more accurate system design [38].

Another option in the project area is to reduce the use of utilities and size of equipment by intensifying operations and processes where the heat necessary to the achievement of operations is transferred directly by the equipment wall. Where the heat is released from certain operation and received in another operation without the need for use of thermal fluids. If energy balance in petrochemical is noted that the use of exchangers with hot thermal fluids (steam) and cold (cooling water) consumes 4000 GJ / h compared effectively with the necessary heat (80 GJ / h) to unit operations work without utilities. With intensive use of energy can significantly reduce power consumption and size of equipment [38].

The rationalization of energy and water can be conducted using techniques which integrate the process with the utility and waste. The identification of water users with the material restrictions for equipment allows the water and energy savings through its reuse in equipment. This technique is applied in sewage networks and networks of heat exchangers.

In the case of wastewater networks (pinch), the contamination levels of waste water streams are analyzed to compose a quality goal that meets the requirements of equipment that can use this water. If exchanger network (pinch), the residual heat through the cooling water reuse allows a better use to reduce energy losses.

The main equipment that consumes water and has high potential for use of waste water is the cooling tower [32]. Thus, it is the integration of processes (sources) with the effluent (waste water standard), and returns to the utilities (water cooling).

Energy recovery from the best thermal performance in cooling tower and application of process integration (pinch) has as a result the reduction of water loss and the reduction of CO2 emissions. To account for this loss reduction in CO2 emission area is seeks to study the integration of the cooling tower, heat exchangers, separation operation (reflux), the heat demand (steam oven), burning fuel and CO2 emissions.

The decrease in water supply from climate change requires research on the feasibility of industrial desalination including comparing other solutions involving energy supply and water (eg lifting deep brackish water). The current technology of reverse osmosis membrane desalination is still not profitable for water treatment industries in the private sector.

4.5 Culture analysis and people

Private and public institutions recognize the importance of investigating the matter culture and people for sustainability, but do not yet know what priority to this issue. To establish an organizational and strong safety culture is required: to know the tasks and risks; analyze the behavior through the routine signals; and establish metrics to identify the stage of safety culture.

An onshore oil industry indicated that the image directly affects the profitability of the company, so often the actions prioritize avoid independent accident effect on production and on revenues. A proposed action was suggested for intervention in three years: (1) Task: To establish indicators of efficiency, formalizing forms of communication in organizational levels to review the inadequate standards establishing best practices; (2) Behavior: analyze routine deviations, establish a causal link with the process losses, see how minor emergencies are handled (decisions) and indicate gaps in relation to skills, knowledge and standard of acceptance; (3) Culture: Establish indicators for technical culture (bottom-up) and measure the difference in perception between the entrepreneurial vision and practices. Set organizational practices (rituals) to neutralize or mitigate biases culture.

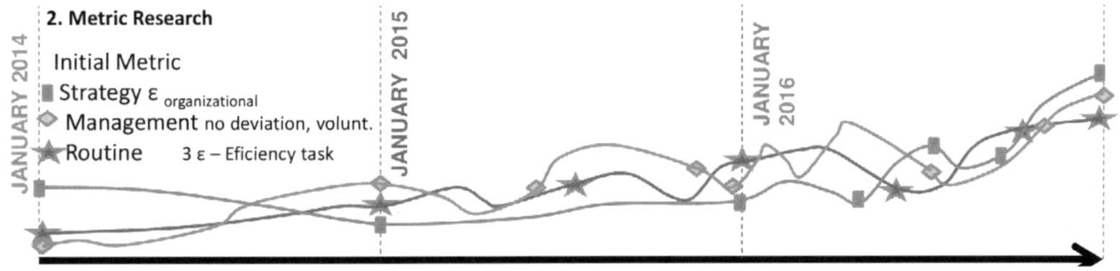

This initiative for the onshore site has been replicated through intervention in the fertilizer industry to analyze events in the routine that involves the effluent, where old vices of operation contribute to the process losses. The suggestion to change behavior and neutralization of decisions that led to contamination of effluent was through the educational program "Friends of emergency pond." The tools used in this program were: lectures in the control room and field, event analysis and hypothesis of the root cause, exchange of information through informative texts and making changes in the decisions of supervision and management.

The process that involves loss materials are more easily perceived as contamination of wastewater, fugitive emissions or the generation of solid waste. Large losses that affect the company's image as an accident or leaks has immediate impact on the stock market. Since energy losses are less visibility difficult behavior change to achieve targets in reducing GHG emissions. So it is important to consider cultural aspects and behavior to take more assertive decisions.

<u>Analysis of elements or human factors [24]</u>

In this GCPS organizational factors related to regional-global culture that are entered in 20 events in the chemical industry were discussed. These factors can initiate failure mode with final impact on energy losses. Thus, conflicts between policies and practices, economic difficulties, poor planning of the task, cultural traits that induce inappropriate decisions, risk management that does not include human factors are important factors to analyze the operational routine which leads to energy losses. The routine in deviant behavior difficult to establish operational discipline to avoid accidents and loss of energy [26].

Cultural analysis based on Archetypes [17]

Cultural phenomena do not exactly follow a logical order and are based on ancient customs rooted through tribal and new rituals introduced by the media. In an attempt to develop cultural barriers to the accident and GHG emissions seeks to analyze the phenomena resulting from the action of primary and derivative archetypes in the organization's tasks. This kind of technique indicates which new rituals should be deployed to vaccinate cognitive impairment.

Analysis of Cognitive Failure [19]

The cultural environment induces undue behaviors in individual or group during the task. Understanding the type of cognitive failure depends on the analysis of executive function in the various stages of cognitive processing: memory operations, environmental awareness for the construction of knowledge and action, attention calibration, job planning, job control and socioeconomic utility affective. It is important to identify the type of human error in the energy losses to prevent the ineffectiveness of action taken.

Reliability Analysis Sociotechnical [27]

The integration of production technology (product, process and control) with its operations on critical equipment indicates the importance of joint analysis of reliabilities. The operation achieves success if the understanding of the task objectives and knowledge of the process conditions (standard and fault). Thus, reliability of operations, processes and equipment depends on human reliability that receives the direct influence of cultural environments.

A propylene pumping operation in the refining industry was analyzed and it was concluded that the redundancy pumps is not enough when cognitive functions are weakened in conflict between policies and practices environment.

4.6 Routine management

The quality brought about the revolution of the administration [39] but did not ensure the sustainability of their results. The complex decisions and the investigation of systems failure mode indicate that to avoid losses should measure the efficiency and effectiveness of the task should be controlled thermal systems, critical procedures and also analyze the level of material losses and energy.

In the area of utilities and processes which circulates high amount of energy it is important to perform measurements and balances to prevent high losses of energy and water industry. The control of thermal systems involves critical steps such as maintaining water quality, purge tower and boiler, the level of control and the use of cells in cooling towers [31].

The production control by type of raw material defines the economic quantities of schedules based on variable cost of energy. In the petrochemical industry this issue has great importance in the definition of production programs knowing input naphtha type (weights and average).

Depending on the input naphtha occur demand for changes in processes, for example, the recycle gas will have another quality and will need to review the burners in the process gas boilers reuse [38]. To carry out the control of these tasks with high impact for the loss of energy is important to apply task analysis techniques in complex processes and have control over the realization schedule these tasks [40].

The mass balance and energy (water, materials and supplies, steam, fuel) enables the detection of losses and control at source [35] [23]. The balance of goods and inputs is closed with the quantification of losses: solid waste; liquid contaminants to the effluent; and fixed and fugitive emissions to the atmosphere.

4.7 Compliance Analysis of water and energy consumption

ISO 50001 [22] are described important procedures for the management and control of energy. It is important to identify sources (equipment and systems) lose water and energy in the production process. It is important to define if the losses are isolated or continuous, which the physical state and properties, the transport fluid and the impact receiving body. The output of thermal systems is measured by measuring the consumed energy and the comparison with expectations. Part of the energy loss and water is recoverable and the cost outweighs the benefit. compliance analysis indicates where the main losses and what the distance from the expected default behavior.

The fixtures and fittings that consume fuel or indirectly affect this consumption are extremely important: boiler, furnaces, heat exchangers, steam traps, cooling towers, water and steam distribution pipes, flash vessels, safety valves, and distillation columns .

It is important to define the consumption profile (baseline), the expected pattern of consumption, benchmark-based target, the difference between the measurement and the expected pattern (loss) and the percentage of impairment. It is also important to describe the operational procedures that can impact on the loss of water and energy consumption, expected standard and goal. These losses may have been caused by misconduct or lack of knowledge.

The criteria for equipment performance analysis are based on rules established by institutions such as ASME (boilers and furnaces) and CTI (cooling tower). Other criteria related to behavior and procedures may be based on API770 and operational discipline guidelines CCPS.

5 Conclusion

Initially considered the normal trend of events without intervention in the provision of management or demand management. On the one hand occur climate change altering the supply of water and energy, on the other hand is the population and industrial growth by increasing demand for water and energy. In this case the industry will be trapped in a situation of lack of water and energy across the demand relative to supply. In our case this will occur in 2022 (for water).

According to the problems already experienced shortages, so there is a balance between supply and water demand and energy are already demonstrated the means and methods that need to be taken into consideration and applied in the productive sectors and society. A case of the fertilizer and petrochemical industry have been discussed.

Already have the knowledge that technologies and techniques have been implemented in the industrial sector considerably reducing the waste of raw materials and fuel and thus environmental damage, but their implementations are not prioritized the way it should, as a general in contrast, there is a cultural influence, behavioral, economic and management that becomes increasingly slow the process of improvement.

With the projection of a future scenario and possible interventions mainly developing countries GHG emissions planned for the year 2050 is approximately 14GTano, very difficult task to achieve.

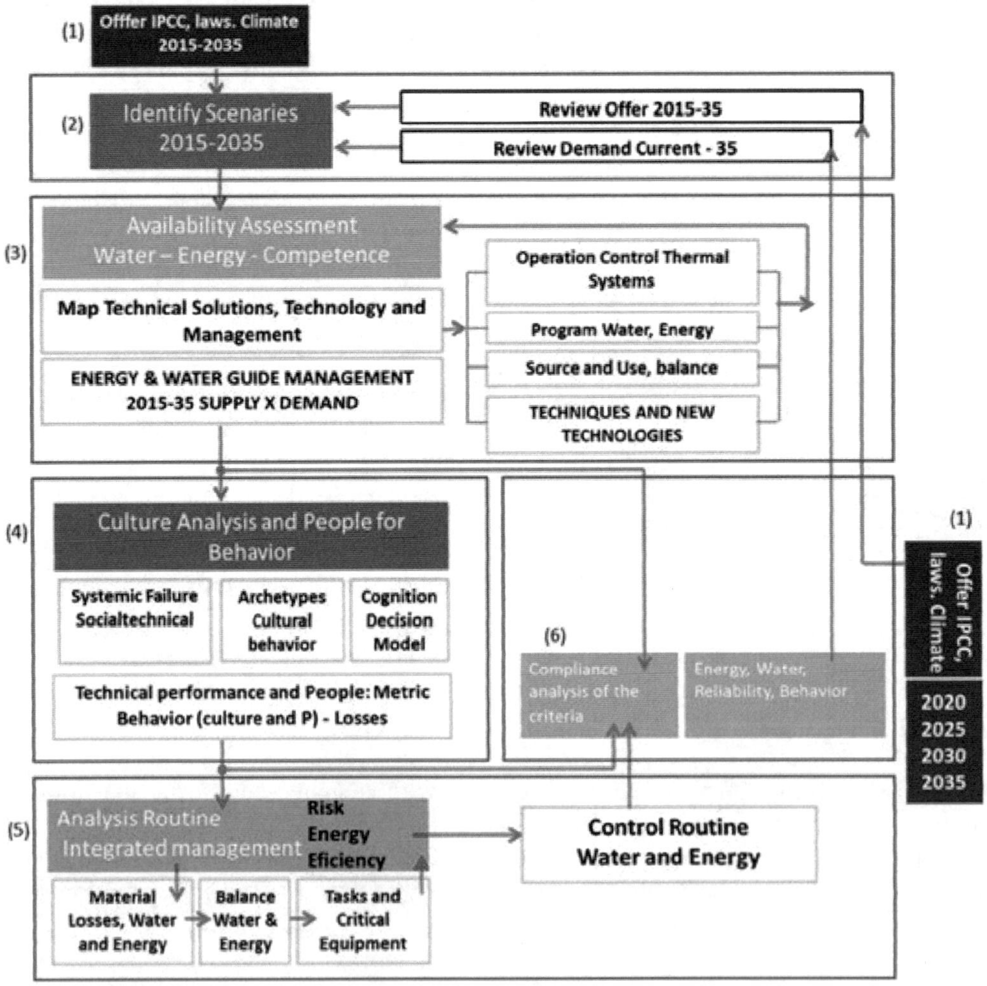

Figure 1. Analysis of Water Availability, Energy and behavior.

6 Bibliography

[1] CARMO, R. L.; DAGNINO, R. S.; FEITOSA, F. F.; JOHANSEN, I. C.; CRAICE, C. População e consumo urbano de água no Brasil: interfaces e desafios. In: SIMPÓSIO BRASILEIRO DE RECURSOS HÍDRICOS, **Anais...** Bento Gonçalves, 2013.

[2] WWDR – The United Nations World Water Development Report 2015: Water for a Sustainable World. Paris, UNESCO, 2015.

[3] FUNDAÇÃO SISTEMA ESTADUAL DE ANÁLISE DE DADOS – SEADE. Disponível em: < http://www.seade.gov.br/pibtrimestral/analise/>. Acessado em: 06 fev. 2016.

[4] PQU. SABESP. Sabesp e Polo Petroquímico lançam o maior projeto de Água de Reuso do País. 2008. Disponível em: <http://www.agenciagabc.com.br/noticias.asp?id=1045>. Acesso em: 06 fev. 2016.

[5] ROCKSTRÖM, J. *et al.* Planetary Boundaries: Exploring the Safe Operating Space for Humanity. Ecology and Society 14(2): 32. 2009. Disponível em: <http://www.ecologyandsociety.org/vol14/iss2/art32/>. Acessado em: 06 fev. 2016.

[6] ANEEL – Agência Nacional de Energia Elétrica. Matriz de Energia Elétrica. 2016. Disponível em: <http://www.aneel.gov.br/aplicacoes/capacidadebrasil/OperacaoCapacidadeBrasil.cfm>. Acesso em: 06 fev. 2016.

[7] MIERZWA, J. C.; HESPANHOL, I. Água na indústria: uso racional e reuso. São Paulo: Oficina de Textos. 2005.

[8] ÁVILA, S. F.; LOPES, J. R. N.; SOUZA, C. R. O. Strategies for industry sustainability. In: WREF 2012 - WORLD RENEWABLE ENERGY FORUM. 2012, Denver.

[9] BONALUME NETO, R. Perda de 30% da água durante distribuição contribui para crise em SP. **Folha de São Paulo**, São Paulo, 05 de jun. 2014. Fórum de Sustentabilidade.

[10] DOORNBOSH, R.; GIELEN, D.; KOUTSTAAL, P. 2008. Mobilising Investments in Low-Emission Energy Technologies on the Scale Needed to reduce the Risks of Climate Change, OECD SG/SD/RT (2008) 1, Paris, France.

[11] IPCC (2007), Metz, B.; Davidson, O. R.; Bosch, P. R.; Dave, R.; Meyer, L. A. Technical Summary in: Climate Change 2007: Mitigation. Contribution of Working Group III to the Fourth Assessment Report of the Intergovernmental Panel on Climate Change, Cambridge University Press, Cambridge, United Kingdom and New York, NY, USA.

[12] FURTADO, M. Água – Clientes exigem resfriamento de baixo custo. Revista Química e Derivados. São Paulo, pp. 32 – 34., 2009.

[13] PARRY, M. L. et al. (Ed.). Climate change 2007: impacts, adaptation and vulnerability; contribution of Working Group II to the Fourth Assessment Report of the Intergovernmental Panel on Climate Change. Cambridge (UK): Cambridge University Press, 2007f. 976p.

[14] ÁVILA, S. F. Análise de falha em processos complexos. In: CONGRESSO BRASILEIRO DE ENGENHARIA QUÍMICA, 19,, 2012. **Anais...** Búzios, Rio de Janeiro.

[15] ÁVILA, S. F.; PESSOA, F. L. P. **Proposition of review in EEMUA 201 & ISO Standard 11064 based on cultural aspects in labor team, LNG case.** Procedia Manufacturing, vol. 3, pp. 6101 – 6108, 2015.

[16] ÁVILA, S. F. **Reliability analysis for socio-technical system, case propene pumping**. Engineering Failure Analysis, vol. 56, pp. 177 – 184, 2015.

[17] ÁVILA, S. F.; MENEZES, M. L. A. Influence of local archetypes on the operability & usability of instruments in control rooms. In: EUROPEAN SAFETY AND RELIABILITY CONFERENCE – ESREL 2015. 25., Zürich, Switzerland.

[18] AMERICAN PETROLEUM INSTITUTE. API 770.2001: A manager´s guide to reducing human errors. Improving Human Performance in the Process Industries. Washington, DC., 2001.

[19] ÁVILA, S. F.; COSTA, C. Cognitive deficit analysis in routine tasks, as a strategy to reduce accidents and increase of industrial production. In: EUROPEAN SAFETY AND RELIABILITY CONFERENCE – ESREL 2015. 25., Zürich, Switzerland.

[20] REASON. J. Human error. New York: Cambridge University Press, 2003. 302p.

[21] ÁVILA, S. F.; DANTAS, E. T.; DUARTE, J. The Requirements & tools to treat process safety risk as result of Organizational Change. In: GLOBAL CONGRESS ON PROCESS SAFETY. 11., 2015. **Anais...** Austin, Texas.

[22] ABNT, Associação Brasileira de Normas Técnicas. **NBR ISO 50001:2011: Sistemas de Gestão de Energia: Requisitos com orientações de uso.** Rio de Janeiro: ABNT, 2011.

[23] ÁVILA, S. F.; KALID, R. A.; KIPERSTOK, A.; CARVALHO, V. S. Program to Avoid Product Leakage at Source, Fertilizer Case. Chemical Engineering Transactions. vol. 26, 2012.

[24] ÁVILA, S. F. Assessment of Human elements to avoid accidents and failures in task perform, cognitive and intuitive schemes. In: GLOBAL CONGRESS ON PROCESS SAFETY. 7., 2011. **Anais...**Chicago, Illinois.

[25] ÁVILA, S. F.; KIPERSTOK, A.; KALID, R. A.; SANTOS, A. C.; CARVALHO, V. S. Environmental Deviation Assessment, Record, Failure, Probability and Investment. Chemical Engineering Transactions. vol. 26, 2012.

[26] CCPS (Center for Chemical Process Safety), Conduct of Operations and Operational Discipline: For Improving Process Safety in Industry, American Institute of Chemical Engineers, New York, NY, 1985.

[27] ÁVILA, S. F. Reliability analysis for socio-technical system, case propene pumping. Engineering Failure Analysis, vol. 56, pp. 177 – 184, 2015.

[28] GERTMAN, D.; BLACKMAN, H.; MARBLE, J.; SMITH, B. C. The SPAR-H Human Reliability Analysis Method. Idaho National Laboratory, Washington DC, 2005.

[29] ÁVILA, S. F.; LOPES, J. R.N.; KALID, R. A.; KIPERSTOK, A. F. Gestão da disponibilidade de água e energia através do Plano Diretor. In: RIO OIL & GAS EXPO AND CONFERENCE 2014. Rio de Janeiro.

[30] ÁVILA, S.; KIPERSTOK, A.; BRAGA, B.; KALID, R. Avoiding loss of energy in a petrochemical industry, operation and design Congress 2011. In: WORLD RENEWABLE ENERGY CONGRESS 2011. Linköping, Sweden.

[31] ÁVILA, S. F.; COIMBRA, H.; SOUZA, C.; GOLÇALVES, A. Iniciativas de eficiência energética para sustentabilidade, oportunidades & desafios. In: RIO OIL & GAS EXPO AND CONFERENCE 2014. Rio de Janeiro.

[32] ÁVILA, S. F.; JESUS, Z. M. R.; KALID, R. A.; NASCIMENTO, V. Good practices in Cooling Towers: Comparison between industrial audits. In: Cooling Technology Institute Annual Conference. Corpus Christi, Texas, 2003.

[33] GENZ, F., TANAJURA, C. A. S., DE ARAÚJO, H. A. Impacto das Mudanças Climáticas nas Vazões do Rio Pojuca sob o Cenário A2 do IPCC - 2070 a 2100. In.: I CONGRESSO BAIANO DE ENGENHARIA SANITÁRIA E AMBIENTAL. Salvador. 2010.

[34] SILVA, S. F. Análise da disponibilidade e demanda para o sistema de abastecimento de água de salvador frente a cenário de mudanças climáticas. Dissertação (mestrado). Programa de pós-graduação em Meio Ambiente, Águas e Saneamento. UFBA. Salvador BA. 2012.

[35] KIPERSTOK, A.; BRAGA, E. A.; CESAR, M.; ÁVILA, S. F.; SAMARA.; KALID, R. A. Eds: 2012. Base de Informações para Plano Diretor de Água Fafen-BA, TECLIM RESEARCH FAFEN – UFBA. Camaçari, Brasil.

[36] SEDUR, SECRETARIA DE DESENVOLVIMENTO URBANO DA BAHIA. Estudos de Aproveitamento dos Mananciais da RMS – Revisão e Atualização das Proposições do Plano Diretor de Abastecimento de Água, Incluindo o rio Pojuca, e Visando o Uso Racional dos Recursos Hídricos Regionais - Estudo de Demandas. Relatório 1100.11-REL. EST. DEMANDAS-00. 73p. Salvador. 2010.

[37] SANTIAGO, P. S., & SANTOS, R. P. Major Improvements in cooling system performance at an ammonia/urea plant. IWC – International Water Conference, Proceedings: Pensylvania: 2000.

[38] ÁVILA FILHO, S., NASCIMENTO LOPES, J.R. Cooling Water Systems: An Overview, in: Z. Amjad, K.D. Demadis (Eds.), Mineral Scales and Deposits: Scientific and Technological Approaches, 2015, pp. 499–531.

[39] DEMING, W. E. Qualidade: A revolução da Administração. Rio de Janeiro: Editora Marques-Saraiva. 1990. 367 páginas.

[40] ÁVILA FILHO S., SOUSA C. R., CARVALHO A. C. Assessment of complexity in the task to define safeguards against dynamic risks. 6th International Conference on Applied Human Factors and Ergonomics (AHFE 2015) and the Affiliated Conferences, AHFE 2015. Procedia Manufacturing 3 (2015) 1772 – 1779.

442403 A Novel Open Source CFD Code for Explosion Modelling

Abstract:

The porosity parametrization of complex geometry is very important in explosion simulation. This project proposes a novel approach to calculate the mesh porosity (volume and area) based on the Gilbert–Johnson–Keerthi (GJK) distance algorithm. The GJK algorithm can be used to check the collision between convex objects. This concept is used in the current work to check the collision between an element of the mesh and an object of the geometrical model. The parametrized geometry is combined with the combustion model based on flamelet formulation in accordance with BML (Bray-Moss-Libby) model. The typical chemical libraries are replaced by the integral length scale of wrinkling based on semi-empirical formulation. One simple step reaction mechanism is considered. The turbulence closure problem is addressed via Boussinesq formulation using two equations model for turbulent kinetic energy and the rate of dissipation of turbulent kinetic energy.

The CFD (Computational Fluid Dynamics) solver relies on an explicit finite-volume code based on a structured Cartesian grid. The classical fourth order Runge-Kutta time marching approach is considered. The PDR (porosity distributed resistance) approach is used to model all geometry under analysis (including the large-scale objects and the small ones). The central differencing scheme is applied to discretize the governing equations and artificial viscosity is used to smooth out numerical wiggling. Additional resistance terms are considered in the momentum equation. Additional generation source term for turbulence is considered in the turbulent kinetic energy equation. The simulation of standard flow across different geometries presented consistency with the literature. Good agreement is found for similar simulations using a commercial CFD software. The developed open CFD code seems to be promising as an alternative to solve complex explosion problems from the engineering view point.

442442 Compliance with Process Safety Management Elements

Abstract:

OSHA promulgated 29 C.F.R. § 1910.119 regulation in 1992 and the fourteen Process Safety Management (PSM) elements. The main objective of defining these elements is to prevent unwanted discharge of highly hazardous chemicals exposing personnel to serious hazards. The elements are linked to each other and have intricate interdependencies. Every element provides information to other element(s) or uses information from other element(s). Given the interdependencies of these elements, an effective Process Safety Management Program (PSM) has to be in place which considers process design, technology, procedures (both normal and emergency), emergency response plans, training programs etc. The presentation talks about all these fourteen elements in details with main emphasis on the relief systems design.

442454 A Fuzzy Logic and Probabilistic Hybrid Approach to Quantify the Uncertainty in Layer of Protection Analysis

Abstract:

Layer of Protection Analysis (LOPA) is a widely used semi-quantitative risk assessment method. It provides a simplified but less precise method to assess the effectiveness of protection layers and the risk of an incident scenario. The outcome frequency and consequence are intended to be conservative by prudently selecting input data, given that design specification and component manufacturer's data are often overly optimistic. There are many influences, including design deficiencies, lack of layer independence, availability, human factors and wear in testing and maintenance, which are not quantified and are dependent on type of process and location. This makes the risk usually overestimated. So, there are different sources and types of uncertainty in the LOPA model that need to be identified and quantified. In this study, a fuzzy logic and probabilistic hybrid approach was developed to quantify the uncertainty of frequency of initiating event and the probabilities of failure on demand (PDF) of Independent Protection Layers (IPLs) based on the available data and expert experience. The method was applied to a distillation system with a capacity to distill 40 tons of flammable n-hexane. The outcome risk of the new method was proven to be more precise comparing to the results of the conventional LOPA approach.

442474 Quantitative Risk Analysis of Nitrogen Fertilizer Plant, of Moron Petrochemical Complex

Abstract:

Quantitative Risk Analysis of Nitrogen Fertilizer Plant, of Moron Petrochemical Complex

Considering the possible risk scenarios associated with new Nitrogen Fertilizer Plants (Ammonia, Urea and Utilities) of the Moron Petrochemical Complex, in general, are common to most petrochemical complexes in the world and are a result of materials hazardous that are manufactured, processed, handled and stored at the facility; the various processing systems and conditions; the different procedures used for the operation and maintenance of the facility; as well as the response of the detection and mitigation of hazards. Previous experiences in similar plants indicate that the most likely hazards are those who identify with the physical-chemical properties of the materials used and the process conditions.

For petrochemical facilities, the most common consequences are:

- Toxic Gas Effects
- Jet Fire
- Flash Fire
- Pool Fire
- Bleve
- Vapor Cloud Explosions.

Based on the premise of "Only what is known can be controlled," This paper aims to develop the Quantitative Risk Analysis (QRA) for new Nitrogen Fertilizer plants (Ammonia, Urea and Utilities) of Moron Petrochemical Complex, which it is located in the jurisdiction of the Municipality Juan Jose Mora, Carabobo State from Bolivarian Republic of Venezuela, considering for it the most severe scenarios identified previously by HAZOP analysis, SIL studies and previous plannings studies for the control and emergency response and contingency, then determining the levels of social and individual risk based on the probability of occurrence of the events of greater impact and its consequences and impact on people, equipment and buildings, comparing the risk levels obtained with tolerance criteria established by the corporation.

The Quantitative Risk Analysis (QRA) are developed from the standard and guidelines of Petroleos de Venezuela, (PDVSA); Center of Chemical Process Safety (CCPS), and The Netherlands Organization (TNO), through the computational tool PHAST RISK 6.7 . from Det Norske Veritas (DNV).

Management of Plant IPL's

Richard Stanley
Tehama Control Systems Consulting Services
Consultant to Mangan Software Solutions
26741 E Clifton Drive
Aurora, CO 80016
rick@tehamacontrolservices.com

Prepared for Presentation at
American Institute of Chemical Engineers
2016 Spring Meeting
12th Global Congress on Process Safety
Houston, Texas
April 11-23, 2016

Management of Plant IPL's

Richard Stanley
Tehama Control Systems Consulting Services
Consultant to Mangan Software Solutions
26741 E Clifton Drive
Aurora, CO 80016
rick@tehamacontrolservices.com

Keywords: IPL, LOPA, Independent Protection Layer, Safety Life Cycle Management, Safety Instrumented System, Safety Instrumented Function

Abstract

Independent Protection Layers (IPL) are usually identified in a Layers of Protection Analysis (LOPA) during a Process Hazards Analysis. The IPL portion of a LOPA identifies Functions or Design Features within a processing facility that serve to prevent or minimize the consequences of a deviation from the expected processing conditions. Many times a LOPA may identify various IPL's, but the facility then has difficulty in clearly associating the functions identified by an IPL with real plant assets. They then have difficulty in assuring that those assets have the necessary operational and maintenance visibility, and are operated and maintained to assure their availability. This paper discusses these issues and describes the characteristics of an effective Management Program for IPL's.

1 Introduction

During Process Hazard Analysis (PHA) for processing facilities, a number of Independent Protection Layers (IPL) may be identified as having the capability to reduce the risk of a given consequence occurring. A given scenario may require that one or more of these IPL's be in place to reduce risk to tolerable levels. Typically, these IPL's are identified during a Layers of Protection Analysis (LOPA), but other methods can be used.

Once these IPL's have been identified, the process of managing them has just started. Designation of an IPL adds Operational and Maintenance requirements that often exceed typical practices. This presents an operating facility with the challenge of adapting Operation and Maintenance practices to assure that the IPL's that are credited are well understood, that their protective functions remain available and meet the performance requirements that were defined for them.

This paper discusses the various type of IPL's and the requirements and methods for assuring their availability. The basic ownership requirements are described along with observations relative to the difficulty that real facilities have had in addressing these requirements. This particularly applies to the less visible IPL types such as those IPL's that are not associated with Safety Instrumented Functions (SIF).

After discussion of the ownership requirements and obstacles to meeting those requirements, the use of a Safety Life Cycle database to overcome these obstacles is discussed. The effective application of a Safety Life Cycle database can enable an operating organization to manage those requirements with maximum effectiveness and value with minimum impact of existing organizational structures and accountabilities.

2 Independent Protection Layers

A key concept in managing Process Hazards is that of an Independent Protection Layer. The formalized concept of an IPL was developed along with the LOPA process. During a PHA the potential consequences of various deviations in plant operation are evaluated, and the potential consequences of those deviations if no preventative measures are taken are defined in qualitative terms.

When an unmitigated risk identified by the PHA is higher than an organization's tolerance, the probability of the event occurring may be further evaluated to better understand the risk and the effectiveness of the various potential safeguards that could prevent consequence from occurring.

The most common technique for performing this analysis is the Layers of Protection Analysis (LOPA). In a LOPA, functions or design characteristics that are credited as being able to prevent the consequence are called Independent Protection Layers (IPL).

2.1 IPL Identification

The measure of effectiveness of identified IPL's is generally assessed against an organization's risk policies using a Target Mitigated Event Likelihood (TMEL). TMEL is usually based upon the severity of the LOPA scenario unmitigated consequences. For example, a severity level may have a required TMEL of 1e-04, or one occurrence of the unmitigated consequence in every 10,000 years.

During the LOPA, various Initiating Causes for a Deviation scenario are identified and each is assigned an expected frequency of occurrence. Some organizations may choose to apply additional frequency mitigation factors, such as the probability of personnel being in the area at the time of the Initiating Cause. Some organizations will choose to sum Initiating Causes, while others do not.

The product of expected frequency of the initiating causes, any additional frequency mitigation and Probability of Failure Upon Demands (PFD's) of the credited IPL's is used to calculate the value of the Total Mitigated Event Likelihood ($TMEL_{act}$). If the $TMEL_{act}$ is below the threshold specified for the consequence level associated with the event, the number of IPL's is considered to be sufficient. If the $TMEL_{act}$ exceeds the threshold the LOPA team recommends corrective actions, often recommending that additional IPL's be added. However, other recommendations such as process modifications to reduce the consequences or reduce the initiating causes may also be recommended. An example of a LOPA TMEL analysis is shown in Figure 1.

Figure 1 – Sample LOPA TMEL Analysis

LOPA TMEL Calculation Example

Consequence TMEL = 1.0E-05 (1 consequence Event per 100,000 years)

IC1 - 0.1 (once per 10 years)
IC2 - 0.1 (once per 10 years)
MF - 0.1 (Personnel present 10% of the time)

IPL1 PFD = 0.1
IPL2 PFD = 0.01

TMELact = (IC1+IC2) * MF * IPL1 * IPL2
TMELact = (0.1+0.1) * 0.1 * 0.1 * 0.01 = 2.0E-06

Actual TMEL (TMELact) is less than TMEL, so the IPL protection is acceptable. No additional protection is required

2.2 What is an IPL?

An Independent Protection Layer is an instrumented or physical function which, if effective, would prevent the unmitigated consequences of a scenario identified in a PHA or LOPA from occurring.

The primary characteristics of an IPL are:

- The proper functioning of the IPL would prevent the consequence associated with a scenario from occurring
- The IPL is independent of any of the scenario initiating causes and is independent of other IPL's cited for the scenario.
- The IPL is reliable enough to perform the expected function with the probability of failure being with the range with which the IPL is credited.

- The physical and functional design of the IPL meets the required independence and availability requirements.
- The IPL is Operated and Maintained in a manner that assures the credited reliability is achieved during plant operation.

CCPS [1] provides guidelines as to what functional, design or physical aspects of a process installation can qualify for being credited as an IPL, and describes the limitations for where multiple IPL credits may be taken. However, it is up to the facility owner or operator to determine specific local requirements. It should be noted however, that a 2015 US OSHA interpretation [2] appears to extend the concept of Recognized and Accepted Good Engineering Practices (REGAGEP) to publications that are not necessarily consensus documents. CCPS publications, such as the LOPA publication cited above, fall into this category. Furthermore, the OSHA interpretation notes that internal standards must represent REGAGEP and will reject standards that do not comply.

2.3 IPL types

Some of the more common IPL types are:

- Operator response to an Alarm
- A Basic Process Control System Automatic Action
- A Safety Instrumented Function (SIF) implemented in a Safety Instrumented System (SIS)
- A non-SIF interlock
- Pressure relief devices
- A physical barrier that would contain process fluids after a loss of primary containment. An example is a tank dike or a drainage system
- Physical equipment that would prevent a scenario from causing the unmitigated consequence associated with it. For example, this could be a valve limit stop, a restriction orifice or other flow restriction that prevents excessive inflow.
- Inherently Safe Design

2.4 IPL's and the Safety Life Cycle

The concept of the Safety Life Cycle is often considered to be associated with Safety Instrumented Systems (SIS) because its prominence in IEC 61508 and IEC 61511. However, the concepts are applicable to any protective functions or systems and are used in other contexts.

An important factor to recognize is that the risk reduction provided by IPL's is the product of their RRF's, and that if any IPL fails to meet the performance credited by a LOPA, the associated risk reduction of that IPL is lost. CCPS recognizes this by requiring that any IPL be Operated and Maintained in manner that assures that the credited Risk Reduction is effective.

The IEC 61511 and ISA 84.00.01-2004 Standards provide a framework for managing of Safety Instrumented Systems (SIS) and the Safety Instrumented Functions (SIF) that were identified by the Hazard Analysis process. However, these Standards are silent on managing of other non-SIF IPL types.

This has led some operating organizations to conclude that there are no corresponding requirements relative to the non-SIF IPL's. This is not the case: non SIF IPL's require sufficient care to assure their availability and cannot just be run to failure or be treated like non-IPL functions. The sections in the remainder of this paper discuss the practicalities of IPL implementation and ownership, the risk gaps that have been observed, and suggested management practices to prevent these gaps from developing.

3 What Should an IPL Management System Look Like

One of the fundamental gaps observed in practice is that there is often a perception that once an IPL is credited in a LOPA, and something exists in the facility that appears to perform the function defined by the LOPA, no additional action is required. This results in a system in which the effectiveness of an IPL is left to chance and the assumed risk reduction becomes an illusion.

An effective IPL Management System must be able to assure that IPL's are visible, that their design meets the functional and independence requirements for an IPL, and that the IPL's are Operated and Maintained to assure the required availability.

IPL's types and their implementation can vary considerably. While SIF IPL's may be paid more attention because of standards such as IEC 61511 and ISA 84.00.01-2004, non-SIF IPL's are usually shared with other basic plant systems such as the Basic Process Control System (BPCS) or blend in with other mechanical systems. Without an effective IPL Management System, non-SIF IPL functions often are either not effective to begin with, or atrophy to the point were they do not deliver the risk reduction with which they were credited.

Table 1 below identifies some of the gaps in IPL Management that have been observed in real operations. Unfortunately, these gaps are more common than not.

Table 1 – IPL Management Gaps

IPL Management Gap	Cause	IPL Management Requirements
Improper IPL credits are taken	Personnel are not properly trained on what the requirements and limitations for IPL qualification are, or local standards do not comply with REGAGEP	LOPA facilitators must be well trained on IPL requirements and advise participants on IPL requirements including IPL Types allowed and restrictions taking multiple credits. LOPA's should be reviewed by qualified and empowered individuals to assure validity.
IPL's as implemented do not meet the effectiveness, independence and reliability requirements defined in the LOPA's	Plant assets are identified of being IPL's, but are not evaluated to verify that they meet the IPL requirements.	Checklists for assessment of IPL qualifications are available and all IPL's are assessed against those requirements.
IPL functions are not understood by Operations and Maintenance personnel	IPL's are identified in LOPA documents and are not clearly identified in documents that are used by Operations and Maintenance personnel. LOPA documents are not readily available to non-PSM personnel and then are difficult to correlate to actual plant installations.	A system is needed that clearly identifies plant assets that are used to implement IPL's. The data should identify real plant assets that perform IPL functions, the Operational and Maintenance expectations for the IPL's, including operator actions, restrictions on disabling or bypassing the IP and Maintenance and Testing Requirements.
IPL's are bypassed or disabled	Assets credited as IPL's are not clearly identified. Operating and Maintenance procedures for IPL's are not prepared, available or enforced	Operating procedures need to exist that define criteria for bypassing or disabling of IPL's and requirements for management or operations while bypassed or disabled.
IPL's are modified or removed without proper MOC	The IPL list is not available, or is out of date. Projects making modifications do not recognize devices which are in IPL service and treat their modification like modification of non-IPL services	A continuously up to date IPL list is provided and linked to all LOPA, IPL design and operations and maintenance data. Projects evaluating proposed modifications have all the data needed to perform a proper MOC available in one place.
IPL's are not tested at the intervals required to maintain performance	IPL's are not identified in maintenance systems or do not have test procedures available. When procedures are available, management is ineffective because of many documents in many places, many of which may not be consistent or up to date	Test intervals, maintenance procedures and testing procedures are integrated in one database. Test records are captured in the Operations and Maintenance Module to allow for identification of next required test interval.

3.1 Why Is Compliance So Hard?

Many organizations have demonstrated difficulty in incorporating the requirements of IPL ownership and operation into their day to day practices. The basic reasons for this are that the importance of IPL's is not always recognized, and that effective management of IPL's generally does not fit within traditional plant management structures.

Effective IPL management requires that information cross traditional organizational boundaries. In traditional operating facilities, these boundaries usually result in many parallel information and documentation systems that have limited accessibility or provide only the functionality that a particular group requires and does not necessarily support the needs of other groups. Some examples of this are:

- Process Safety Groups have proprietary software that often has highly restrictive incensing terms that effectively lock up HAZOP and LOPA data access to just a few people.
- Maintenance Management Systems have historically been developed for management of major rotating equipment by and for rotating equipment personnel. Other functions are focused upon parts inventory management and work order tracking. These systems perform poorly when other systems are involved. They are particularly bad at handling Instrumented Systems. Resistance to making them effective for Instrumented Systems, or handling safety related non-rotating equipment has typically been quite high.
- Much of other traditional information management is done with discrete paper files or electronic files that are often not coordinated or readily available. These records are held by various organizations and are difficult coordinate and keep up to date. Often these records do not survive changes in personnel.

Management of IPL related data across these multiple platforms and formats is extremely inefficient and easily broken. Most facilities do not have the personnel or skill sets required to manage these problems. The Systems, even when well intentioned, tend to fail under the burden the highly labor intensive requirements of maintaining this data.

3.2 Safety Life Cycle Management

Effective management of IPL's across the entire Safety Life Cycle requires Safety Life Cycle Management systems that address the cross discipline nature of the data required and supports the following types of information:

- Defining relationships between Hazard Identification, Functional Requirements and Design of Protective Functions and Equipment, and Operation and Maintenance requirements.
- Capturing the Operational and Maintenance Events that are related to the IPL functions and equipment and using that data to provide Operation and Maintenance Management a clear picture of compliance and performance.

- Allowing users from various organizations to readily access IPL related data including the Hazard associated with IPL Assets, IPL functional requirements and equipment specifications and IPL Asset events and performance

Figure 2 illustrates a high level view of the relationships that should be defined and maintained and their general relationship to the Safety Life Cycle.

- HAZOP scenarios that have a high consequence are related to LOPA's
- LOPA's and the IPL's that were credited by the LOPA are identified
- Plant Assets that are used to perform the functions identified by a LOPA IPL are identified and their relationships to to the LOPA IPL's are shown. Note that in some cases, multiple LOPA IPL's may refer the the same Plant Asset.
- The Plant Asset data contains definition of the functions required, performance requirements, testing requirements, expected operator responses and physical equipment used to implement the functions.
- Operation and Maintenance Events are recorded and associated with the Plant Assets or Plant Equipment.
- KPI's are derived from the Operation and Maintenance Events and used to manage such things as period testing and assess performance such as demands, failures, bypasses.

Figure 2 – IPL Asset Data and Relationships

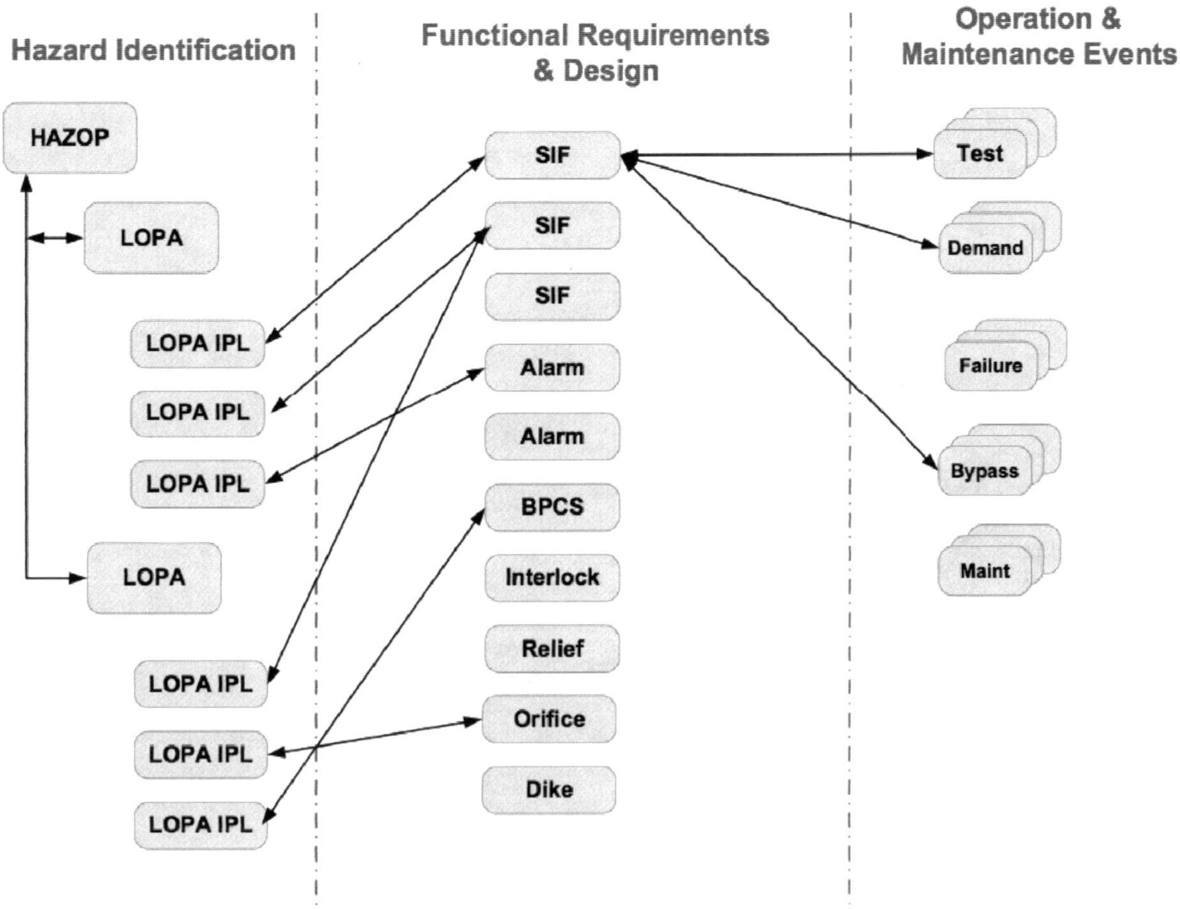

Figure 3 shows a more detailed set of data and relationships that focuses on one Plant Asset that is used to perform IPL functions.

- The Asset is related to one or more LOPA IPL's and to the equipment used to perform the Asset functions
- Plant Equipment that is used to implement an IPL Asset, such as instrumentation or other physical equipment is associated with the IPL Asset
- Assessment checklists and results are associated with the LOPA IPL's. Design Assessment checklists are associated with each Plant Asset and its associated equipment.
- Important documents such as Operating and Maintenance procedures or design documents are associated with the Asset for quick access.
- Operating and Maintenance Events are recorded and associated with their related equipment or Plant Assets

Figure 3 – IPL Asset Data and Relationship Details

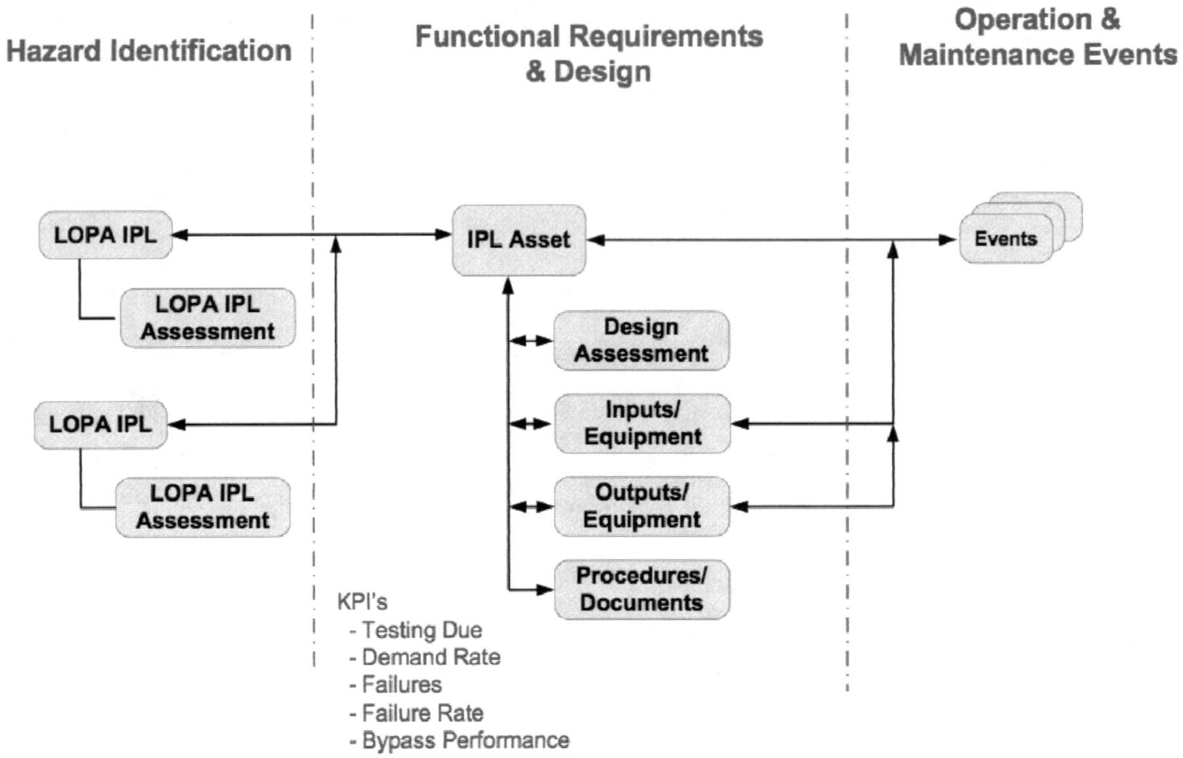

4 Using a Safety Life Cycle Management Data Base

A Safety Life Cycle Management System is best implemented in a database that is constructed specifically to manage the data and events that apply to the Safety Life Cycle. The data should be highly linked, so that a User can readily access related data. For example, an operator would be able to view the functional requirements for an IPL Asset, using a tag or other ID that is used within the Unit, and then directly access the Hazard data associated with the IPL. Appropriate Operations, Maintenance or Engineering personnel should then be able to associate Events with the IPL Assets and view KPI's associated with those Assets.

4.1 Data Relationships

A Safety Management Life Cycle Management Data Base should contain data objects that provide relationships among the various items shown in Figure 3. Among these are

- LOPA IPL's that define basic definitions of the IPL function and performance requirements and relate the LOPA IPL to the appropriate LOPA and HAZOP scenarios

- IPL Assets that define the actual plant assets that are used to implement the IPL functions. These should be identified using the same identifiers used in other plant documentation and by which plant operators, engineers and maintenance personnel know the asset. The IPL Asset data should define the design requirements such as function, expected response, accuracy and test and maintenance requirements
- Assessment Check Lists that define the items relative to performance, design, documentation or testing that is required of the IPL
- Input and Output Asset and Device objects that define the functional requirements for the Asset, Inputs or Outputs that are used to implement the IPL function and the specification data for the Devices that are installed in those Asset locations
- If IPL Performance is to be tracked, Event data objects that allow for tracking of events such as Demands, Failures, Testing, Bypasses, etc.
- Reference Documents such as procedures or design or specification data

4.2 Data Content

The data for each of the objects shown in Figures 2 and 3 should capture information necessary to support IPL Management. Some examples are:

- LOPA IPL's define the type of IPL, the performance requirements - typically in terms of a Risk Reduction Factor (RRF), and a functional description of the IPL. If operator response to an alarm is involved, the expected operator action and response time should be specified
- SIF Asset data should define the functional requirements, performance expectations and testing and maintenance requirements. In some implementations, data for a SIF could include all data necessary for definition of the Safety Requirements Specification (SRS) as defined in IEC 61511 and ISA 84.00.01-2004.
- Non-SIF IPL Asset data should define define the functional requirements, performance expectations and testing and maintenance requirements appropriate to the type of IPL.
- A LOPA IPL Assessment is a check list of items and their assessment results necessary to validate that the IPL will be effective and independent of other IPL's credited for the LOPA scenario
- An IPL Design Assessment is a check list of items and assessment results necessary to verify that the IPL Asset as designed meets the functional, performance and independence requirements for the IPL.
- Input/Equipment and Outputs/Equipment should define the functional requirements for Inputs and Outputs used by the IPL. The Equipment sections should contain or be linked to specification data for the specific equipment performing the functions.
- Procedure/Documents data should identify the specific Operating Procedures, Maintenance and Test Procedures or other Design Documents that apply to the IPL.

- Events – Events are used to describe the date, time and results of various types of Operation and Maintenance Events that occur during the IPL Life Cycle. The data for each event should be appropriate to the Event Type, such as as-found and as-left data for Functions and Equipment, start and end of bypasses, etc.

5 AN EXAMPLE IPL

The Figures below illustrate how a Safety Life Cycle data base application can be used to manage an IPL. An Alarm IPL is shown as an example.

5.1 IPL Functional Requirements

Figure 4 shows an example of an IPL Functional Specification for an Alarm with Operator Action IPL. This allows a User to define the functional requirements for the Alarm including such data as

- The Alarm ID – Typically this is the Tag Number of the Alarm as it appears on P&ID's, Operator Interfaces and other documentation
- Alarm functions, set points and basic implementation requirements
- Expected operator response and the time within which the response is expected
- Alarm operating restrictions such as when the alarm may be disabled or restrictions on changing set points or other configuration data
- Alarm Testing Requirements – type of testing and frequency of testing.
- LOPA Associations including the LOPA scenario for which the Alarm was cited as an IPL, the LOPA IPL and other IPL's that were cited by same LOPA scenario
- Associated Documentation such as Test Procedures, Operating Procedures

Figure 4 – Alarm IPL Functional Requirements Data

Alarm O-M Data | Alarm Functional | Alarm Devices | Revisions

Print | 1 Document | Attach Tasklist | Subscription | Bookmark | Admin Tools

Alarm Functional Requirements - FCC-FAL-138, Feed to East Riser

Alarm Requirements

Alarm ID:	FCC-FAL-138
Alarm Description:	Feed to East Riser Low Flow
Functional Description:	Feed flow to the East Riser is indicated on the BPCS by the analog input poing FCC-FI-138. This point has a low flow alarm confirgued to alert the operator of low riser feed flow
Asset Protection Category:	IPL
Operator Repsonse Time:	10 Min
Operator Repsonse Time Notes:	Nominal time to responsd to a high priorty alarm
Defined Corrective Action:	The operator shall place the East Riser Assist Steam flow controller FCC-FC-282 into automatic with a set point of 1100 PPH. Verfiy that the controller is properly proving steam to the riser

Protective Category:

Health and Safety	Environmental	Commercial	Reputation
Yes	No	No	No

Performance Requirements:	RRF 10 PFD 1.0 E-01
Alarm Operating Requirements:	Alarm must be enabled at all times when feed is in the unit. If the alarm must be bypassed or disabled prepare Safety Bypass Authorization. Set point is not to be changed without an MOC being performed
Alarm Response Time:	5 sec
Alarm Response Time Notes:	BPCS must activate alarm within 5 seconds of when the transmitter senses low flow
Alarm Location:	FCC Operator's Console in Main Control Room
Alarm Implementation:	Soft alarm on analog input tag
Alarm Priority:	High
Alarm Priority Rationale:	The risler low feed flow alarm indicates that conditios may exist the contribute to a riser flow upset but the situation is no yet crtical. Additional SIF layers of protection exist
Alarm Testing Requirements	Test transmitter, calibrate and test alarm activation and verify set points at turnarounds
On-Line Testing Interval	N/A
Off-Line Testing Interval	60 Months

Alarm Set Point (Software Based)

Set Point ID	Value	Type	Tolerance	Time Delay	Basis
FCC-138-SPL	115	Decreasing	5%	1 sec	Minimum HC flow w/o assist steam

Alarm Set Point (Device Settings)

Set Point ID	Value	Type	Tolerance	Time Delay	Basis
N/A	N/A	N/A	N/A	N/A	N/A

LOPA References

LOPA IPL ID	LOPA ID	LOPA IPL Description	LOPA Hazard	LOPA Consequnces	Required RRF
FCC-LOPA-15-1	FCC-LOPA-15	Low Feed Flow Alarm	Increase in riser catalyst density and loss of flow stability. Potential loss of flow and flow reversal	Hydrocarbon in regenerator. Very high regenerator temperataures and potenial hydrocarbon in preciptator. Potential explosion	10

Associated IPL's

ID	Type	Description
FCC-SIF-02	SIF	East Riser Emergency Steam
FCC-SIF-05	SIF	FCC Catalyst Reversal Protection System

Alarm Testing Procedures

Document ID	Title	Description
CE-TEST-7764	Alarm Testing Procedure	General Refinery Alarm Testing Procedures

5.2 IPL Operating and Maintenance History

Figure 5 shows an example of a view of the Alarm IPL's Operating and Maintenance history for the same Alarm as illustrated above. This view repeats the basic Alarm Functional Requirements data and the provides data specific to the Alarm operating status and the history of the events which have been entered into the database for the Alarm. This view provides the following types of data.

- Basic Alarm Function Requirements
- Alarm Testing and Performance Data – this section lists things such as the testing status for the alarm, the last test date and next test data, commissioning data, hours in service and a summary of performance statistics such as the number of demands and failures and the in-service failure rate
- Lists of the events that have been entered into the database for the alarm listed by type of event. In the example, this includes the Alarm Service Status history, Test Events, Demand Events, Bypass Events, etc.

In this view, many data fields are linked so the User can click on things such as Event ID's to go directly to the event detail view.

Figure 5 – Alarm Operation and Maintenance Data

Alarm O-M Data | Alarm Functional | Alarm Devices | Revisions

Print | 1 Document | Attach Tasklist | Subscription | Bookmark | Admin Tools

Alarm O-M View - FCC-FAL-138, Feed to East Riser

Alarm Data

Alarm ID	FCC-FAL-138
Alarm Description	Feed to East Riser Low Flow
Functional Description:	Feed flow to the East Riser is indicated on the BPCS by the analog input poing FCC-FI-138. This point has a low flow alarm confirgued to alert the operator of low riser feed flow
Operator Repsonse Time:	10 Min
Defined Corrective Action:	The operator shall place the East Riser Assist Steam flow controller FCC-FC-282 into automatic with a set point of 1100 PPH. Verfiy that the controller is properly proving steam to the riser

Testing and Performance Data

Alarm Testing Requirements	Test transmitter, calibrate and test alarm activation and verify set points at turnarounds
On-Line Testing Interval	N/A
Off-Line Testing Interval	60 months
Test Status	Current
Last Test Date	3/26/12
Next Test Due Date	3/26/17
Due Date Based Upon	Turnaround Schedule

Determining Devices	ID	Type	Description
	FCC-FT-138	Flow Transmitter	Feed to East Riser

Alarm Service Status	Commissioned
Service Hours	108,624 hrs
Alarm Commission Date	4/28/08
Alarm Commission Notes	None
Alarm Decommission Date	N/A
Alarm Decommission Notes	N/A

Alarm Performance	Life of Plant	5 Years	1 Year	Year to Date
No. of Failures	1	1	0	0
No. of Demands	6	6	0	0

In Service Failure Rate 7.0 E-02

Alarm Service Status History

Event ID	Date	From	To
FCC-FAL-138-Serv-001	4/28/2008	Created	Commissed

Test Groups

Test Group ID	Test Group Description	Date Last Run	Test Event ID
FCC-Alarms-001	FCC Safety Related Alarm Testing at Turnaround	4/4/2008	FCC-Alarms-001-Test-002
FCC-Alarms-002	FCC Safety Related Alarm Testing at Turnaround	3/24/2012	FCC-Alarms-001-Test-001

Functional Test Events

Event Date	Event ID	Parent Event ID	As Found	As Left	Event Notes	Event Status
4/4/2008	FCC-FAL-138-Test-001	FCC-Alarms-001	Failed	Passed	Pre-Commissioning Test	Approved
3/4/2012	FCC-FAL-138-Test-002	FCC-Alarms-002	Passed	Passed	Routine Turnaournd Tst	Approved

Demand Events

Demand Date	Demand Event ID	As Found	As Left	Event Notes	Event Status
4/17/2008	FCC-FAL-138-Demand-001	Passed	Passed	Startup	Approved
9/8/2008	FCC-FAL-138-Demand-002	Passed	Passed	Feed Pump Trip and Spare Start	Approved
12/17/2008	FCC-FAL-138-Demand-003	Passed	Passed	Feed Pump Trip and Spare Start	Approved
7/22/2011	FCC-FAL-138-Demand-004	Passed	Passed	Feed control valve sticking	Approved
10/9/2012	FCC-FAL-138-Demand-005	Passed	Passed	Feed Pump Trip and Spare Start	Approved
11/30/2013	FCC-FAL-138-Demand-006	Passed	Passed	Feed Pump Trip and Spare Start	Approved

5.3 IPL Assessment

As mentioned above, a critical part of the Safety Life Cycle is verification of an IPL's qualification to perform as an IPL. Check Lists can be created and tracked to document this process. Examples for an IPL Independence Check List and an Alarm Design IPL Check List are shown below.

In Figure 6 illustrates a Check List that would be completed near the time an IPL is cited in a LOPA is shown. This task list is intended to verify that the basic requirements to qualify the function as an IPL have been satisfied.

Figure 6 – LOPA IPL Check List

Item No.	Description	Finding	Comments
1.1	Verify that the IPL function, if effective, will prevent the consequence of the initiating cause from occurring	Passed	
1.2	Verify that the IPL is independent of the LOPA scenario imitating cause	Failed	The Alarm appears to use the same transmitter whose failure was the initiating cause.
1.3	Verify that the IPL is independent of any other IPL's that have been credited for the LOPA scenario	Passed	

Figure 7 shows a Check List summary that appears in the LOPA IPL view. This provides the user with a quick overview of the verification status and typically appears on the IPL Functional Requirements view.

Figure 7 – LOPA IPL Check List Status

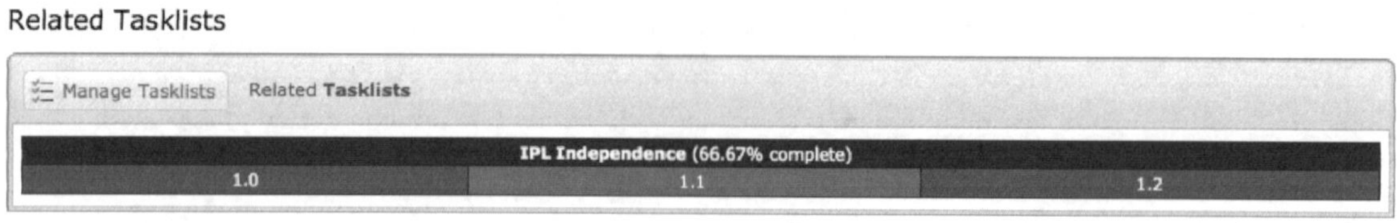

Figure 8 illustrates an IPL Design Check List for an Alarm. This check list addresses design requirements and operational preparedness items such as operation and maintenance preparedness items.

Figure 8 – Alarm IPL Checklist

Item No.	Description	Finding	Comments
3.1	Verify that the Alarm and its associated operator actions are capable of preventing the consequence for which the alarm has been cited as an IPL	Passed	
3.2	Verify that the alarm location is continuously attended	Passed	
3.3	Verify that the initiating or sensing device is independent of the initiating cause of the LOPA scenario	Failed	This alarm is derived from FCC-FT-138 whose failure is the LOPA scenario IC. Need an independent transmitter
3.4	Verify that the initiating device or sensor is independent of other protection layers claimed as IPL credits for the same LOPA scenario	Passed	
3.5	Verify that an operator action has been defined for this alarm, that the action can be accomplished and the result effective within the time between the initiating cause and the consequence occurring	Passed	
3.6	Verify that a written procedure for the operator response to this alarm has been prepared, that operators have been trained to locate the procedure and that the procedure is readily available to the operator	Failed	Operating procedures require updating.
3.7	Verify that this alarm has a suitable priority and it is identified as an IPL Alarm on Operator Graphics and Alarm views	Failed	Alarm needs to be added to operating graphics and IPL Alarm graphic

Item No.	Description	Finding	Comments
3.8	Verify that the settings for this alarm and disabling, inhibiting or bypassing of the alarm is under control of a Bypass Control Procedure and suitable approvals are required for the above actions and that return to operation is managed by this procedure.	Failed	The alarm needs to be added to the list of critical items that require Safety Bypass Authorization.
3.9	Verify that modifications to this alarm are under Management of Change Controls	Passed	Site Standard MOC Procedure
3.10	Verify that Maintenance and Testing Procedures have been prepared which are applicable to this alarm, and that the Procedures include capture of all testing records	Passed	Site Standard Maintenance procedure for testing of alarms
3.11	Verify that the P&ID; correctly shows this alarm	Failed	P&ID has not yet been updated to show this alarm as an IPL
3.12	Verify that the Alarm has been included in the Unit IPL List	Passed	

5.4 KPI's from Data Base

Once IPL data is captured in a data base, KPI's associated with the IPL Operation and Maintenance can be readily created and displayed. Some of the KPI's and summary data that can be developed are

- Identification of IPL Testing that is upcoming or overdue
- Identification on IPL's that have had high Demand or Failure rates
- Calculation of In-Service IPL Failure rates
- Reporting on IPL Assessment status and and completion of open items
- IPL quantities for Units, Sites or Enterprise-wide
- Identification of LOPA IPL's that are not linked to a Plant IPL Asset
- Identification of Plant Assets that are associated with multiple LOPA IPL's

Figure 9 shows a few examples of the types of IPL KPI data that can be provided to Users

Figure 9 – Typical IPL KPI's

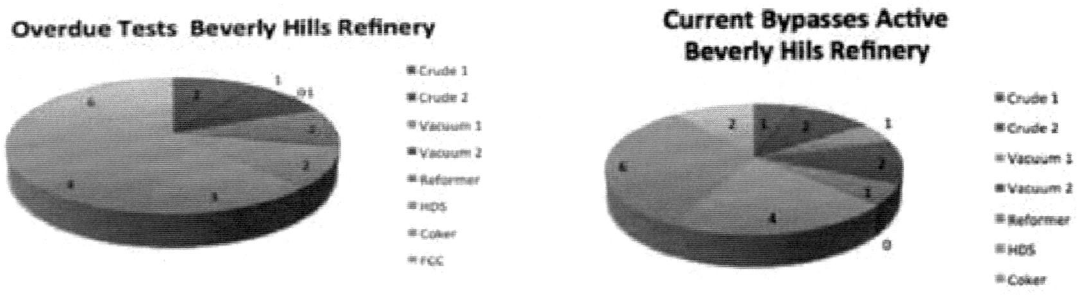

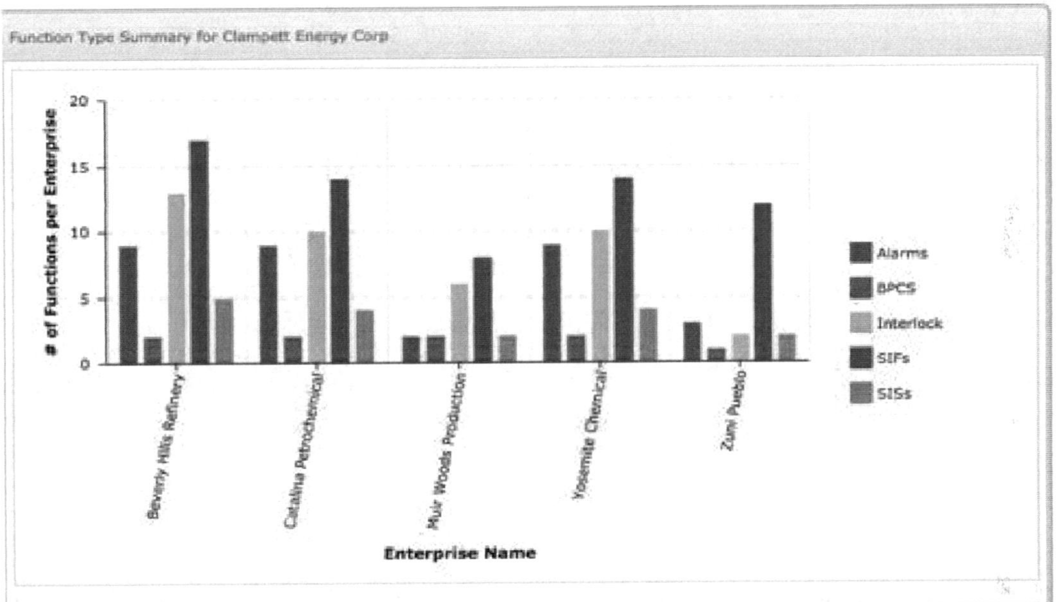

Enterprise Name	Alarms	BPCS	Interlock	SIFs	SISs
Beverly Hills Refinery	9	2	13	17	5
Catalina Petrochemical	9	2	10	14	4
Muir Woods Production	2	2	6	8	2
Yosemite Chemical	9	2	10	14	4
Zuni Pueblo	3	1	2	12	2

6 Conclusions

The Management of IPL's within an Operating Facility presents distinct challenges to Facility Management and trickles down to many organizations within the Facility. Many organizations are having signification problems in managing these requirements due to lack of the clarity of requirements and the cross discipline aspects that do not fit within the practical restrictions of management tools traditionally used. Force fitting IPL Safety Life Cycle management requirements as an overlay to these tools has generally been highly inefficient and generally ineffective.

The compliance gaps that have been observed as operating facilities struggle with management of IPL's and the Safety Life Cycle are a result of these problems. A practical and effective means of compliance is to apply a Management Process that is designed specifically for support of the Safety Life Cycle. This Management Process can maximize effectiveness of compliance and assure that the risk reduction that has been assumed for a facility is actually realized.

7 References

[1] – Layer of Protection Analysis, Center for Chemical Process Safety, ISBN 0-8169-0811-1, 2001

[2] – OSHA Memorandum – June 5, 2015 – Thomas Galassi, RAGAGEP in Process Safety Management Enforcement.

Rick Stanley has over 40 years' experience in Process Control Systems and Process Safety Systems with 32 years spent at ARCO and BP in execution of major projects, corporate standards and plant operation and maintenance. Since retiring from BP in 2010, Rick has consulted with Mangan Software Solutions (MSS) on the development and use of MSS' ProSys SLM Safety Lifecycle Management software and has performed numerous Functional Safety Assessments for both existing and new SISs.

Rick has a BS in Chemical Engineering from the University of California, Santa Barbara and is a registered Professional Control Systems Engineer in California and Colorado. Rick has served as a member and chairman of both the API Subcommittee for Pressure Relieving Systems and the API Subcommittee on Instrumentation and Control Systems.

442704 Jack Rabbit II: Experimental Large-Scale Chlorine Release Trials

Abstract:

In August and September of 2015, the U.S. Department of Homeland Security (DHS) Science & Technology (S&T) Chemical Security Analysis Center (CSAC) conducted the Phase 1 field trials of the Jack Rabbit II (JR II) program, a series of 5 large-scale outdoor chlorine release experiments with a collaborative team of partners from government, industry, and academia. Sponsored by the CSAC, the Defense Threat Reduction Agency (DTRA), and Transport Canada, these field trials were performed to fill critical knowledge, data, and capability gaps to support and improve chemical release modeling, industrial safety, and emergency response procedures. This paper summarizes the execution of the 5 successful chlorine release trials ranging from 5 to 9 tons, the instrumentation deployed, the data collected, and preliminary analyses of the high-definition video footage, cloud movement, and behavior and concentration/time profiles as a function of distance from the release site. This paper will also present plans for the JR II Phase 2 field trials scheduled for late 2016 which will include additional chlorine releases up to 16 tons.

445083 Novel Methodology for Inherent Safety Assessment in the Process Design Stage

Abstract:

Various physical and chemical hazards are intrinsic to the material itself or its manner of storage and use. Inherently safer design of the chemical process and equipment has been increasingly used by the design engineers to minimize or eliminate these hazards. This approach aims to mitigate hazards by design instead of add-on measures. This presentation describes a new design approach, introducing comprehensive inherent safety index (CISI), for use in the early stages of process development. Chemical, process and connectivity scores are calculated for individual units in the process, based on which the CISI assigns safety scores for the equipment. Each chemical in the unit has a weighted severity score. Reactivity score is calculated for mixture of chemicals in that unit. Connectivity score in introduced because the units can be highly interconnected, leading to compounding of hazards. Biodiesel and methyl methacrylate processes were studied to demonstrate the new safety assessment methodology. The results of the assessment are used to clearly identify key areas for improvement in a root-cause analysis.

AUTHOR INDEX

AUTHOR INDEX